PLASMA ASTROPHYSICS

ASTROPHYSICS AND SPACE SCIENCE LIBRARY

VOLUME 279

PLASMA ASTROPHYSICS

Kinetic Processes in Solar and Stellar Coronae

SECOND EDITION

by

ARNOLD O. BENZ

Institute of Astronomy,
ETH Zürich, Switzerland

KLUWER ACADEMIC PUBLISHERS

DORDRECHT / BOSTON / LONDON

A C.I.P. Catalogue record for this book is available from the Library of Congress.

ISBN 1-4020-0695-0

Published by Kluwer Academic Publishers,
P.O. Box 17, 3300 AA Dordrecht, The Netherlands.

Sold and distributed in North, Central and South America
by Kluwer Academic Publishers,
101 Philip Drive, Norwell, MA 02061, U.S.A.

In all other countries, sold and distributed
by Kluwer Academic Publishers,
P.O. Box 322, 3300 AH Dordrecht, The Netherlands.

Printed on acid-free paper

Printed in the Netherlands.

It is through wonder that men now begin and originally began science; wondering in the first place at obvious perplexities, and then by gradual progression raising questions about the greater matters too, e.g. about the changes of the Moon and of the Sun, about the stars and about the origin of the universe.

Aristotle
384 - 322 B.C.

CONTENTS

PREFACE

This textbook is intended as an introduction to the physics of solar and stellar coronae, emphasizing kinetic plasma processes. It is addressed to observational astronomers, graduate students, and advanced undergraduates without a background in plasma physics.

Coronal physics is today a vast field with many different aims and goals. Sorting out the really important aspects of an observed phenomenon and using the physics best suited for the case is a formidable problem. There are already several excellent books, oriented toward the interests of astrophysicists, that deal with the magnetohydrodynamics of stellar atmospheres, radiation transport, and radiation theory.

In kinetic processes, the different particle velocities play an important role. This is the case when particle collisions can be neglected, for example in very brief phenomena – such as one period of a high-frequency wave – or in effects produced by energetic particles with very long collision times. Some of the most persistent problems of solar physics, like coronal heating, shock waves, flare energy release, and particle acceleration, are likely to be at least partially related to such processes. Study of the Sun is not regarded here as an end in itself, but as the source of information for more general stellar applications. Our understanding of stellar processes relies heavily, in turn, on our understanding of solar processes. Thus an introduction to what is happening in hot, dilute coronae necessarily starts with the plasma physics of our nearest star. The main-sequence stars from about F5 to M, including the Sun, appear to have comparable atmospheres (in particular a magnetized, hot, low density corona) with similar plasma processes. Additionally, A7 to F5 main sequence stars, T Tau, and other pre-main sequence objects, giants and subgiants, close binaries, Ap and Bp stars may have 'solar-like' outer atmospheres.

The theory of coherent emission and absorption processes, as peculiar to radio astronomy, has here been limited to the most useful and successful concepts. Given the wealth of recent results from X-ray and gamma-ray observations, and from kinetic processes correlating with optical and UV phenomena, the idea of separating fields according to wavelength seems to be outdated. Nevertheless, I have found solar radio astronomy particularly instructive, since it involves basic principles of general interest.

A beginning graduate student (or advanced undergraduate) is confronted with a confusing amount of work on kinetic plasma processes published in a widely dispersed literature. On the other hand, an astronomer's education hardly includes the basic concepts of plasma physics. Yet a student may be asked to interpret observations. In such a dilemma we learn best by example. Accordingly, the purpose of this book is not to review observations comprehensively, nor to unfold theory in all its details, but to introduce concisely some basic concepts and to show how kinetic plasma physics can be applied in astronomy. Some exemplary observations particularly relevant to the physics of coronae are presented and compared with theory. Such a partial synthesis is urgently needed as a basis for assimilating the flood of information expected from future space instrumentation.

The book has two main parts: a basic introduction to kinetic theory and applications to coronae and similar plasmas. After a brief overview of the phenomenology of solar and stellar coronae and the fundamental equations of plasma physics, the next four chapters provide the physical background and a minimum introduction suitable for every course in plasma astrophysics. Chapter 2 discusses some elementary properties of plasmas. A statistical approach is used in Chapter 3 to introduce magnetohydrodynamics and its fundamental characteristics. Chapter 4 gives an overview of the most important high-frequency waves. Finally, in Chapter 5 kinetic plasma theory is developed for the important example of a particle beam. The second part of the book is the substantial core, presented in Chapters 6-10. They are more or less independent of each other and need not be covered in sequence or *in toto*. These chapters provide the tools for a proper investigation of kinetic phenomena, such as electron beams (Chapter 6), ion beams (Chapter 7), and magnetically trapped particles (Chapter 8). Thus equipped, the reader can proceed to the important, energy related topics of electric currents and fields in Chapter 9. Chapter 10 presents the foundations for kinetic (collisionless) shock waves with emphasis on particle acceleration. Plasma processes interfere drastically with the propagation of electromagnetic waves at low frequencies. This is discussed in Chapter 11 from a pragmatic point of view, primarily of interest to radio astronomers.

A complete list of the literature would fill a book. References are only given for a few outstanding historical discoveries and – in rare cases – for results quoted without derivation. For detailed lists of the literature, the reader is referred to the reviews and other further reading listed at the end of each chapter.

This textbook has grown out of my experience teaching introductory courses on plasma astrophysics at the Swiss Federal Institute of Technology (ETH) for two decades. I wish to thank my colleagues and students for countless helpful remarks. Special mention must be made of Dr. D.G. Wentzel, who has carefully and critically read preliminary versions of the text and supplied thoughtful advice. Drs. M.J. Aschwanden, B.R. Dennis, S. Krucker, M.R. Kundu, D.B. Melrose, and E.R. Priest have read the manuscript and have contributed valuable comments. Dr. J. Gatta, I. Priestnall, and K. Siegfried have patiently corrected my English, and Mrs. M. Szigeti has helped with typing. Finally, I would like to thank my son Pascal whose generous donation of a dozen erasers has considerably helped to improve this book. I carry the full responsibility for any oversights resulting from my neglecting to use them. There is no way to adequately express how much I owe to my wife, Elisabeth, and children, Renate, Christoph, Pascal, and Simon, for their love and patience.

Institute of Astronomy *Arnold O. Benz*
ETH Zurich
Easter, 1993

PREFACE TO THE SECOND EDITION

The program outlined in the first edition has been found satisfactory by beginning graduate students, although the distant mountain tops, on which the reader should gain a view, have grown and multiplied. Climbing them involves problems in theory and observations that cannot be presented in detail by an introductory text. The reader should understand the basics of the approach and how the results could be derived.

The text of the first edition was complemented in the sections on collisional processes and stochastic acceleration. Misprints and errors were corrected and the literature up-dated.

Zurich, March 2002 *Arnold O. Benz*

CHAPTER 1

INTRODUCTION

People who have the chance to see the solar corona during a total solar eclipse are fascinated by its brilliant beauty. Appropriately, the word 'corona' is the Latin root of the English 'crown'. Its optical luminosity is less than a millionth of the Sun's total flux and is normally outshone by atmospheric stray light. In a total solar eclipse, on average about once every 18 months, a narrow strip of the Earth's surface is shielded completely by the Moon from the brilliant disk of the Sun, and the corona appears. The phenomenon has been known from antiquity, and is described by Philostratus and Plutarch with considerable astonishment. Almost identical prehistoric paintings in Spanish caves and on rocks in Arizona, USA, seem to represent the Sun surrounded by coronal streamers and to testify to human emotions of even older times. Nevertheless, solar eclipses are rare, and the corona escaped scientific scrutiny and even interest until the middle of the 19th century, when it became clear that it is a solar phenomenon, not related to the Moon, nor an artifact produced by the Earth's atmosphere. It entered the limelight in 1939 when W. Grotrian and B. Edlén identified coronal optical lines as originating from highly ionized atoms, suggesting a thin plasma of some million degrees. The temperature was confirmed in 1946 by the discovery of thermal radio emission at meter waves. This unexpectedly high temperature is still an enigma today: it is one of the reasons why the solar corona will be a primary target of solar research in the coming years.

Stellar flares and activity observed at optical and radio wavelengths have long been taken as hints for the existence of solar-like coronae of other stars. The recent advent of X-ray and UV observing satellites, and the advancement of optical spectroscopy and radio interferometers, have finally made it possible to detect and study the quiescent emissions of stellar coronae and transition regions. They have shown persuasively that main sequence stars with less mass than about 1.5 times the mass of the Sun (astronomically speaking 'later' than about spectral type F5, and beyond type M) probably all have coronae. The extremely hot and massive coronae of young, rapidly rotating stars are of particular interest.

1.1. The Solar Corona

1.1.1. Brief overview of the Sun

The Sun is a gaseous body of plasma structured in several, approximately concentric layers. In the central *core*, hydrogen nuclei are fused into helium nuclei, providing the energy for the solar luminosity and for most of the activity in the outer layers. The size of the core is about a quarter of the radius of the visible disk, $R_\odot$. From there, the energy is slowly transported outward by radiative diffusion. Photons propagate, are absorbed and remitted. The *radiation zone* extends to 0.71 $R_\odot$, as revealed by the seismology of surface oscillations. The temperature decreases from $1.55 \cdot 10^7$ K in the center to some $2 \cdot 10^6$ K. At this point, the decrease becomes too steep for the plasma to remain in static equilibrium, and convective motion sets in, becoming the dominant mode of energy transport. The *convection zone* is the region where the magnetic field is generated by the combination of convection and solar rotation, called a *dynamo* process.

The solar atmosphere is traditionally divided into five layers. The reader should not imagine these as spherical shells, but as physical regimes with different characteristics. The boundaries follow the spatial structures and are extremely ragged. Figure 1.9 shows the schematic changes of density and temperature with height. Note that in reality the product of density and temperature, i.e. the pressure, does not decrease smoothly. In the outer layers, the local pressure of the magnetic field and dynamic phenomena dominate and control the pressure.

The lowest atmospheric layer, called the *photosphere*, emits most of the energy released in the core. It is the region where the solar plasma becomes transparent to optical light. The thickness of the photosphere is little more than one hundred kilometers and so forms an apparent 'optical surface'. It is just above the convection zone. Convective overshoots are observed in the photosphere as a constant, bubbling *granulation*, a coarse cell structure with a mean size of 1300 km. At larger size, supergranules suggest convective structure of a superior order. In spite of its thinness, the photosphere has been the major source of information on the Sun. The analysis of global oscillations in photospheric lines has revealed the structure of the interior. The magnetic field can be accurately measured and mapped by observing the Zeeman effect. The photosphere comprises the footpoints of the field lines extending into the regions above. Sunspots are observed as groups of dark dots having a large magnetic flux.

After reaching a minimum value of about 4300 K, the temperature surprisingly rises again as one moves up from the solar surface. The region of the slow temperature rise is called *chromosphere* after the Greek word for 'colour'. Its light includes strong lines which produce colourful effects just before and after totality of an eclipse. There are more than 3500 known lines, mostly in the optical and ultraviolet (UV) range. Of particular relevance is the Balmer-alpha line of hydrogen (Hα) at 6563 Å, which is sensitive, for example, to temperature, to plasma motion, and to impacts of energetic particles from the corona. The chromospheric

Fig. 1.1. White-light photograph of the corona taken during the eclipse of 1980, February 16 (courtesy Dr. J. Dürst). The moon blocks the bright photosphere and most of the chromosphere. Streamers (A), coronal holes (B), coronal loops (C), and prominences (D) become visible.

plasma sometimes penetrates into the upper atmosphere in the form of arch-shaped prominences (visible as bright dots at the limb of the Moon in Fig. 1.1).

Within a narrow *transition region* of only a few hundred kilometers, the temperature rises dramatically from 20 000 K at the top of the chromosphere to a few million degrees in the corona. It is the region where many emission lines in the extreme ultraviolet (EUV) originate, as the temperature is *already* high, but the density is *still* high. Most important are the Lyman-alpha line of hydrogen at 1215.7 Å and the lines of partially ionized heavier elements (e.g. O IV and C III). Another source of information is thermal bremsstrahlung emitted by free electrons as a continuum in microwaves (radio waves having frequencies in the range 1–30 GHz, introduced in Section 1.1.4). The transition region is far from being a static horizontal layer, but rather a thin envelope around relatively cool chromospheric material protruding (such as prominences) or shooting out into the corona (as observed in spicules and surges).

1.1.2. Optical observations of the corona

The solar *corona* can be observed at many wavelengths including soft X-rays, atomic lines, optical, and radio waves. Figure 1.1 shows an image in the integrated optical range (called *white light*) during an eclipse. A radial filter was used to correct for the orders of magnitude decrease in brightness with radius. The white light is produced by Thomson scattered photospheric photons on free electrons in the coronal plasma. (Another cause of white light, particularly at large angular distance, is scattering by interplanetary dust particles. It will not be discussed here.) High density structures therefore appear bright. They outline myriads of loop structures in the inner corona up to half a photospheric radius in height (1.5 $R_\odot$ from the center of the Sun). The density enhancement is a factor of 3 to 10 and outlines the magnetic field (Section 3.1.3.C). At higher altitude the high density structures consist of roughly radial streamers extending beyond 10 $R_\odot$. At their base, they blend into loop structures and resemble the helmets of old Prussian soldiers. Beyond about 1.5 $R_\odot$ the remaining magnetic field lines do not return within the heliosphere. Here the plasma is not bound and flows outwards, forming the *solar wind*. When followed back to the inner corona, one notes that most of the solar wind originates from coronal holes, regions of low density, nearly vertical magnetic fields, and conspicuous darkness in soft X-ray pictures (see Fig. 1.2). This outermost layer extends far beyond the Earth's orbit, gradually changing into the interplanetary medium, until it meets the interstellar medium at the heliopause located at some 100 to 200 astronomical units (AU).

1.1.3. Soft X-rays and EUV lines

Most of the thermal energy directly radiated by the coronal plasma is in the form of soft X-rays (photons having energies from 0.1 to 10 keV) and EUV line emission. As the contribution from the lower atmosphere is negligible, the coronal X-ray emission shown in Figure 1.2 appears on a dark background. The emission includes both a continuum emitted by free electrons and lines from highly ionized ions. The bright structures outline magnetic loops of high density (typically 10^9 electrons cm^{-3}) and high temperature ($2 - 3 \cdot 10^6$ K). Their ends are rooted in the photosphere and below.

Loops appear in different intensities and various sizes. The brightest species arch above sunspot groups and constitute the third dimension of so-called *active regions*. Whereas sunspots are cooler than the ambient photosphere, since the magnetic field inhibits convection, the active regions above sunspots – or at least some loops – are heated much more than the quiet regions of the corona. This fact strongly suggests that the heating process involves the magnetic field.

Interconnecting loops join different active regions, sometimes more than a photospheric radius apart. Systems of interconnecting loops may exist for several rotations, although individual loops often last less than one day. *Quiet region* loops are not rooted in active regions, but in magnetic elements of the quiet photosphere.

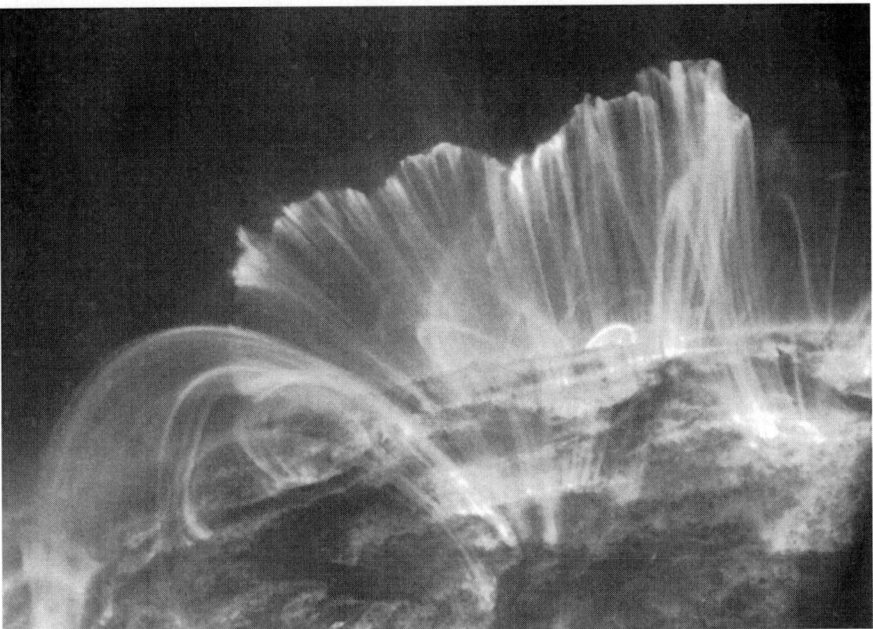

Fig. 1.2. Picture of the solar corona in the Fe IX line at 171 A taken with the TRACE satellite. The horizontal size of the frame is 0.37 $R_\odot$ ($2.6 \cdot 10^{10}$ cm). The solar limb is visible in the middle of the image. High density, hot loops are seen to outline the magnetic field of the corona.

Such loops are somewhat cooler ($1.5 - 2.1 \cdot 10^6$ K) and less dense. They are visible in projection above the limb as a forest of overlapping structures (Fig. 1.2).

1.1.4. THERMAL RADIO EMISSIONS

The coronal plasma emits thermal radio waves by two physically different mechanisms. It is important to distinguish between these different processes. (*i*) The usually dominant process is the *bremsstrahlung* ('breaking radiation' or free–free emission) of electrons experiencing collisions with other electrons or ions. (*ii*) In active regions, the enhanced magnetic field strength increases the gyration frequency of electrons in the field (Section 2.1.1) and makes *gyroresonance emission* the dominant thermal radiation process between roughly 3 GHz to 15 GHz. This process opens a possibility to measure the coronal magnetic field. High-frequency bremsstrahlung originates usually at high density in the atmosphere. Similarly, the higher the magnetic field, the higher the observed frequency of gyroresonance emission. Therefore, thermal high-frequency sources are generally found at low altitude.

The large range of radio waves is conveniently divided according to wavelength into metric, decimetric, centimetric, etc. radiation (30 – 300 MHz, 0.3 – 3 GHz, 3

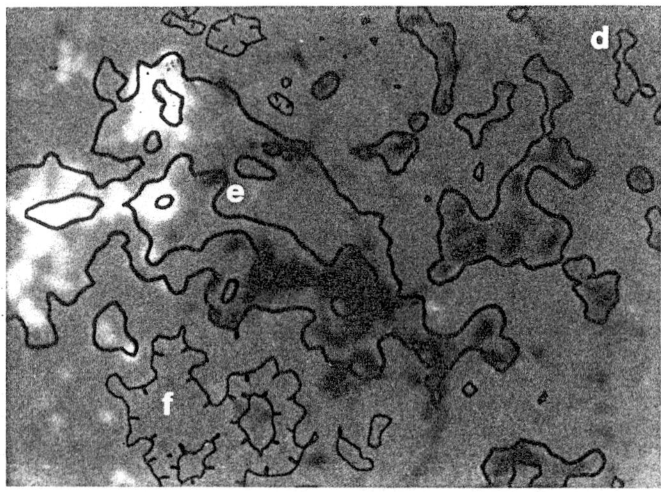

Fig. 1.3. A magnetogram of a decayed (spotless) active region in the left part and quiet photosphere at right is presented as a grayscale image. Positive polarity (where the magnetic field vector points out of the picture) is represented as bright, and negative polarity as dark. Superposed are the contours of the radio intensity at 8.5 GHz observed at the VLA. The size of the total image is 0.36 $R_\odot$. Some particular locations are labelled and discussed in the text (from Gary *et al.*, 1990).

– 30 GHz, etc.). Alternatively, the technical term *microwaves*, meaning the range 1 – 30 GHz, is also frequently used.

Figure 1.3 compares the radiation intensity at 8.5 GHz with the underlying magnetic field strength along the line of sight. The peak brightness temperature of $4.1 \cdot 10^4$ K suggests that radiation originates from the transition region. Model calculations assuming bremsstrahlung emission show that this is the case. Enhanced radio emission correlates with patches of an enhanced photospheric magnetic field (called network) a few 1000 km below, as first noted by M.R. Kundu *et al.* in 1979. The network of the quiet part (d) outlines the boundaries of supergranules, formed by large-scale convection cells underneath. The depression in brightness (f) coincides with a filament seen in absorption of the Hα line. The magnetic field at the height of the radio source cannot be measured in general, but may be estimated using the photosphere as a boundary surface. The peak radio intensity is above the neutral line of the decayed active region (marked e in Fig. 1.3), where the longitudinal field component vanishes and (just below the letter e) a weak loop structure seems to bridge the two polarities. The marvellous correlation between magnetic field and transition region suggests that the field also influences the heating of quiet regions.

1.2. Dynamic Processes

1.2.1. Processes in the Upper Corona

The solar corona is remarkably variable on all time scales. Even the early observers of eclipses noted the different shape of the corona, year after year. The 11-year magnetic cycle, conspicuous in the sunspots of the photosphere, has a striking effect on the corona. During the minimum years the corona shows few streamers, and these have enormous 'helmets' and are located near the equator. In solar maximum years the average density increases, and numerous streamers are more regularly distributed (Fig. 1.1).

Pictures taken by coronagraphs (producing an artificial eclipse) on board spacecraft show more rapid changes. As a rule, structures do not persist long enough to be followed during the few days it takes to traverse the limb. Bright and dark rays appear and disappear within hours. Major events, called *coronal transients*, occur daily during the solar maximum. They are connected (though not always) to coronal mass ejections or flares, and sometimes result in disrupting or dissolving streamers.

1.2.2. Processes in the Lower Corona

Soft X-ray motion pictures show ceaseless variability in active regions. Loops brighten up and disappear within a few hours. Also, interconnecting loops sometimes light up within hours. Loops that connect an active region to an old remnant remain stable for several days.

Also fascinating are the incessant brightenings in *quiet regions and coronal holes*, as discovered 1997 by S. Krucker *et al.* in soft X-rays and radio emission. They consist of loops, typically 10 000 km long and 2000 km wide, flaring up and fading on time scales of a few minutes. The footpoints of the loops have been found in tiny bipolar regions in photospheric magnetograms. More than a million such microevents can be observed on the whole Sun per hour in the very sensitive coronal EUV lines of Fe IX and Fe XII. Interestingly, the fluctuations do not indicate much heating of already hot plasma, but rather the addition of new plasma presumably of chromospheric origin. The events constitute also an energy input of a sizeable fraction of what is needed to heat the quiet corona.

Duration and size become even smaller in the sources located in the transition region. The radio and UV sources in the quiet regions (Fig. 1.3) vary within less than one hour, sometimes within one minute. Jets of a few thousand kilometers in diameter have been observed in a line of C IV to accelerate upward to speeds as high as 400 km s^{-1} (Brueckner and Bartoe, 1983).

The plasma in active-region loops is in continual motion along the magnetic field lines. Upflows in chromospheric *spicules* and the more dramatic *surges* move typically at a speed of 20 to 30 km s^{-1} and reach heights of 11 000 km and 200 000 km, respectively. *Coronal rain* is a relatively cool plasma observable in the Hα line, flowing down at nearly the free-fall speed of 50 to 100 km s^{-1}.

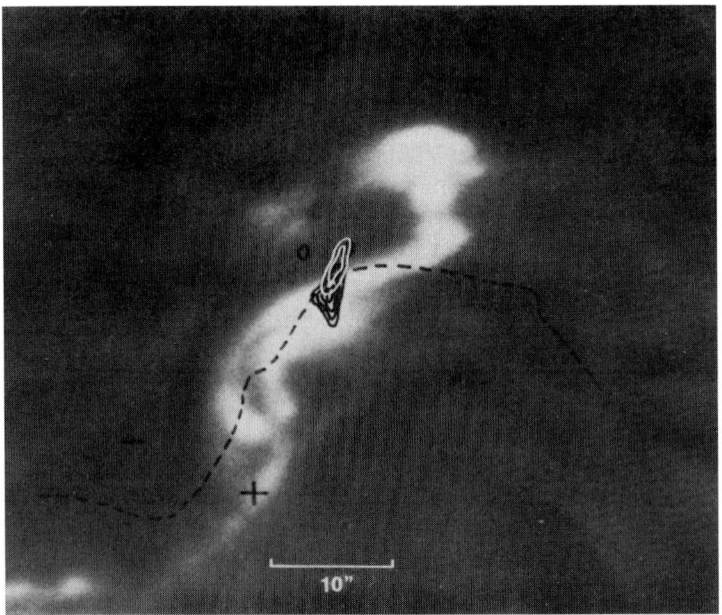

Fig. 1.4. A flare appears bright in a picture taken in the optical Hα line at the Big Bear Observatory. The flare partially follows the line (dashed) of zero longitudinal magnetic field measured in the underlying photosphere. The contours outline the microwave intensity at 15 GHz observed by the Very Large Array. White contours show left circular polarization, black contours show right circular polarization. The scale of 10$''$ corresponds to 7260 km (from Hoyng *et al.*, 1983).

1.2.3. SOLAR FLARES

The *flare* is a process of dramatic and truly remarkable dimensions. Its energy release spans many orders of magnitude, from the smallest microflares observable in EUV lines and radio emission of the solar corona (less than 10^{24} erg) to the most energetic events in young or rapidly rotating late-type stars in stellar clusters (10^{37} erg). The first solar flare was observed as a brightening in white light by R.C. Carrington and R. Hodgson in 1859. The emission originated from a small area in an active region and lasted only a few minutes. When, a few years later, the Sun was studied extensively in the chromospheric Hα line, the reports of flares became much more frequent, but also bewilderingly complex. Variations of source size, ejections of plasma blobs into the solar wind, and blast waves were noted. Meterwave radio emissions, detected serendipitously in 1942 by J.S. Hey during military radar operations, revealed the presence of non-thermal electrons in flares. During a radio burst, the total solar radio luminosity regularly increases by several orders of magnitude.

In the late 1950s it became possible to observe the Sun in hard X-rays by balloons and rockets. L.E. Peterson and J.R. Winckler discovered the first emission during a flare in 1958. Later, the enhancements observed in microwave and hard X-ray emissions have led to the surprising conclusion that these energetic particles contain a sizeable fraction (up to 50%) of the initial flare energy release. The broadband microwave emission results from the gyration of mildly relativistic electrons in the magnetic field (gyrosynchrotron emission, Section 8.1.2). Hard X-rays ($\geq$ 10 keV) are caused by electrons of similar energies having collisions (Section 6.4). In 1972 the γ-ray emission of heavy nuclei excited by MeV protons was discovered. This emission indicates accelerated ions (Section 7.1.1). Finally, EUV and soft X-ray ($<$ 10 keV) emissions have shown that the flare energy heats the plasma of coronal loops to temperatures from 1 to $3 \cdot 10^7$ K. Within minutes, some active-region loops become brilliant soft X-ray emitters, outshining the rest of the corona. Obviously, such temperatures suggest that the flare is basically a coronal phenomenon.

One may thus define the flare observationally as a brightening within minutes of any emission across the electromagnetic spectrum. The different manifestations seem to be secondary responses to the same original process, converting magnetic energy into particle energy, heat, and motion.

The timing of the different emissions of the same flare is presented schematically in Fig. 1.5. In the *preflare phase* the coronal plasma in the flare region slowly heats up and is visible in soft X-rays and EUV. The large number of energetic particles is accelerated in the *impulsive phase*, when most of the energy is released. Some particles are trapped and produce the extended emissions in the three radio bands. The thermal soft X-ray emission usually reaches its maximum after the impulsive phase. The heat is further distributed during the *flash phase*. In the *decay phase*, the coronal plasma returns nearly to its original state, except in the high corona ($\gtrsim 1.2\ R_\odot$), where plasma ejections or shock waves continue to accelerate particles causing meter wave radio bursts and interplanetary particle events (Section 10.2).

1.2.4. Other dynamic processes

Somehow related to flares in their physics, but even more powerful, are *coronal mass ejections (CME)* which will be discussed in Section 10.2.3. The first observers of CMEs in optical lines (using ground-based coronameters) describe the events as 'coronal depletions' (e.g. Hansen *et al.*, 1971). The full scope of CMEs was not realized until 1973, when white-light observations were possible from space by the Skylab mission. CMEs occur when a coronal structure – a system of loops or a prominence – loses its equilibrium and starts to rise. Such events carry away remarkable amounts of coronal mass (up to 10^{16} g) and magnetic flux; most of their energy does not show up in the brightening of some radiation or flare. CMEs reach velocities of several hundred km s^{-1} and can be followed on their path throughout the heliosphere.

Radio *noise storms* are also relatives of flares (Section 9.5.2). They were identified soon after the discovery of the solar radio emission. Noise storms consist

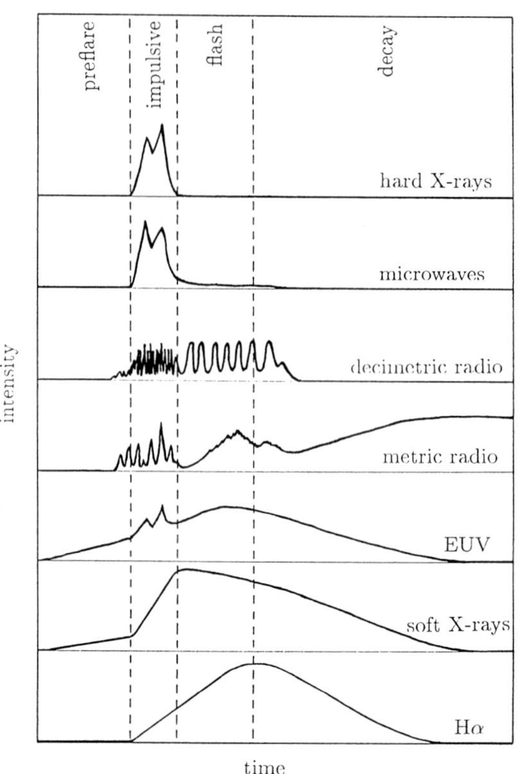

Fig. 1.5. A schematic profile of the flare intensity at several wavelengths. The various phases indicated at the top vary greatly in duration. In a large event, the preflare phase typically lasts ten minutes, the impulsive phase one minute, the flash phase five minutes, and the decay one hour.

of myriads of tiny meter wave bursts of duration about a second or less, and a broadband, slowly varying continuum ranging from 60 to 400 MHz. The storms last for days and are not directly associated to flares, although the start of noise storms may be initiated by a large flare or a CME. More frequently, storms appear in the upper corona ($> 1.2\ R_\odot$) when a new active region forms or rapidly grows. This suggests that they represent the adjustment of the previous coronal plasma and magnetic field to the newly emerged magnetic flux.

The various dynamic processes discussed in this section represent energy inputs in different forms. Are they enough to heat the solar corona? The answer seems to become more difficult the more we know about the corona. Regular flares in active regions release most energy by the largest events. They are not frequent enough to account for the quasi-steady heating of the entire corona. Microflares similar to noise storm sources but at low height having energies below 10^{26} erg may provide the energy in active regions. It has been proposed that microevents and network sources heat the quiet corona and coronal holes. Note that additional forms of energy input, in particular by waves and stationary electric currents, must also be considered. The heating problem of the solar corona can only be solved by a better understanding of all the various processes.

1.3. Stellar Coronae

1.3.1. Soft X-ray Emission

The Sun was the only star known to possess a high temperature corona until X-ray telescopes became feasible. There was no great hope of detecting other coronae with the first soft X-ray instruments. Capella, a nearby binary system of two late-type giants, was discovered in 1975 when the X-ray detector of a rocket payload was briefly pointed at the object (Catura *et al.*, 1975), a detection soon confirmed by satellite (Mewe *et al.*, 1975). Capella's corona was found to be a thousand times more luminous than the Sun!

The discoveries of other coronae, and in particular of coronae of other types of stars, followed slowly until the launch of the Einstein satellite. Its greatly superior sensitivity revealed that particular stars of nearly all types are X-ray emitters (Vaiana *et al.*, 1981). Figure 1.6 gives an overview of different types of prominent X-ray stars. Note that X-ray emissions from stars earlier than type A have a different origin. The strong X-ray emission of O and B stars is caused by hot stellar winds. Late-type (F to M) stars on the main sequence have coronae similar to the Sun with a wide range of luminosities for each spectral type. The dependence on photospheric temperature and – even more surprisingly – on photospheric radius is small.

The post-T Tau stars indicated in Figure 1.6 are of particular interest. They are a class of young stars that have evolved beyond the accretion phase (being so-called weak-line T Tau stars), but have not yet arrived on the main sequence. BY Dra and RS CVn stars are binary systems with no significant mass exchange, but close enough so that the tidal effects of their gravitational interaction synchronize the rotation periods of the individual stars to the orbital period of the system. Since the orbital periods range from less than a day to a few days, the stars are forced to rotate rapidly. In BY Dra systems both components are late-type, main-sequence stars; in RS CVn systems at least one component is a subgiant. FK Com stars are even more rapidly rotating objects and may be the remnants of collapsed binary systems.

The X-ray luminosity given in Figure 1.6 indicates a minimum requirement on the coronal heating process. The actual energy input is higher if there are energy losses other than radiation. Since we do not yet have good stellar models, the Sun may serve as an example. Conduction losses of the solar corona to the underlying chromosphere have been estimated as 50% in quiet regions, but less in active regions. In coronal holes, about 90% of the input energy ends in solar wind expansion (Withbroe and Noyes, 1977).

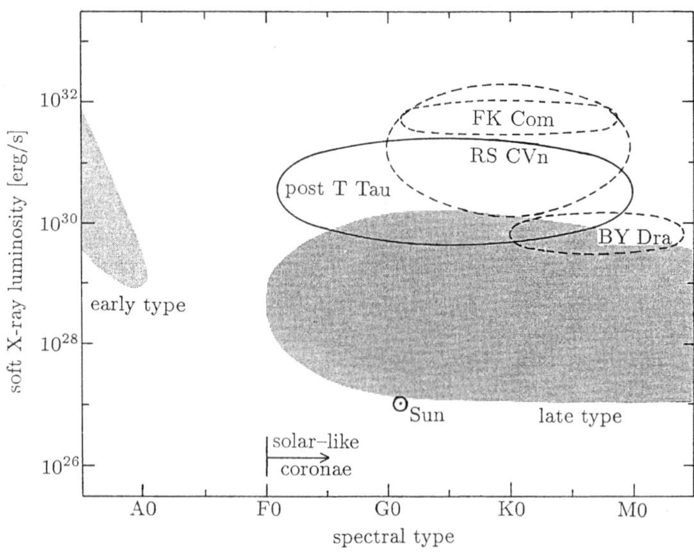

Fig. 1.6. Thermal soft X-ray emission of hot stellar plasma has been observed from nearly all types of stars. The spectral type characterizes the temperature of the photosphere and is also an indication of stellar mass (decreasing to the right). X-ray emitting main-sequence stars are represented by shading, young main-sequence stars (post T Tauri type) by full contours, and binary systems of BY Draconis and RS Canem Venaticorum types, as well as 'post-binaries' of FK Draconis type, by dashed curves.

The range of solar-like coronae overlaps approximately with the stellar types that have a convection zone below the photosphere. Another necessary condition is stellar rotation: the faster a star rotates, the more luminous the corona (Pallavicini *et al.*, 1981).

In open field regions the hot coronal plasma expands thermally and forms a stellar wind. The loss of angular momentum carried away by the wind is the likely cause of the present day slow equatorial rotation of the Sun in 26 days, compared to ten times faster rates in young main-sequence stars. Observing star clusters of different age, O. Struve and others found in 1950 that a braking process operates for young main-sequence stars of type F and later. The existence of coronae seems to be important in slowing down the rotation of late-type stars.

1.3.2. Stellar flares

The existence of stellar coronae could have been guessed from stellar flares, known for half a century. As in the case of the Sun, stellar flares were first detected in white light. In K and M main-sequence stars having low intrinsic optical luminosity, the effect of a flare on the layers below is more easily observable than in the Sun. Flares temporarily increase the total stellar luminosity. In UV light (U-band) this may reach a factor of ten.

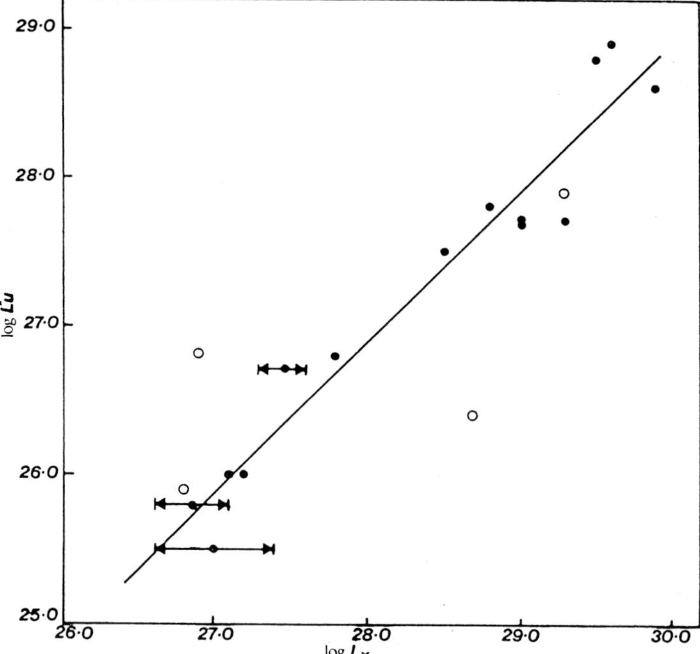

Fig. 1.7. The time averaged luminosity of flare emission in the ultraviolet, L_u^*, is compared to the quiescent soft X-ray emission, L_x, for nearby main-sequence M stars (from Doyle and Butler, 1985).

Stellar flares have also been detected in radio, EUV, and soft X-ray emissions (Section 8.4.4). Reliable and frequent measurements became possible around 1980 with the Very Large Array (a radio interferometer in New Mexico) and the Einstein satellite, respectively. The radio flares exist as at least two types: short ($\lesssim$ 10 minutes), circularly polarized events not associated with optical or X-ray emissions; and long-duration, weakly polarized events ($\lesssim$ 40%) coincident with optical and X-ray flares.

Flaring has been reported in all spectral classes along the main sequence, both on single stars and on members of multiple systems. Flare activity starts at an early phase of stellar evolution, and flaring is particularly frequent in pre-main-sequence stars. Most importantly, stars with considerable coronae are likely to have frequent flares. The integrated luminosity in X-rays of one flare ranges from the detection limit of 10^{27} erg in nearby M stars to 10^{37} erg in star clusters.

Figure 1.7 shows an important discovery, made independently in 1985 by J.G. Doyle and C.J. Butler, D.R. Whitehouse, and A. Skumanich. They found a tight correlation between the time averaged luminosity of U-band flares and the quiescent X-ray luminosity in dMe, a subclass of main-sequence (*dwarf*) M stars with chromospheric emission (*e*) lines. A relationship between quiescent and flaring activity is not unexpected from the solar observations. They show that (*i*) the heating of coronal regions is connected to the magnetic field and that (*ii*) the flare energy is drawn from the same magnetic field. Thus both originate from the same dynamo process in the stellar interior that generates the magnetic field.

1.3.3. QUIESCENT RADIO EMISSION

The discovery of a low-level microwave emission ($\gtrsim$ 1 GHz) from apparently non-flaring, single dMe stars in 1981 by D.E. Gary and J.L. Linsky has added another mystery to coronal physics. The radiation is 2–3 orders of magnitude more luminous than the quiet Sun. Spectral investigations and VLBI measurements of the source size have shown that the radio emission must originate from more energetic electrons than the X-rays. It is probably non-thermal. In contrast, the quiet and slowly varying solar radio emissions are caused thermally and by the same plasma that emits soft X-rays. The different word 'quiescent' has been chosen for stars to indicate the difference from the solar non-flare radio emissions and to emphasize the possibility that the observed low-level variability (on time scales longer than ten minutes) may, in fact be a superposition of many flare-like events. The typical quiescent stellar radiations have a brightness temperature of some 10^9 K and are often weakly polarized. This is compatible with an interpretation by gyrosynchrotron emission of a population of energetic ($\gtrsim$ 100 keV) electrons spiraling in the coronal magnetic field. The presence of such particles is ubiquitous during solar flares (Section 1.2.3), but does not noticeably contribute to the quiet solar radio emission. The gyrosynchrotron emission, however, permanently dominates the microwave emission of dMe stars, surpassing the thermal solar radio luminosity by orders of magnitude.

Figure 1.8 is the gist of a decade of painstaking measurements by many observers and instruments. The quiescent soft X-ray and microwave emissions are compared not only for different stars and types of stars, but also with the temporary sources of solar flares. The correlation between thermal emission and radiation of non-thermal particles is obvious and stretches over many orders of magnitude. It tells us that the continuous presence of accelerated particles and coronal heating have something in common. Figure 1.8 supports the conjecture that a common process in the stellar interior drives the coronal heating and flares. Moreover, the similarity of flares and coronae suggests a more direct relation between the thermal constituent of coronal plasmas and powerful, possibly violent acceleration processes in the corona. The heating of the coronae of rapidly rotating stars and flares both accelerate electrons and may be similar processes!

Are coronal plasmas heated by flares? The answer is not simple, as it has become clear in Section 1.2 that there are different types of flare-like, dynamic processes even in a relatively inactive corona like the Sun's. There is substantial variability in the X-ray and radio emissions of most stellar coronae over a variety of time scales. The observed variability is in the form of individual flares and slow variations of the quiescent background. Flares show a wide range of amplitude (up to a factor of 10), whereas the variations of the quiescent component on time scales of hours do not exceed amplitudes of 50%. Continuous low-amplitude, short-period

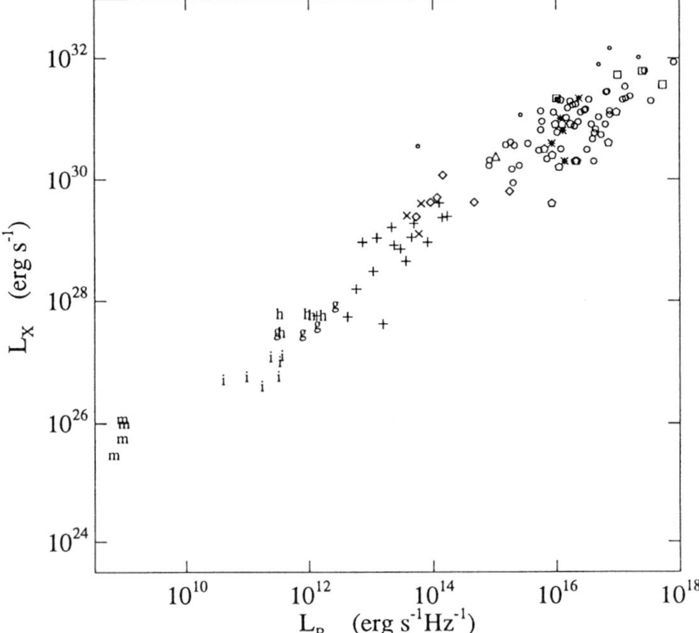

Fig. 1.8. The soft X-ray luminosity is plotted vs. the radio luminosity at a frequency of 5–10 GHz. Different types of stars are compared with the peak flux of solar flares. Key to the symbols: *m* solar microflare; *i* intermediate impulsive solar flare; *h* gradual solar flare with dominating large impulsive phase; *g* pure gradual solar flare; + dM(e) stars; × dK(e) stars; ◇ BY Dra binaries; ◯ RS CVn binaries; ○ RS CVn binaries with two giants; △ AB Dor; ∗ Algols; ☐ FK Com stars; (pentagon) post-T Tau stars (from Benz and Güdel, 1993).

variability (microflares) has been proposed for the heating of stellar coronae, but the observational evidence is still meager.

In conclusion, observations – both solar and stellar – strongly suggest that it is not possible to comprehend coronae without understanding their dynamic phenomena. Apart from the heating problem, coronal processes offer numerous and exciting challenges for plasma astrophysics in general. Some of them will be introduced as applications throughout the book.

1.4. Fundamental Equations

When a gas like a corona is heated beyond its atomic binding energy, collisions strip off electrons from the atoms. Freely moving electrons and ions eventually recombine, but may become ionized again. The two rates cancel in equilibrium. This fourth state of matter behaves peculiarly in the presence of electric and magnetic fields. We shall call an ionized gas a *plasma*, if the free charges are abundant enough to influence the relevant properties of the medium. The equilibrium fraction of neutral atoms is given by the Saha-Boltzmann ionization equation. As an example, the ratio of neutral to ionized atoms is $5 \cdot 10^{-17}$ for a hydrogen plasma with a density of 10^8 cm^{-3} at a temperature of 10^6 K. These values are typical of the solar and some stellar coronae. We shall neglect the influence of the neutrals in the following, and thus consider only fully ionized plasmas.

Let us consider the motion of a particle i with charge q_i, mass m_i, velocity $\mathbf{v}_i$, and Lorentz-factor γ_i. In electrodynamics the particle is influenced by the Coulomb force due to the electric field $\mathbf{E}$ and the Lorentz force due to the magnetic field $\mathbf{B}$ (sometimes also called magnetic flux density or magnetic induction). The equation of particle motion is Newton's second law, which equates the temporal derivative of the momentum with the forces acting on the particle. This basic equation provides a first connection between particles and fields,

$$\frac{\partial}{\partial t}(m_i \gamma_i \mathbf{v}_i) = q_i(\mathbf{E} + \frac{1}{c}\mathbf{v_i} \times \mathbf{B}) \quad . \tag{1.4.1}$$

We are using Gaussian (cgs) units. A conversion table to mks units can be found in Appendix B. Vector variables are signified by bold face or underbars. Tensor variables are, in the following, indicated by hats. The product of two vectors is defined in Appendix A in three ways: scalar product (indicated by $\cdot$), vector product (indicated by $\times$), and tensor product (indicated by $\circ$). We consider the particles of a plasma to move in a vacuum, and therefore the dielectric tensor equals the unity tensor ($\hat{\epsilon} = \hat{1}$), and similarly the magnetic permeability $\hat{\mu} = \hat{1}$. The electromagnetic fields are given by Maxwell's equations, which then reduce to:

$$\nabla \times \mathbf{B} = \frac{4\pi}{c}\mathbf{J} + \frac{1}{c}\frac{\partial \mathbf{E}}{\partial t} \tag{1.4.2}$$

$$\nabla \times \mathbf{E} = -\frac{1}{c}\frac{\partial \mathbf{B}}{\partial t} \tag{1.4.3}$$

$$\nabla \cdot \mathbf{B} = 0 \tag{1.4.4}$$

$$\nabla \cdot \mathbf{E} = 4\pi\rho^* \quad . \tag{1.4.5}$$

∇ is the operator $(\partial/\partial x, \partial/\partial y, \partial/\partial z)$, which can be treated like a vector. We shall refer to it as the *nabla* operator, after the Phoenician stringed instrument having the same shape. ∇^2 is the scalar product of two such operators, $\partial^2/\partial^2 x + \partial^2/\partial^2 y + \partial^2/\partial^2 z$; it is generally termed the Laplace operator. $\nabla\times$ is the *curl* (or rotation), and $\nabla\cdot$ is the divergence (cf. definitions A.4 and A.5 in Appendix A). The electric charge density ρ^* and the current density $\mathbf{J}$ are defined by the sums over all particles i in an elementary volume $\Delta\mathcal{V}$,

$$\rho^* := \frac{1}{\Delta\mathcal{V}} \sum_{i,\Delta\mathcal{V}} q_i \quad , \tag{1.4.6}$$

$$\mathbf{J} := \frac{1}{\Delta\mathcal{V}} \sum_{i,\Delta\mathcal{V}} q_i \mathbf{v}_i \quad , \tag{1.4.7}$$

providing a second link between particles and fields.

Equation (1.4.2) is Ampère's law complemented by the displacement current. It states that either electric currents or time-varying electric fields may produce magnetic fields and *vice versa*. Equation (1.4.3) is credited to Faraday and Equation (1.4.5) to Poisson, implying the origin of electric fields by time-varying magnetic fields and electric charges, respectively. Equation (1.4.4) expresses the absence of magnetic monopoles.

Since each particle obeys Equation (1.4.1), the system of Equations 1.4.1) – (1.4.7) is complete. It can, in principle, be solved if the particle masses, charges, initial positions, and velocities are known. The charged particle motions create the electromagnetic fields, which in turn influence the motion of the particles. A coupled system of this kind is obviously highly non-linear.

The large number of equations of the kind of (1.4.1) inhibits the solution of the system (1.4.1) – (1.4.7) for real cosmic plasmas, even using the fastest computer available. One has the choice of reducing the particles to a tractable number, or of using appropriate statistical methods. Numerical simulations generally overcome the non-linearities. Analytical and statistical calculations render a better overview and understanding of the physics.

1.4.1. Magnetohydrodynamic approach

There are two major statistical approaches to plasma physics. One, generally termed *magnetohydrodynamics (MHD)*, assumes that the plasma can be sufficiently described by a fluid, i.e. by velocity averaged parameters such as density n, mean velocity $\mathbf{V}$, temperature T, etc.,where

$$n := \int f(\mathbf{v}) d^3v \quad , \tag{1.4.8}$$

$$\mathbf{V} := \frac{1}{n} \int \mathbf{v} f(\mathbf{v}) d^3v \quad , \tag{1.4.9}$$

$$T := \frac{m}{3k_B n} \int (\mathbf{v} - \mathbf{V})^2 f(\mathbf{v}) d^3v \ , \tag{1.4.10}$$

and $f(\mathbf{v})$ is the velocity distribution given by the mean number of particles per volume element in space and velocity. Equation (1.4.8) can be considered as the normalization of f. In Equations (1.4.8) – (1.4.10) we have simplified to the non-relativistic case. Conservation of particle number, total momentum and energy, combined with Maxwell's equations and an equation of state – after some simplifying approximations – form a complete system of a more humanly manageable number of equations (to be discussed in Chapter 3.1).

Gradual plasma processes in coronae do not require a description at the detail of the particle level. MHD is widely used to study equilibrium conditions and the onset of global instabilities. Ideal MHD does not include electric fields parallel to the magnetic field, and neglects populations of energetic particles cospatial with the thermal background of rapidly colliding particles. We shall find MHD extremely useful if the velocity distributions are close to Maxwellian. This is certainly the case in thermal equilibrium and small deviations from it. Collisions (and other randomizing processes) tend to bring any distribution toward a Maxwellian shape, the maximum entropy distribution.

MHD becomes questionable for processes which are faster than the collision time. Such a period may seem too short to be relevant. However, in certain plasma phenomena *collisionless* processes (occurring much faster than the time between collisions) have turned out to control the situation. Let us illustrate by three examples how this can happen. (*i*) High-frequency waves can grow within a fraction of the collision time, absorb a considerable fraction of the total energy, and change drastically the plasma resistivity. (*ii*) Electric fields, both stationary or in waves, can rapidly accelerate charged particles so that their velocity distribution deviates considerably from a Maxwellian form. (*iii*) The collision time of energetic particles increases with the third power of velocity (Section 2.6). High energy particles thus have long collision times. They may spread over a large volume and form a population of particles that can be considered 'collisionless' over time scales of minutes to hours.

1.4.2. Kinetic approach

The other statistical approach, *kinetic plasma physics*, starts at a much lower level, considering particle orbits and the properties of $f(\mathbf{v})$ without integration in velocity. The basic equation, named after L. Boltzmann, will be derived from particle conservation in Chapter 3,

$$\frac{\partial f}{\partial t} + \mathbf{v}\frac{\partial f}{\partial \mathbf{x}} + \frac{q}{m}(\mathbf{E} + \frac{1}{c}\mathbf{v} \times \mathbf{B})\frac{\partial f}{\partial \mathbf{v}} = \left(\frac{\partial f}{\partial t}\right)_{\rm coll} \quad . \tag{1.4.11}$$

The electromagnetic fields have been split into macroscopic components and microscopic components. The macroscopic part is denoted $\mathbf{E}$ and $\mathbf{B}$, the spatial averages over a volume much larger than the Debye shielding length (Chapter 2.4). The rapidly fluctuating fields of the neighboring particles are taken care of by the *collision term* on the right hand side. We do not need to apply quantum mechanics as long as the interparticle distance exceeds the de Broglie wavelength of the particles (but note quantum effects in close collisions, Eq. 2.6.17) and the photon energies are smaller than the rest mass of the particles. For a more complete introduction and discussion of this fundamental equation the reader is referred to Section 3.1.1.

The full scope of kinetic plasma physics of the universe cannot be estimated yet. In the realm of stars, it is mainly the hot, dilute coronae where collisionless plasma processes occur. As an example, let us look at the collision time of thermal electrons as a function of height in the solar atmosphere (Figure 1.9). It increases more than six orders of magnitude from the bottom of the chromosphere to the corona. In the chromosphere, collisions of a charged test particle are mostly with neutral atoms. The collision time increases with height as the material becomes less dense, but decreases again as the temperature (and mean velocity of the electrons) increases. It has a narrow minimum at the bottom of the transition region where the dominant collision partners change from neutrals to charged particles, and it becomes orders of magnitude larger in the hot corona.

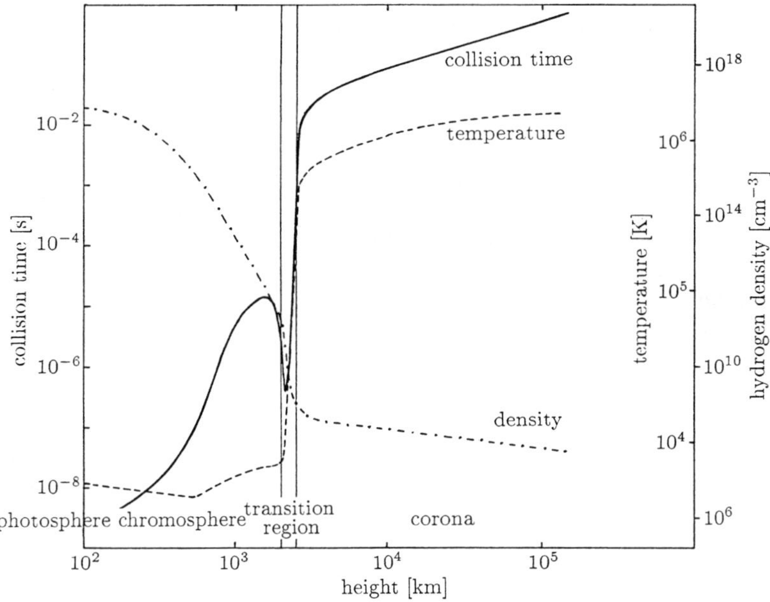

Fig. 1.9. Schematic drawing of the collision time vs. height of an electron moving at the mean thermal velocity in the quiet solar atmosphere. The relevant plasma parameters are from standard models and are also shown (dashed). The hydrogen density includes both neutral atoms and ions.

Other fields for applying kinetic plasma physics include magnetospheres of planets and neutron stars, extended atmospheres of stars in formation, accretion discs, supernova remnants, and cosmic rays. Even some massive (early-type) stars show signs of energetic particles and magnetic activity.

Plasma physics at the particle level may be required in very localized regions during times of rapid change. Such processes, however, may be at the heart of some of the classical, unsolved coronal phenomena. Let us consider ten key examples of kinetic plasma processes in coronae:

- High energy particles are widely observed in the universe; *particle acceleration* has various origins. The most basic mechanism is simply a quasi-static electric field. In turn, accelerated particles drive collisionless plasma waves to grow unstably and may alter the acceleration process (Section 4.6).

- Interactions of plasma waves with particles in resonance cause various phenomena where energy is exchanged between waves and particles (Chapter 5). The result is either *heating* (more general: energization) of certain particle species or unstable growth of some other wave modes.
- Wave-particle interactions cause *anomalous (i.e. additional) resistivity* in an electric current. It can rapidly release the free energy of a strong current or, in other words, of its associated magnetic field (Chapter 9).
- Wave-particle interactions also increase viscosity and generally change the *transport coefficients* of a plasma. They have the effect of additional collisions.
- *Collisionless shock waves* can also accelerate particles and heat. They are sites of enhanced turbulence of various wave modes (Chapter 10).
- *Particle beams* are frequently observed in the solar corona and in interplanetary space cospatial with the background plasma. Beams can be signatures of some cosmic accelerator, or can form simply by evaporation from an impulsively heated source (Section 2.3).
- Beams are often unstable to *growing plasma waves* in the background plasma. The wave turbulence thrives on the beam energy and accompanies the beam until exhaustion (Chapters 6 and 7).
- Energetic particles having long collision time can be *trapped in magnetic mirror* configurations that are abundant in the form of coronal loops or dipole fields. A large number of fast particles accumulate near the acceleration region (Chapter 8). The velocity distribution of magnetically trapped particles, having the form of a *loss-cone*, gives rise to several instabilities of growing collisionless waves. They may control the trapping time (Section 8.3).
- Some kinetic processes cause *observable electromagnetic radiation.* Plasma wave turbulence of various origins can emit electromagnetic radiation by wave-wave coupling (Section 6.3). Such emissions yield information on plasma processes in cosmic sources. Another example is one of the loss-cone instabilities driving electromagnetic waves. This very efficient emission corresponds to a *maser* process, whereby electrons change gyration orbits (Section 8.2).
- Large amplitude waves develop *non-linear structures*, such as solitons. Their energy density is a considerable fraction of the thermal background. Furthermore, they can be powerful emitters of coherent radiation.

One sometimes notes in plasma astrophysics, and particularly in coronal physics, a gap between observation and theory. This disparity, which hinders progress in the field, exists for several reasons. (*i*) Plasma phenomena – in particular kinetic processes – tend to have many parameters, and they are often not observable. Therefore, theories are difficult to test by observations. (*ii*) Scientists in the field separate into observers and theoreticians early in their career and so lose the capability to combine the two tools. This book intends to bridge this gap.

Further Reading and References

Introductory texts into solar physics

Bray, R.J., Cram, L.E., Durrant, C.J., and Loughead, R.E.: 1991, *Plasma Loops in the Solar Corona*, Cambridge Astrophysics Series, Vol. **18**, Cambridge University Press.

Durrant, C.J.: 1988, *The Atmosphere of the Sun*, Hilger, Bristol UK.

Stix, M.: 1989, *The Sun*, Springer, Berlin.

Tandberg-Hanssen, E. and Emslie, A.G.: 1988, *The Physics of Solar Flares*, Cambridge University Press, Cambridge UK.

Zirin, H.: 1987, *Astrophysics of the Sun*, Cambridge University Press, Cambridge UK.

Reviews and articles on stellar coronae

Güdel, M.: 1994, 'Non-Flaring Stellar Radio Emissions', *Astrophys. J. Suppl.* **90**, 743.

Linsky, J.L.: 1985, 'Non-radiative Activity Across the HR-diagram: Which Types of Stars are Solar-like?', *Solar Phys.* **100**, 333.

Greiner, J., Duerbeck, H.W., and Gershberg, R.E. (eds.): 1995, 'Flares and Flashes', IAU Coll. 151, *Lecture Notes in Physics* **454**.

Haisch, B., Strong, K.T., and Rodonó, M.: 1991, 'Flares on the Sun and Other Stars', *Annual Rev. Astron. Astrophys.* **29**, 275.

Narain, U. and Ulmschneider, P.: 1991, 'Chromospheric and Coronal Heating Mechanisms', *Space Sci. Rev.* **57**, 199.

Schmitt, J.H.M.M.: 1990, 'X-ray Astronomy', *Adv. Space Res.* **10**, 115.

References

Benz, A.O. and Güdel, M.: 1993, 'Radio and X-Ray Observations of Flares and Coronae', *Astron. Astrophys.* **285**, 621.

Benz, A.O. and Krucker, S.: 2002, 'Energy Distribution of Micro-events in the Quiet Solar Corona', *Astrophys. J.*, in press.

Brueckner, G. and Bartoe, J.: 1983, 'Observations of High-Energy Jets in the Corona above the Quiet Sun, the Heating of the Corona, and the Acceleration of the Solar Wind', *Astrophys. J.* **272**, 329.

Catura, R.C., Acton, L.W., and Johnson, H.M.: 1975, 'Evidence for X-Ray Emission from Capella', *Astrophys. J.* **196**, L47.

Doyle, J.G. and Butler, C.J.: 1985, 'Ultraviolet Radiation from Stellar Flares and the Coronal X-Ray Emission for Dwarf-M_e Stars', *Nature* **313**, 378.

Gary, D.E., Zirin, H., and Wang, H.: 1990, 'Microwave Structure of the Quiet Sun at 8.5 GHz', *Astrophys. J.* **355**, 321.

Hansen, R.T., Garia, C.J., Grognard, R.J.-M., Sheridan, K.V.: 1971, 'A Coronal Disturbance Observed Simultaneously with a White-Light Coronameter and the 80 MHz Culgoora Radioheliograph', *Proc. Astron. Soc. Australia* **2**, 57.

Hoyng, P., Marsh, K.A., Zirin, H., and Dennis, B.R.: 1983, 'Microwave and Hard X-Ray Imaging of a Solar Flare on 1980 November 5', *Astrophys. J.* **268**, 879.

Krucker, S., Benz, A.O., Acton, L.W., and Bastian, T.S.: 1997, 'X-ray Network Flares of the Quiet Sun', *Astrophys. J.* **488**, 499.

Kundu, M.R., Rao, A.P., Erskine, F.T., and Bregman, J.D.: 1979, 'High Resolution Observations of the Quiet Sun at 6 Centimeters Using the Westerbork Synthesis Radio Telescope', *Astrophys. J.* **234**, 1122.

Mewe, R. *et al.*: 1975, 'Detection of X-Ray Emission from Stellar Coronae with ANS' *Astrophys. J.* **202**, L67.

Pallavicini, R., Golub, L. Rosner, R., Vaiana, G.S. Ayres, T., and Linsky, J.L.: 1981, 'Relation between Rotation and Luminosity', *Astrophys. J.* **248**, 279.

Vaiana, G.S. *et al.*:1981, 'Results from an Extensive Einstein Stellar Survey', *Astrophys. J.* **245**, 163.

Withbroe, G.L. and Noyes, R.W.: 1977, 'Mass and Energy Flow in the Solar Chromosphere and Corona', *Ann. Rev. Astron. Astrophys.* **15**, 363.

CHAPTER 2

BASIC CONCEPTS

With the ever more detailed and profound observations of the universe, we have become exposed to a vast field of coronal phenomena in the formation, activity, and late phases of stars. However, coronal physics poses some unexpected, recondite difficulties to the beginner and to the uninitiated astronomer in particular. The fundamental principles are simple: Maxwell's equation of electrodynamics and Boltzmann's equation of statistical mechanics (Chapter 1). On the other hand, the complexity of plasma phenomena is bewildering. The pioneers of plasma physics have had similar experiences; they have subdued some of the problems with elegant approximations, clever tricks, and deep insights. Their ideas have been checked in the laboratory, the solar wind and the magnetosphere, and the results have initiated new developments. The beginner should try to understand the fundamental concepts and the range of their applicability, even if they later seem to be outshone by brilliant mathematics or covered up by cumbersome algebra.

2.1. Single Particle Orbit

It will be shown in Section 2.6 that the collision time increases with particle velocity. Therefore, the assumption of a single particle moving in a vacuum, permeated by magnetic and electric fields, is particularly relevant for energetic particles, which can reach large distances before being deflected by collisions with the thermal background particles. In other words, they feel primarily the large scale electric and magnetic fields created by background particles, but have little stochastic interactions with individual particles.

The orbit of just one single particle in time-varying, spatially inhomogeneous electromagnetic fields is already a problem of considerable complexity. In this section we only consider the more elementary principles.

2.1.1. Homogeneous magnetic field

Magnetic fields are ubiquitous in stellar coronae (Figure 1.2). The fields are related to currents driven by sub-photospheric motions. Hindered by the high conductivity of a plasma atmosphere, coronal magnetic fields can change only slowly. The field lines guide the motion of charged particles and have an important physical meaning.

Let us start with a particle moving in a stationary, homogeneous magnetic field. We assume that no other charges and fields interact with our particle. The equation of motion (1.4.1) then reduces to

$$\frac{\partial(m\gamma\mathbf{v})}{\partial t} = \frac{q}{c}(\mathbf{v}\times\mathbf{B}) \quad . \tag{2.1.1}$$

As the Lorentz force is perpendicular to the velocity, the motion can be regarded as a superposition of a circular orbit perpendicular to the magnetic field and a straight, inertial movement parallel to the field. The circular motion is called *gyration*. The gyroradius R follows immediately from Equation (2.1.1),

$$R = \frac{m\gamma c}{|q|B}\, v_\perp \quad , \tag{2.1.2}$$

which we define more generally as a vector,

$$\mathbf{R} := -\frac{mc\gamma}{qB^2}(\mathbf{v}\times\mathbf{B}) \quad . \tag{2.1.3}$$

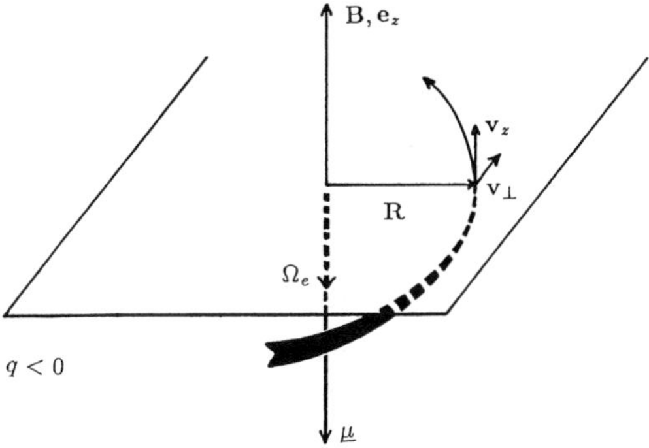

Fig. 2.1. The orbit of an electron in a homogeneous magnetic field.

The subscripts $\perp$ and z denote components perpendicular and parallel to the magnetic field, respectively. Throughout the book the local coordinates are chosen such that the z-axis is parallel to $\mathbf{B}$.

The gyration resulting from Equation (2.1.1) is counterclockwise for a positive charge when viewed along $\mathbf{B}$, and clockwise for an electron. As a consequence, the radius vector $\mathbf{R}$ always points from the center of the circular motion to the particle, independently of the sign of q. Furthermore, the gyrofrequency is defined in vector notation as

$$\mathbf{\Omega} := \frac{q}{m\gamma c}\mathbf{B} \quad . \tag{2.1.4}$$

In this chapter we limit ourselves to non-relativistic phenomena, and we shall put $\gamma = 1$ in the following. We note that the gyrofrequency (Eq. 2.1.4) then becomes independent of velocity and equal for all particles of a species. The gyrofrequencies of the various species are characteristics of a plasma and will play an important role in the theory of waves (Chapter 4). In Equations (2.1.3) and (2.1.4) vectors have been defined to simplify the mathematics later. They are drawn in Figure 2.1, which depicts the spiraling orbit of an electron, the combination of parallel and circular motion.

No work is done by the Lorentz force, as it is perpendicular to the orbit. An important result of the gyration is the electric ring current due to the circular motion of the particle. It is charge per time, thus

$$< I >= \frac{|q\Omega|}{2\pi} \quad . \tag{2.1.5}$$

This ring current is the cause of a magnetic moment of a particle defined by

$$\underline{\mu} := \frac{q}{2c}(\mathbf{R} \times \mathbf{v}_\perp) \quad . \tag{2.1.6}$$

Its absolute value is

$$\mu = \frac{\pi R^2}{c} < I > = \frac{\frac{1}{2}mv_\perp^2}{B} \quad . \tag{2.1.7}$$

The magnetic moment induces a secondary magnetic field, $\mathbf{B}^{\rm ind}$, which has the form of a dipole in the far field (at distances $r \gg R$). For example, in the axis of the particle orbit, the induced field has the value $\mathbf{B}^{\rm ind} = 2\underline{\mu}r^{-3}$. There are two important points to note in Equation (2.1.6):

- The magnetic moments of positive and negative charges are pointing in the same direction. The induced magnetic fields of all particles add constructively.
- The directions of $\mathbf{B}^{\rm ind}$ and $\mathbf{B}$ are antiparallel. The induced field counteracts the primary field and reduces the total field in the plasma. This diamagnetism of a plasma has far-reaching consequences (see example below).

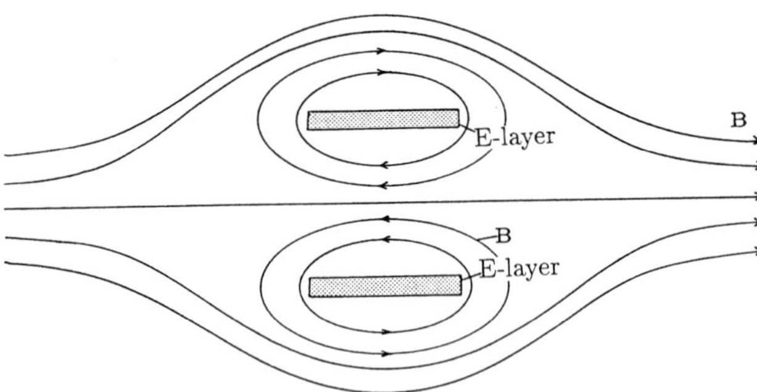

Fig. 2.2. Monoenergetic electrons accelerated in a magnetic field form a cylindrical shell (called E-layer, shaded) and deform the field.

The Astron machine in fusion research may serve as an illustration (Fig. 2.2). Energetic electrons are injected into initially parallel magnetic field lines. These electrons circle in a cylindrical shell, termed the E-layer. Their diamagnetism reduces the field inside the E-layer and adds to the field outside. Thus the magnetic field forms a 'magnetic bottle', shown in Section 2.2, which is suitable for confinement. The Astron configuration makes use of the diamagnetism of the injected electrons to confine ions. Heating or acceleration of particles in coronal loops may lead to similar effects.

2.1.2. Inhomogeneous Magnetic Field

We now look at a particle in a converging magnetic field as sketched in Figure 2.3. Similar field geometries may exist near the footpoints of coronal loops or in the polar regions of stars and planets. The magnetic field at the particle's instantaneous position can be decomposed into a component, $\mathbf{B}_z$, parallel to the field at the center of the gyration, and a perpendicular component, $\mathbf{B}_r$. The Lorentz force on the particle can also be written in two components:

$$\mathbf{F}_{L,r} = \frac{q}{c}(\mathbf{v}_\perp \times \mathbf{B}_z) \quad , \tag{2.1.8}$$

$$\mathbf{F}_{L,z} = \frac{q}{c}(\mathbf{v}_\perp \times \mathbf{B}_r) \quad . \tag{2.1.9}$$

The first, Equation (2.1.8), is the radial force component analogous to the homogeneous case. Equation (2.1.9) describes a force along the axis of the spiraling particle motion. It can be evaluated from $\nabla \cdot \mathbf{B} = 0$ (Eq. 1.4.4), which in cylindrical coordinates becomes:

$$\frac{1}{r}\frac{\partial}{\partial r}(rB_r) + \frac{\partial B_z}{\partial z} = 0 \quad . \tag{2.1.10}$$

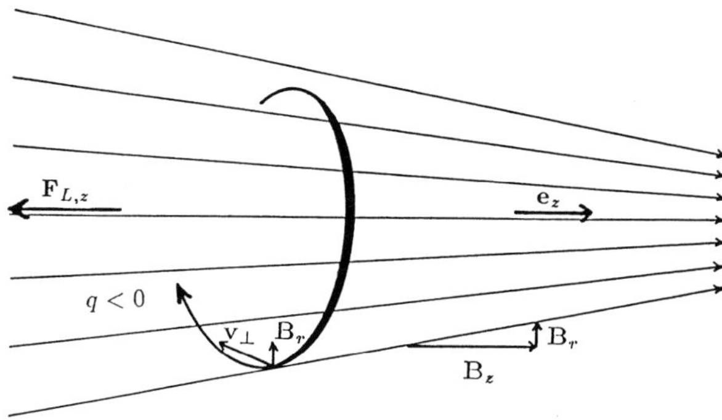

Fig. 2.3. Particle motion in a converging magnetic field produces a Lorentz force $\mathbf{F}_{L,z}$ in the opposite direction.

Let us consider the case where r is much smaller than the scale length of the derivative of B_z ($r \ll |B'_z/(\partial B'_z/\partial r)|$, where $B'_z := \partial B_z/\partial z$). In such a mildly inhomogeneous field, Equation (2.1.10) can be integrated approximately to

$$B_r \approx -\frac{1}{2} r \frac{\partial B_z}{\partial z} \quad . \tag{2.1.11}$$

Inserting in the gyroradius (Eq. 2.1.2) for r, Equation (2.1.9) yields

$$F_{L,z} = -\mu \frac{\partial B_z}{\partial z} \quad . \tag{2.1.12}$$

For example, $\partial B_z/\partial z > 0$ (a converging field as in Fig. 2.3) produces an F_z in the negative z-direction. It slows down the parallel velocity of a particle moving along the converging field and can reflect the particle into the opposite direction. In Section 2.2 we shall take a first look at the possibility of trapping particles by such 'magnetic mirrors'.

2.1.3. Conservation of the Magnetic Moment

We shall show in this section that in a mildly inhomogeneous, stationary magnetic field the magnetic moment of a particle is a constant of motion. It is generally sufficient to require that $|(\mathbf{R} \cdot \nabla)\mathbf{B}| \ll |\mathbf{B}|$ and $B_r \ll B_z$. We neglect collisions, electric fields and other forces. With $F_z = m\dot{v}_z$ we derive from Equation (2.1.12) the rate of change of the parallel energy component,

$$m v_z \dot{v}_z = -\mu \frac{\partial B}{\partial z} \frac{dz}{dt} = -\mu \frac{dB}{dt} \quad . \tag{2.1.13}$$

The forces are evidently perpendicular to the velocity components (Eqs. 2.1.8 and 2.1.9); thus the particle energy is conserved,

$$\frac{d}{dt}\left(\frac{1}{2} m v_z^2 + \frac{1}{2} m v_\perp^2\right) = 0 \quad . \tag{2.1.14}$$

Inserting (2.1.7) and (2.1.13), one obtains

$$-\mu \frac{dB}{dt} + \frac{d}{dt}(\mu B) = 0 \quad . \tag{2.1.15}$$

Since $B \neq 0$ by assumption, it immediately follows that

$$\frac{d\mu}{dt} = 0 \quad . \tag{2.1.16}$$

It is also instructive to consider the conservation of the magnetic moment from a different point of view. Since the transverse motion is periodic, an action integral $\mathcal{J} = \int p_\perp ds_\perp$ can be defined, where $p_\perp$ is the transverse momentum and $ds_\perp$ is the perpendicular component of the gyration path. It is easy to show that $p_\perp$ and $s_\perp$ satisfy the conditions of canonical coordinates. The integration is over a complete cycle of $s_\perp$, i.e. over one gyration. The action integral for the transverse part of the motion is

$$\mathcal{J} = \frac{4\pi mc}{q}\mu \quad . \tag{2.1.17}$$

Theoretical mechanics proves that, if a system changes slowly compared to the period of motion (and is not in phase with the period), an action integral remains constant. Such a variation of a system is termed *adiabatic*, and action integrals become *adiabatic invariants*.

In our case the system consists of the magnetic field and the particle. If the particle moves through a region where the magnetic field varies slowly, Equation (2.1.17) confirms that the magnetic moment is conserved. In addition to spatial variations, the magnetic moment is also conserved during temporal changes of magnetic field strength. This property can accelerate particles: Consider an increase of B, say by compression, at a rate well below the gyrofrequency. The perpendicular part of the particle energy must then increase in proportion to B according to the definition of μ (Eq. 2.1.6). Such an energy increase is generally referred to as *betatron acceleration* and will be applied in Section 10.3.3.

The magnetic moment μ is the 'most invariant' action integral of charged particles. The obvious analogue to Equation (2.1.17) for periodic motion in a magnetic mirror is the *longitudinal invariant*,

$$\mathcal{L} := p_z L \quad , \tag{2.1.18}$$

where L is the mirror length. It may be used in connection with Fermi acceleration (Section 10.3.1).

2.1.4. Particle drifts

Our next step in understanding single particle orbits is to include some other stationary force, $\mathbf{F}$, in addition to the Lorentz force of the magnetic field. As its parallel component to the magnetic field, $\mathbf{F}_z$, accelerates the particles into a direction where they are freely moving, which is trivial, we concentrate on the effect of $\mathbf{F}_\perp$. In the non-relativistic limit, the perpendicular component of Equation (2.1.1) is

$$\frac{d\mathbf{v}_\perp}{dt} = \frac{q}{mc}(\mathbf{v}_\perp \times \mathbf{B}) + \mathbf{F}_\perp/m \quad . \tag{2.1.19}$$

This is a linear, inhomogeneous, first-order differential equation in $v_\perp$. The theory of such equations prescribes that its general solution is simply the superposition of one particular solution, say $\mathbf{w}^d$, of the full equation and of the general solution, $\mathbf{u}$, of its homogeneous part (Eq. 2.1.19 without the $\mathbf{F}_\perp/m$ term). Thus

$$\mathbf{v}_\perp = \mathbf{w}^d + \mathbf{u} \quad . \tag{2.1.20}$$

Using the general vector identity (A.9),

$$\frac{(\mathbf{F} \times \mathbf{B}) \times \mathbf{B}}{B^2} = \frac{(\mathbf{F}_\perp \cdot \mathbf{B})\mathbf{B} - (\mathbf{B} \cdot \mathbf{B})\mathbf{F}_\perp}{B^2} = -\mathbf{F}_\perp \quad , \tag{2.1.21}$$

it can be shown that the form

$$\mathbf{w}^d = \frac{c}{q} \frac{\mathbf{F}_\perp \times \mathbf{B}}{B^2} \tag{2.1.22}$$

is a solution of Equation (2.1.19). The homogeneous equation

$$\frac{d\mathbf{u}}{dt} = \frac{q}{m\,c}(\mathbf{u} \times \mathbf{B}) \tag{2.1.23}$$

describes the action of the Lorentz force on the particle, but in the frame of reference moving with velocity $\mathbf{w}^d$. Equation (2.1.23) represents a motion without the additional force. Its solution, $\mathbf{u}$, is the circular velocity around what is called the *guiding center*. The velocity of the guiding center is simply

$$\mathbf{v}^{gc} = \mathbf{w}^d + \mathbf{v}_z \quad . \tag{2.1.24}$$

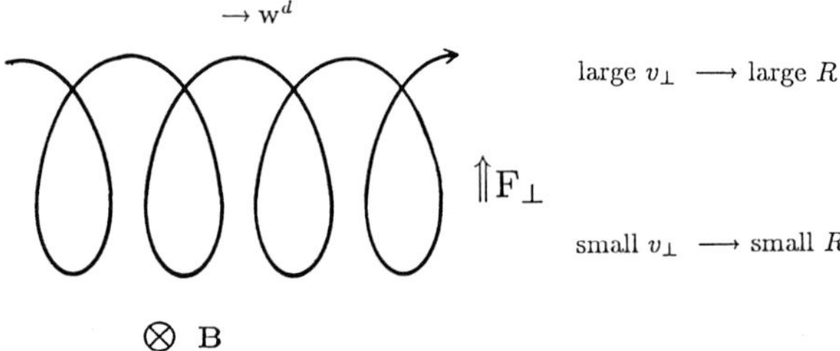

Fig. 2.4. A perpendicular force $\mathbf{F}_\perp$ on a gyrating particle causes a drift perpendicular to $\mathbf{F}_\perp$ and $\mathbf{B}$.

This tells us that the perpendicular force results in a *drift* $\mathbf{w}^d$ perpendicular to the magnetic field, in addition to the previously introduced gyration and parallel motions. The reason for this drift can be visualized in Figure 2.4. The force $\mathbf{F}_\perp$ accelerates the particle in the upward direction in Figure 2.4. It is therefore faster in the upper part of its orbit. The gyroradius increases with velocity (Eq. 2.1.2). The opposite occurs in the lower part of the figure. The effect of the variation of the gyroradius is a drift to the right (assuming a positive charge). An analogous effect, called *gradient drift*, occurs when the magnetic field strength increases in the downward direction of Figure 2.4 and reduces the gyroradius in the lower part of the orbit. In homogeneous magnetic fields and in the absence of other forces, collisionless particles are bound to a particular field line like pearls on a string. Drifts, however, can move the particles across field lines and spread them out from their original line to a larger volume. In the following, we discuss some key examples of particle drifts.

A. *Electric Field*

Let there be a stationary electric force

$$\mathbf{F}_E = q\mathbf{E} \quad . \tag{2.1.25}$$

The drift velocity resulting from Equation (2.1.22) is

$$\mathbf{w}_E = c\frac{\mathbf{E} \times \mathbf{B}}{B^2} \quad . \tag{2.1.26}$$

This is generally known as $E \times B$ drift. We note that $\mathbf{w}_E$ is independent of charge, mass, and velocity. As it is identical for all particles, the electric field can be 'transformed away' by a Lorentz transformation into a suitable frame of reference. This coordinate system moves with velocity $\mathbf{w}_E$, it is the rest frame of the $E \times B$ drifting plasma. Since $\mathbf{B}$ is perpendicular to $\mathbf{w}_E$, the magnetic field in the moving frame of reference is the same. The $E \times B$ drift can be viewed as the moving plasma carrying along its embedded magnetic field.

B. *Gravitational Field*

Let gravity be given by a gravitational acceleration $\mathbf{g}$ and

$$\mathbf{F}_g = m\mathbf{g} \quad . \tag{2.1.27}$$

According to Equation (2.1.22) it causes a drift

$$\mathbf{w}_g = \frac{mc}{q}\frac{\mathbf{g} \times \mathbf{B}}{B^2} \quad . \tag{2.1.28}$$

Note that ions and electrons drift in opposite directions. Gravitational drift produces a current! As the drift velocity, $\mathbf{w}_g$, is proportional to mass, the current consists mainly of perpendicularly moving ions. A well-known example is the ring current in the terrestrial ionosphere, where the observed ion drift velocity is of the order of 1 cm s^{-1}, in agreement with Equation (2.1.28).

C. *Curved Field Lines*

In a curved magnetic field a 'centrifugal force' may be defined, simulating the effect of particle inertia. The centrifugal drift becomes

$$\mathbf{w_c} = \frac{mc\ v_z^2}{qB^4}[\mathbf{B} \times (\mathbf{B} \cdot \nabla)\mathbf{B}] \tag{2.1.29}$$

(Exercise 2.1, below) and depends on the ratio of particle energy to charge. It has different signs for electrons and ions. The resulting current is perpendicular to the magnetic field and is directly related to the field curvature by Ampère's equation (1.4.2).

2.2. Particle Trapping in Magnetic Fields

A considerable fraction of the energy released in solar flares (estimates range from 0.1 to 50%) resides temporarily in electrons with energies far above thermal. Such particles could easily escape from the corona without collisions if they were not hindered by the magnetic field. In fact, only 0.1 to 1% of these electrons are later found in interplanetary space. This evidently reflects the predominance of loop-shaped magnetic field lines in the corona and, particularly, in active regions where flares generally occur. 'Open' field lines, which connect the active region to interplanetary space, seem to be rare or extremely well shielded from the site of acceleration. (Note that the widely used term 'open field line' is a misnomer, since in the absence of magnetic monopoles and in a finite universe all magnetic field lines must eventually close and return from space to the Sun. Nevertheless, such field lines can easily have a length exceeding the diameter of the Galaxy, and they are quantitatively different in this respect from 'closed' field lines, returning within the corona, or within about one solar radius: see Fig. 2.5)

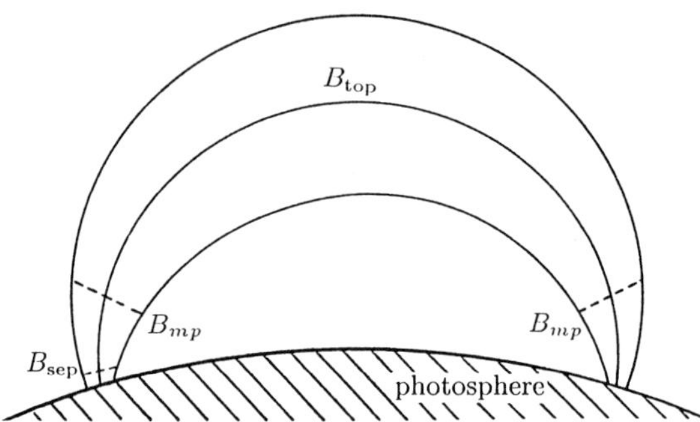

Fig. 2.5. Schematic drawing of 'closed' magnetic field lines forming loops between photospheric spots of opposite polarity.

Most energetic particles are apparently guided back to the Sun by 'closed' field lines. Some of them immediately penetrate denser plasma or even the chromosphere, where they rapidly lose their energy by collisions, emitting bremsstrahlung in hard X-rays. Others, however, remain in the corona up to several minutes, as some microwave emissions (around 3 GHz) and decimetric continuum bursts (0.3 – 3 GHz) indicate. The flare microwaves originate primarily from relativistic electrons (Section 1.2.3), and trapped electrons seem to play a role in long duration events. Decimetric continuum radiation is generally interpreted as coherent radiation of trapped electrons (Chapter 8).

Magnetic trapping can be understood as the result of conservation of the magnetic moment. The condition of smooth and only slowly varying magnetic field lines is easily satisfied in practically all we know about solar and stellar atmospheres. According to Equation (2.1.16), the magnetic moment

$$\mu = \frac{\frac{1}{2} m v_\perp^2}{B} = \text{const} \quad . \tag{2.2.1}$$

Assuming a stationary magnetic field, no work is done by the Lorentz force. Thus the particle energy is also conserved,

$$v^2 = v_\perp^2 + v_z^2 = \text{const} \quad . \tag{2.2.2}$$

Let us consider a system of looped magnetic field lines, typical of solar and stellar spots (or magnetic poles of opposite polarity), sketched in Figure 2.5. The field strength has a minimum near the top of the loop and is assumed to increase toward the photospheric footpoints. If a charged particle spirals in the direction of increasing field strength, it experiences an opposing component of the Lorentz force (given by Eq. 2.1.9), reducing v_z. This may continue until $v_z = 0$, when the particle changes the sign of its parallel velocity and is reflected. The magnetic field strength at the mirror point, B_{mp}, can readily be calculated from Equation (2.2.1),

$$B_{mp} := B_0 \left(\frac{v}{v_\perp^0} \right)^2 \quad . \tag{2.2.3}$$

The subscript 0 refers to a given point in the orbit, for instance to the top of the loop, where the particle has a perpendicular velocity component $v_\perp^0 = v_\perp^{\text{top}}$. Equation (2.2.2) requires the perpendicular velocity at the mirror point to equal the initial total velocity v.

Let B_{sep} be the magnetic field strength at the critical height below which particles are not mirrored but lost by collisions. The altitude of the separation between collisional and collisionless behavior is somewhere in the upper chromosphere or transition region. If $B_{mp} < B_{\text{sep}}$, the particle is reflected before entering the region of rapid collisions. Provided that this is also the case at the other footpoint (where the field strength may be different), the particle remains trapped in the corona. The coronal collision times are about five orders of magnitude longer than in the chromosphere (Figure 1.9, but note that super-thermal particles have a much longer collision time, as will be shown in Section 2.6). Then the particle bounces between the mirror points. The full bounce period τ in a parabolic field (like the far field of a dipole, $B = B_{\text{top}}(1 + s^2/H_B^2)$, where s is the path length measured from the top of the magnetic loop), is independent of the initial parallel velocity, since the larger v_z^{top} is, the farther away is the mirror point. One derives

$$\tau = \frac{2\pi H_B}{v_\perp^{\text{top}}} \quad . \tag{2.2.4}$$

The proof of Equation (2.2.4) is left to an exercise (2.2). The geometric parameter H_B is related to the loop length L (say from chromosphere to chromosphere) by

$$H_B \approx \frac{L/2}{\sqrt{\frac{B_{\text{sep}}}{B_{\text{top}}} - 1}} \quad . \tag{2.2.5}$$

$B_{\text{sep}}/B_{\text{top}}$ is called the *mirror ratio*. It has been estimated from soft X-ray observations to be in the range of 2 – 10 for solar coronal flare loops.

If $B_{mp} \geq B_{\text{sep}}$, the particle penetrates the chromosphere and is lost from the trap. Equation (2.2.3) then states that trapping or precipitation depends only on the ratio $v/v_\perp^{\text{top}}$, or the particle's pitch angle, α_{top}. The pitch angle is defined by the angle of the orbit to the magnetic field,

$$\alpha_0 := \arcsin(\frac{v_\perp^0}{v}) \quad . \tag{2.2.6}$$

The particle is trapped if the initial pitch angle is larger than the critical value given by $\arcsin(B_{\text{top}}/B_{\text{sep}})^{1/2}$. A critical pitch angle α_c can be determined from Equation (2.2.6). It is given at every point in the loop by the local magnetic field strength B_0 and amounts to

$$\alpha_c(B_0) = \arcsin\sqrt{\frac{B_0}{B_{\text{sep}}}} \quad . \tag{2.2.7}$$

If the pitch angle is below this value, the particle will get lost with high probability before reflection.

The velocity distribution of magnetically trapped particles has characteristic cones with half-angle α_c and axes in the positive and negative v_z-directions, where the number of particles is strongly reduced. They are known as *loss-cones*. Figure 2.6 shows a typical observation of protons trapped in the Earth's dipole field with a clearly developed loss-cone. The thermal, collisional background plasma forms a nearly isotropic distribution in the center. Velocity distributions with a loss-cone can be expected whenever particles are mirrored in a converging magnetic field within less than a collision time. Particles outside of the loss-cone are trapped if their mean free path exceeds the size of the magnetic configuration. This is well-known to occur in the tenuous plasmas of planetary magnetospheres and the solar and stellar coronae, but can also be expected in the atmospheres of white dwarfs and neutron stars, in galactic magnetic fields, etc.

Loss-cone features are more than identification tags of trapped particle populations. They are an important deviation from thermal equilibrium. Even if the rest of the velocity distribution is Maxwellian, loss-cones constitute free energy. In Chapter 8 it will become clear that this free energy can be tapped by plasma instabilities causing various types of observable emission.

2.3. Generation of Beams

The previous section introduced the imprint of the magnetic field's spatial structure on the velocity distribution. Here we consider temporal changes of the distribution and study how they propagate in space. In the absence of collisions and any other forces, the particle distribution at a place $\mathbf{x}+\Delta\mathbf{x}$ and time t is determined by the distribution at $\mathbf{x}$ and $t - \Delta t$, where $\mathbf{v} = \mathbf{\Delta x}/\Delta t$ is the particle velocity. Thus

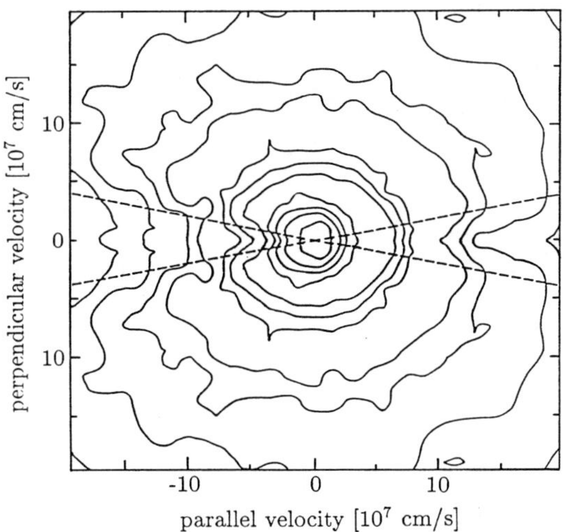

Fig. 2.6. Contour plot of proton velocity distribution observed by the VIKING satellite in the Earth's magnetosphere. The theoretical loss-cone angle given by Equation (2.2.7) is indicated (dashed). Contours are separated by a factor of 3.2 (5 dB) (after Boström, Koskinnen, and Holback, 1987).

$$f(\mathbf{x}+\mathbf{\Delta x},\mathbf{v},t) \;=\; f(\mathbf{x},\mathbf{v},t-\Delta t) \quad . \tag{2.3.1}$$

Let us use this simple model to outline the evolution of a local disturbance in velocity space. Example: Consider the rapid heating of a fraction of the particles to a temperature $T_h \gg T_c$, the temperature of the ambient medium. We take only one hot particle species for simplicity, which we assume to have a Maxwellian distribution. Since charged particles propagate along field lines (neglecting gyration and drifts), the problem can be reduced to one dimension (the z-direction). Let the number of hot particles at z_0 now increase exponentially with a time constant τ. Their distribution f at the heating site is a Maxwellian function multiplied by an exponential function,

$$f(z_0,v,t) \;=\; \frac{n_h}{\sqrt{2\pi}v_{th}}\exp\left(-v^2/2v_{th}^2\right)\cdot\exp\left(\frac{t}{\tau}\right) \quad . \tag{2.3.2}$$

With this expression, the mean thermal velocity in one velocity component and the temperature T_h are related by $v_{th} =: (k_B T_h/m)^{1/2}$ (k_B is the Boltzmann constant). The subscript t stands for thermal, h stands for the hot population. Using Equations (2.3.1) and (2.3.2) it is straightforward to derive the hot particle distribution at a site $z_0 + \Delta z$ outside the heating region from the distribution at z_0 shifted by $-\Delta t =-\Delta z/v$,

$$f(z_0 + \Delta z, v, t) \;=\; \frac{n_h e^{t/\tau}}{\sqrt{2\pi} v_{th}} \exp\left(-v^2/2v_{th}^2 - \frac{\Delta z}{v\tau}\right) \quad . \tag{2.3.3}$$

The velocity dependent part of Equation (2.3.3) – the exponential function – is plotted in Figure 2.7 for various normalized distances ξ, where

$$\xi \;:=\; \frac{\Delta z}{v_{th}\tau} \quad . \tag{2.3.4}$$

The curve $\xi = 0$ represents the initial distribution in the heating region given in Equation (2.3.2). At $\xi > 0$ a hump appears in velocity space. It has a maximum at

$$v_{\max} \;=\; \xi^{1/3} v_{th} \quad . \tag{2.3.5}$$

The evolving particle beam is simply an effect of spatial dispersion or, plainly, of fast particles arriving first. The average velocity increases with distance at the expense of beam density. The beam amplitude decreases, since the larger ξ, the earlier in the heating event the particles have originated. The gap at $v = 0$ for $\xi > 0$ expresses that the slowest particles have not yet arrived.

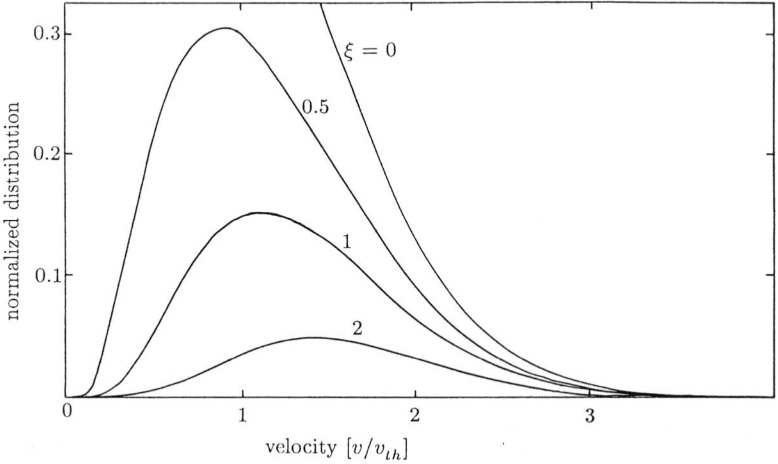

Fig. 2.7. Particle velocity distributions due to a local heating at $\xi = 0$ for various distances ξ from the heating region at any given time $t > 0$.

The simple model lacks many problems encountered in reality. Equation (2.3.1) obviously cannot describe a situation where forces act on the particles. Particle conservation in the presence of electromagnetic forces is expressed by the Boltzmann equation (1.4.11). It can be shown (Exercise 2.3) that Equation (2.3.1) follows from the Boltzmann equation under simplifying conditions. There are two major shortcomings of the simple model, which are briefly mentioned here and which will later be discussed in greater detail:

- Most importantly, an electric field builds up if the evaporating particles carry a charge. The electric field decelerates the particles of the beam and accelerates background electrons to form a *return current.* In the absence of friction (collisions) between background electrons and ions, the electric field is minute, and return current and beam current cancel each other. For large beam currents the frictional loss of the return current becomes an important energy drain of the beam.
- If the peak velocity, $v_{\max}$, of the beam exceeds about three times the thermal velocity of the ambient population, interactions of beam particles with Langmuir waves come into play (Section 5.2). Kinetic particle energy is transferred to oscillating electric fields and substantial energy is lost. Beam-wave interactions have received great interest since electromagnetic emission of the enhanced waves has been observed in solar radio bursts (Section 5.1).

We note that particle beams, as developed in this section, constitute an energy loss of the hot region not included in first-order heat transfer. First-order thermal conductivity is calculated from small deviations of a Maxwellian distribution in the presence of a temperature gradient. Collisionless particles can shorten the cooling time far below first-order calculations.

2.4. Debye Shielding

Up to now we have singled out a particle and calculated its orbit, neglecting interactions with other particles. In this and the following section we consider two fundamental *collective properties* of a plasma. They are both consequences of the presence of free charges of opposite sign.

How do the other particles react to the presence of a charge? Let us calculate the disturbance using classical statistics (see Fig. 2.8). In thermodynamic equilibrium the kinetic particle energy ε is distributed according to Maxwell (Eq. 2.3.2),

$$f_0(\varepsilon) = \frac{n_0}{k_B T}\exp(-\varepsilon/k_B T) \quad . \tag{2.4.1}$$

Now we introduce into the plasma a particle (referred to as the 'test particle') with charge Q creating a potential $\Phi(r)$. If another particle ('field particle') with mass m and charge q approaches the first particle, its total energy is conserved. It remains at the value ε_∞ it had at large distances,

$$\varepsilon + q\Phi(r) = \varepsilon_\infty = \text{const} \quad . \tag{2.4.2}$$

The energy distribution of the field particles also remains constant, but now kinetic and potential energy have to be summed. The energy distribution for field particles with charge q becomes

$$f(\varepsilon, r) = \frac{n_0}{k_B T}\exp[-(\varepsilon + q\Phi)/k_B T] \quad . \tag{2.4.3}$$

The density is given by the normalization

$$n(r) = \int_0^\infty f(\varepsilon, r) d\varepsilon = n_0 \exp(-q\Phi(r)/k_B T) \quad . \tag{2.4.4}$$

Equation (2.4.4) describes how the density of the undisturbed field particles, n_0, is modified by the potential of the test particle. Elementary electrostatics relates Φ to the electric charge density ρ^* by Poisson's equation,

$$-\nabla^2\Phi = \nabla \cdot \mathbf{E} = 4\pi\rho^* \quad . \tag{2.4.5}$$

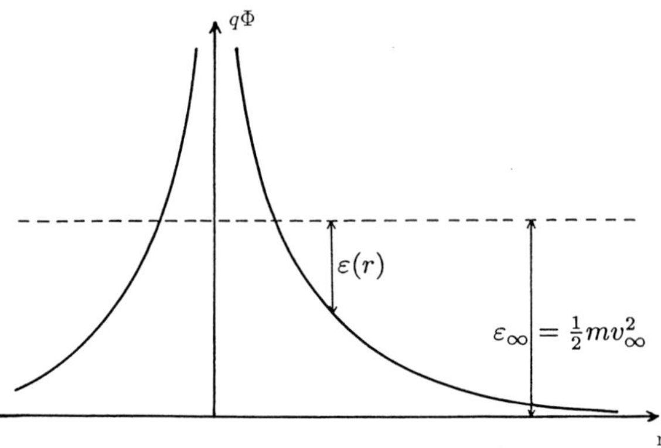

Fig. 2.8. A field particle with charge q experiences the potential Φ of the test particle and (assuming the same sign of the two charges) loses kinetic energy. The total energy of the field particle (dashed line) is conserved.

The charge density ρ^* has been defined in Equation (1.4.6) and is given by the sum over all species α,

$$\rho^* = \sum_\alpha q_\alpha n_\alpha \quad . \tag{2.4.6}$$

As an example we now look at a hydrogen plasma with equal electron and proton temperatures. The undisturbed electron density equals the proton density. Now we disturb this equilibrium with the test charge. Using Equation (2.4.4) and $\rho^* = -e\, n_e + e\, n_p$, the charge density becomes

$$\rho^* = en_0[\exp(-e\Phi/k_B T) - \exp(+e\Phi/k_B T)] \quad . \tag{2.4.7}$$

A Taylor expansion of the exponential for $e\Phi/k_B T \ll 1$ in the region of interest immediately yields for Equation (2.4.5)

$$\nabla^2\Phi = 2\frac{4\pi n_0 e^2}{k_B T}\Phi \quad . \tag{2.4.8}$$

It is easy to generalize Equation (2.4.8) for a plasma of composition different from hydrogen. $\sum_\alpha Z_\alpha^2 n_\alpha$ then takes the place of $2n_0$, where the sum is over all particle species α. Equation (2.4.8) is a second-order differential equation in Φ. It can be solved in spherical coordinates by trying the form $\Phi(r) = g(r)/r$, and yields

$$\Phi(r) = \frac{a}{r} \exp(-\sqrt{2}r/\lambda_D) \quad , \tag{2.4.9}$$

where $a \approx Q$ follows from the boundary condition on the test charge (assuming this charge to reside on a sphere with radius r_0 and $r_0^2 \ll \lambda_D^2$). In Equation (2.4.9) we have defined the 'Debye-Hückel shielding distance', λ_D, usually termed the *Debye length* for short. It is one of the characteristic lengths in a plasma and amounts to

$$\lambda_D := \sqrt{\frac{k_B T}{4\pi \, n_0 e^2}} = 6.65\sqrt{\frac{T}{n_e}} \quad [\mathrm{cm}] \ , \tag{2.4.10}$$

where T is in K, and the electron density n_e is in cm^{-3}. A fully ionized multicomponent plasma with solar abundances (27% helium by mass) has been assumed for the numerical expression in Equation (2.4.10).

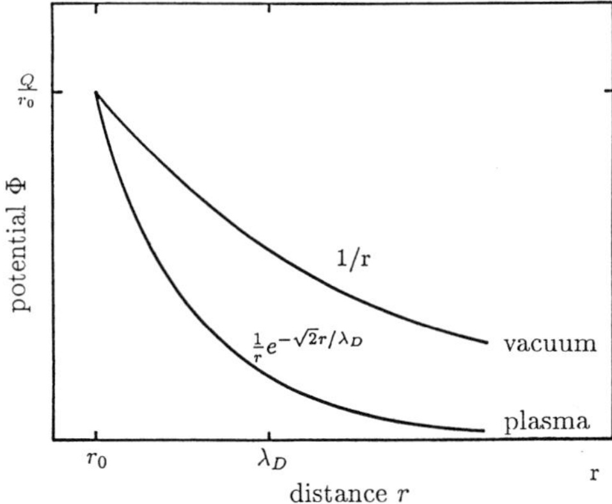

Fig. 2.9. The electric potential of a test charge Q is reduced by the ambient plasma. Here the positive test particle is shielded by field particles of negative charge and by a reduced number of ions in its vicinity.

Equation (2.4.9) means that the electric force of a charge is limited in a plasma to about the Debye length. Particles with opposite charge slightly predominate in its environment, and neutralize its effect (see Fig. 2.9). Only particles separated by less than about λ_D feel the test charge. Of course, the test charge can be any arbitrary particle, and the conclusion is more general: only particles within λ_D are directly influenced by each others' electric fields. The limitation of the direct influence of a charge has far reaching consequences and makes the behavior of a plasma entirely different, for example, from a cluster of gravitationally interacting stars. Since gravitational attraction is not shielded by repulsive forces, it is unlimited in its reach.

With $Q = e$ and $r_0 \ll \lambda_D$ our approximation of Equation (2.4.7) at $r \geq \lambda_D$ is satisfied with

$$\frac{4\pi}{3} n_0 \lambda_D^3 \gg 1 \quad . \tag{2.4.11}$$

The left-hand side is the number of particles in a sphere of radius λ_D (termed the *Debye sphere*) in the undisturbed plasma. It is intuitively clear that this number must exceed unity, as required in Equation (2.4.11), to make Debye shielding work. Assumption (2.4.11) is called the *plasma approximation*.

Debye shielding is usually very efficient in a plasma. For example, in a corona with $n_0 = 10^8 \text{cm}^{-3}$ at a temperature $T = 10^6$ K, $\lambda_D \approx 0.7$ cm; and the number of particles in the Debye sphere is about 10^8, amply justifying our condition (2.4.11).

This process also provides an effective mechanism to prevent local charge accumulation. A concentration of charges of one sign (say electrons) in a region of a plasma would create local electric fields, acting toward the restoration of homogeneity. For $e\Phi \ll k_B T$ the effect of the electric field on the velocity of the field particles is small. The Debye shield builds up mostly by thermal motion: some particles (with opposite charge) remain near the test charge slightly longer, while others (with the same sign of charge) pass by a bit faster. Charge inhomogeneity, as introducing an additional charge, is eroded in this way within a thermal propagation time.

2.5. Charge Oscillations and the Plasma Frequency

In addition to spatial shielding, the plasma species of opposite signs can also lead to charge oscillations around the homogeneous equilibrium discussed in the previous section.

Fig. 2.10. In a 'thought experiment' the electrons are shifted to the right (dashed boundary) and released to oscillate at the plasma frequency.

We perform the following 'thought experiment' drawn in Figure 2.10: The electrons of a plasma are moved slightly to the right by a distance x. Ions are assumed to be fixed (or infinitely inert). This produces a volume $\mathcal{V} = Fx$ to the right where only electrons exist, and an equal volume to the left where only ions remain. Let the number density of electrons in $\mathcal{V}$ be n_e. The electric field produced by this artificial charge separation is given by a simple integration of Poisson's equation (2.4.5). Neglecting edge effects, the problem is one-dimensional and yields, as for an infinitely extended capacitor,

$$E = 4\pi e n_e x \quad . \tag{2.5.1}$$

Now we release the electrons and they are accelerated in the negative x-direction by the electric force. Using Equation (2.5.1) and neglecting thermal motions and collisions,

$$\ddot{x} = -\frac{eE}{m_e} = -\frac{4\pi e^2 n_e}{m_e} x := -(\omega_p^e)^2 x \quad , \tag{2.5.2}$$

where m_e is the electron mass. A characteristic frequency is defined in Equation (2.5.2):

$$\omega_p^e := \left(\frac{4\pi e^2 n_e}{m_e}\right)^{1/2} = 2\pi \cdot 8.977 \cdot 10^3 \sqrt{n_e} \quad \text{[Hz]}. \tag{2.5.3}$$

The solution of Equation (2.5.2),

$$x = a\ cos(\omega_p^e t + b) \quad , \tag{2.5.3}$$

describes the restoring motion of the electrons, their overshooting of the equilibrium position ($x = 0$), and oscillation. It is the eigenmode of charge oscillations of electrons around infinitely inert ions. Its frequency, ω_p^e, is called the *electron plasma frequency*. Here we have derived it under very artificial conditions (which however contain and show the essence). We shall later encounter it again in various, more general circumstances (Sections 4.2 and 5.2). We shall find more eigenmodes due to the gyration of charged particles in a magnetic field or due to a particle beam. They are fundamental plasma properties.

In Section 2.4 we mentioned the elimination of charge inhomogeneities by thermal motion and Debye shielding. To avoid this problem in the above experiment, we have to require that the separation of inhomogeneities is larger than the thermal diffusion per oscillation period. This condition amounts to

$$\lambda > 2\pi \frac{v_{te}}{\omega_p^e} = 2\pi\lambda_D \quad . \tag{2.5.5}$$

Equation (2.5.5) suggests that the wave vector $k := 2\pi/\lambda$ of a space-charge wave has to satisfy $k < 1/\lambda_D$, a result that will be derived in a different way in Section 5.2. The identity relation between the fundamental plasma parameters v_{te}, ω_p^e, and λ_D in Equation (2.5.5) simply follows from the definitions (2.4.10) and (2.5.3). It reveals that Debye shielding is an equilibrium between thermal motion (expressed by v_{te}) and electric acceleration ($e^2 n/m$, implied by ω_p^e). The time to establish plasma shielding, $2\pi\lambda_D/v_{te}$, is the electron plasma period.

2.6. Collisions

So far we have neglected interactions between single particles. What is the range of applicability of this approximation? In addition to particle orbits, collisional interactions are also of interest in the broader context of a gas being not in thermodynamical equilibrium. Collisions are likewise important for the thermalization of a super-thermal population or a high energy tail of some particle species, as well as the transport coefficients, such as resistivity, conductivity, or viscosity in some steady non-equilibrium state.

The concept of particle collisions in a plasma is by far the most complex introduced in this chapter and deserves careful study. The word 'collision' generally evokes an image like that of two billiard balls bumping against each other. While not in touch, there is no interaction to speak of. During the extremely short time they are pushed against each other, they feel a strong repulsive force, which practically ceases at the moment the two balls are separated from each other. A collision between two charged particles, however, is a very different process since they interact through the long-range Coulomb force. Thus the collision between two charged particles takes place as a motion of the two charges on hyperbolic paths in each other's electric field.

2.6.1. Particle encounters in a plasma

Let us follow a particle, again to be called a 'test particle', moving simultaneously in the fields created by many other particles (the 'field particles') in a plasma. These fields add up to a stochastically changing force on the test particle; and its orbit is rarely a hyperbola, characteristic of a two particle encounter, but a jittery motion. Two test particles with similar, but not identical starting points and initial velocities will, after some time, diverge into completely different directions with different velocities (Figure 2.11).

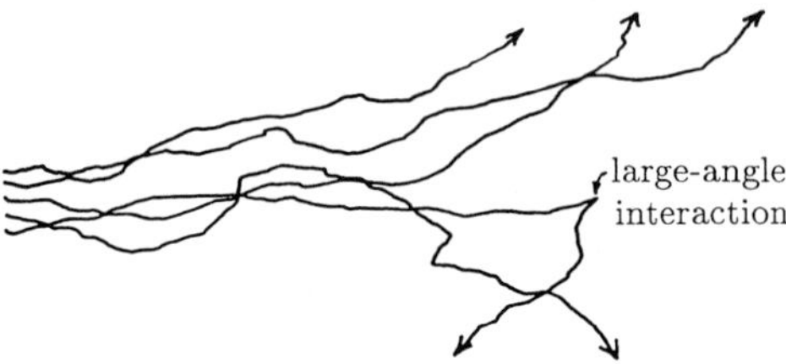

Fig. 2.11. Particles with nearly identical starting conditions diverge in space and velocity due to stochastic collisional interactions.

Collisions in a plasma are therefore of a statistical nature. By a 'collision' we mean the combined effect of a multitude (millions or billions!) of simultaneous interactions with field particles. The collision time in plasma physics is defined as the average time needed for a change of direction, energy, or momentum by an amount to be specified.

It is important for the understanding of collisions in plasmas to distinguish clearly three ranges for field particles at different distances from the test particle. A first dividing line has already been encountered in Section 2.4. The Debye effect shields the test particle from the electric force of charges beyond the Debye length λ_D. We note here that a test particle moving faster than the mean thermal speed of the field particles cannot build up appreciable shielding, being an effect of thermal motion. However, moving in the Debye spheres of field particles, the test particle experiences reduced electric fields.

In fact, particles beyond the Debye length are not without any influence. They may cause a magnetic field or even an electric field that is not completely canceled by shielding (such a situation occurs in electric currents). For such long-range fields the same procedure is adopted as in classical electrodynamics. The local fields of individual charges is neglected in so-called *macroscopic* electric and magnetic fields. The deviations from the smoothed field are *microscopic* fluctuations due to single particles.

The Debye length is about the range of microscopic fields in plasmas, and is a relatively well defined boundary separating two populations of field particles. (*i*) The more distant charges influencing a test particle by macroscopic fields yield smooth test particle orbits, drifts, and oscillations as discussed in Sections 2.1 – 2.3, and 2.5. (*ii*) The charges within the Debye sphere causing microscopic fields lead to stochastic motion, termed *collision*.

A second distinction of field particle distances at the microscopic level is useful to evaluate the process of collision. The criterion is the deflection angle of the path of the test particle in the potential of a given field particle. The deflection is obviously small if the energy of the test particle, $\frac{1}{2}m_T u^2$ (where the subscript T refers to the test particle and u is the velocity relative to the field particle), is large compared to the electric potential, $q_T q_F / r_1$ (where $q_T := Z_T e$ and $q_F := Z_F e$ are the charges of the test particle and the field particle, respectively). The distance of closest approach, r_1, refers to the original orbit of the test particle in the absence of forces. It is generally called the *impact parameter*. The particle is deflected by a small angle only if $r_c < r_1 < \lambda_D$, where

$$r_c := \frac{q_T q_F}{m_T u^2} \tag{2.6.1}$$

is the impact parameter for a 90° deflection. For $r < r_c$ one obtains a large-angle deflection, changing the direction of the orbit by 90° or more by one single encounter. The ratio of small-angle to large-angle interactions varies as the respective impact areas,

$$\frac{\lambda_D^2 - r_c^2}{r_c^2} \lesssim \frac{\lambda_D^2}{r_c^2} \approx \left(\frac{9}{Z_T Z_F}\right)^2 (\frac{4\pi}{3} n_e \lambda_D^3)^2 =: \Lambda^2 \quad . \tag{2.6.2}$$

For the second equation in (2.6.2) we have used Equations (2.6.1) and (2.4.10), and replaced mu^2 by $3k_B T$. The second expression in parentheses is the number of particles in a Debye sphere (derived in Section 2.4) and is assumed to be a large number. Equation (2.6.2) states that small-angle scattering generally outnumbers large-angle deflections by an enormous factor. It will become clear below (Eq. 2.6.18) that this multitude of distant encounters is about two orders of magnitude more efficient in deflecting the test particle than close binary collisions.

The computational simplification of neglecting large-angle deflections follows immediately. Small, simultaneous interactions become additive, and the scattering of a particle by small-angle deflection is a random walk in angle and velocity.

2.6.2. FOKKER-PLANCK METHOD

The statistical method to describe small-angle deflections has been developed by A.D. Fokker and M. Planck. Let $P(\mathbf{v}, \mathbf{\Delta v})$ be the probability that a test particle changes its velocity $\mathbf{v}$ to $\mathbf{v}+\mathbf{\Delta}v$ in the time interval Δt. Provided that the particle number is conserved, the velocity distribution at time t can be written as

$$f(\mathbf{v}, t) = \int f(\mathbf{v} - \mathbf{\Delta}v, t - \Delta t) P(\mathbf{v} - \mathbf{\Delta}v, \mathbf{\Delta}v) d^3\Delta v \quad . \tag{2.6.3}$$

Noting that for small-angle deflections $|\Delta v| \ll |v|$, the product fP in Equation (2.6.3) can be expanded into a Taylor series,

$$\begin{aligned} f(v,t) = \int \{ fP - \Delta t[\frac{\partial}{\partial t} f]P - \mathbf{\Delta v}[\nabla_v fP] \\ + \frac{1}{2}\Delta v_i \Delta v_j [\frac{\partial}{\partial v_i}\frac{\partial}{\partial v_j} fP] + \ldots \} d^3\Delta v \end{aligned} \quad . \tag{2.6.4}$$

The Einstein convention has been introduced that the sums over the indices i and j have to be used if they appear together in the numerator and denominator, or as subscripts and superscripts. Since the probability that some transition takes place is unity, P is normalized to

$$\int P d^3\Delta v = 1 \quad . \tag{2.6.5}$$

We define the average velocity change per time interval Δt:

$$\int \mathbf{\Delta v} P d^3\Delta v := < \mathbf{\Delta v} > \quad , \tag{2.6.6}$$

$$\int \Delta v_i \Delta v_j P d^3\Delta v := < \Delta v_i \Delta v_j > \quad . \tag{2.6.7}$$

Exchanging integration and differentiation, the integral in Equation (2.6.4) is readily evaluated. The first term in the integrand cancels with the left hand side of the equation. The remaining terms form the important *Fokker-Planck equation*,

$$\left(\frac{\partial f(\mathbf{v},t)}{\partial t} \right)_{\text{coll}} = \frac{\partial^2}{\partial v_i \partial v_j} \left(f \frac{< \Delta v_i \Delta v_j >}{2\Delta t} \right) - \frac{\partial}{\partial v_i} \left(f \frac{< \Delta v_i >}{\Delta t} \right) \quad . \tag{2.6.8}$$

The possibility of neglecting the higher-order terms in the expansion (2.6.4) is a property of inverse-square law particles having multiple collisions. Equation (2.6.8) shows that the motion of particles in velocity space then can be visualized as a diffusion process. Its right hand side describes the temporal change of a distribution of test particles by multiple, small-angle collision processes. It corresponds to the right hand side of the Boltzmann equation (1.4.11). The first term in Equation (2.6.8) represents the three-dimensional diffusion of the test particle in velocity space; the second term is a friction, slowing down the test particle and moving it radially toward the origin of velocity space.

2.6.3. COLLISION TIMES

The collision time in a plasma is not uniquely defined as in the case of neutral atoms. A charged particle in the Coulomb potential of another particle experiences two effects: (*i*) It is deflected from its initial direction, and (*ii*) it accelerates the field particle. The latter constitutes an energy loss and a drag force (friction) on the motion of the test particle. The relative importance of the two effects depends on the mass ratio of the two particles. If the field particle is much heavier, its acceleration is small, and the direction of motion of the test particle is changed before it loses its energy. The velocity distribution of a set of test particles with the same initial velocity approaches isotropy before the energy is lost. If, on the contrary, the field particle is very mobile, the drag on the test particle slows it down before deflection. The time when a test particle changes its direction is generally different from the time it loses its energy or momentum. Depending on the particular problem, the collision time has to be defined accordingly. Several relaxation times have been introduced by L. Spitzer, some of which we evaluate in the following.

A. *Angular Deflection*

The scattering of a particle from its initial direction by distant encounters is a diffusion process expressed by the first term of the Fokker-Planck equation (2.6.8). It is natural to define the mean *deflection time*

$$t_d := \frac{v^2}{< \Delta v_{\rm perp}^2/\Delta t >} \quad , \tag{2.6.9}$$

where $< \Delta v_{\rm perp}^2/\Delta t >$ is the average diffusion in velocity perpendicular to the initial velocity $\mathbf{v}$ of the test particle in the time interval Δt. We denote by $v_{\rm perp}$ and $v_{\rm par}$ (needed later) the velocity components perpendicular and parallel, respectively, to the initial velocity. They are generally different from v_z and $v_\perp$, the velocity components relative to the magnetic field. The collision time, as defined in Equation (2.6.9), is the average time in which the test particle is deflected through an angle of order 90°.

The calculation of $< \mathbf{\Delta v} \circ \mathbf{\Delta v} >$ is lengthy but straightforward, and is omitted here. The interested reader is referred to the classical textbooks of plasma physics listed at the end of the chapter. It is based on the simple physical idea that the test particle's velocity changes gradually under the influence of many field particles, whose motions are approximated by straight lines. The electric force on the test particle is given by adding the combined effects of the field particles. To average the effect, the sum is replaced by integration over the distribution in space and velocity. For field particles with an isotropic, Maxwellian distribution, the integral in velocity space reduces to the part $|v_F| < |v_T|$ and yields the error function Ψ and the combination G with its derivative, where

$$\Psi(x) := \sqrt{\frac{2}{\pi}} \int_0^x \exp(-\frac{y^2}{2})dy \quad , \tag{2.6.10}$$

$$G(x) := \frac{\Psi(x) - x\Psi'(x)}{x^2} \quad . \tag{2.6.11}$$

The limiting values are

$$\lim_{x \to \infty} \Psi(x) = 1 \quad , \tag{2.6.12}$$

and

$$\lim_{x \to \infty} G(x) = \frac{1}{x^2} \quad . \tag{2.6.13}$$

The general result of the integration in velocity space is

$$t_d = \frac{v^3}{A_d[\Psi(v/v_{tF}) - G(v/v_{tF})]} \quad , \tag{2.6.14}$$

where the diffusion constant A_d has been defined by

$$A_d := \frac{8\pi n_F q_T^2 q_F^2 \ln \Lambda}{(m_T)^2} \quad . \tag{2.6.15}$$

Λ is the ratio of the boundaries on the integral over space around the test particle where interactions take place. A finite lower limit of the impact parameter has to be introduced to guarantee approximate rectilinear motions. Since small angle deflections dominate, we put $r_{\rm min} = r_c$, and Λ is therefore given by Equation (2.6.2). As it enters only as a natural logarithm, the rough approximations of the boundaries are greatly alleviated. For solar abundances and electrons or protons as test particles,

$$\Lambda \approx \begin{cases} 1.24 \cdot 10^4 (T^{3/2}/n_e^{1/2}), & \text{for } T \lesssim 4.2 \cdot 10^5 \text{ K} \qquad (2.6.16) \\ 8.0 \cdot 10^6 (T/n_e^{1/2}), & \text{for } T \gtrsim 4.2 \cdot 10^5 \text{ K} \quad . \qquad (2.6.17) \end{cases}$$

Equation (2.6.17) includes quantum-mechanical effects on electrons at large velocities where r_c falls below the de Broglie wavelength of the particles. For collisions between ions, the classical expression, Equation (2.6.16), should be used. In the solar and stellar coronae and outer envelopes, $\ln\Lambda$ is about 20 and decreases to around 10 in the chromosphere. The theory breaks down for extremely high densities and low temperatures, where Λ approaches unity.

It is instructive to consider the deflection of a fast particle having $v \gg v_{tF}$ (the thermal velocity of the field particles). According to Equations (2.6.12) and (2.6.13), G can then be neglected and $\Psi \approx 1$. The deflection time becomes

$$t_d \approx \frac{1}{\pi n v r_c^2} \frac{1}{8 \ln\Lambda} := t_{ce} \frac{1}{8 \ln\Lambda} \quad , \tag{2.6.18}$$

where $t_{ce} = (\pi n v r_c^2)^{-1}$ is the collision time of close encounters. In the plasmas of interest here, Equation (2.6.18) is consistent with the assumption of the Fokker-Planck method that close encounters can be neglected. Since a test particle feels both electrons and ions, the total deflection time combines the effect of all field particles, and

$$t_d \approx \frac{m_T^2 v_T^3}{8\pi e^4 (n_e + \sum_i Z_i^2 n_i) Z_T^2 \ln\Lambda} \quad . \tag{2.6.19}$$

A useful approximation for a non-relativistic, super-thermal particle with kinetic energy $\varepsilon_{\rm keV}$ (in keV) in a fully ionized plasma with solar abundances is

$$t_d = 9.5 \cdot 10^7 \frac{\varepsilon_{\rm keV}^{3/2}}{n_e} \left(\frac{m_T}{m_e}\right)^2 \frac{1}{Z_T^2} \left(\frac{20}{\ln\Lambda}\right) \quad [\mathrm{s}] \ , \tag{2.6.20}$$

where collisions with both electrons and ions have been taken into account. The deflection time (2.6.20) has been numerically evaluated in Table 2.1 for electrons and protons moving with $v \gg v_{tF}$. We note that this deflection time, *increasing* with velocity in proportion to v^3, is different from neutral atoms and billiard balls (where collision time decreases as $1/v$). The faster a charged particle moves through a plasma, the smaller its frictional drag. It is this property of longevity which makes super-thermal particles a distinct population in some cosmic plasmas.

The magnetic field has been omitted in this discussion. It does not change the basic process of deflection nor the derived collision times. Collisions, however, allow charged particles to diffuse across magnetic field lines. The diffusion is simply caused by random steps of about one gyroradius within a collision time, t_d.

B. *Energy Exchange*

Another collision time, t_{ex}, is related to the energy change between the test particle and the field particles. In analogy to the deflection time we define the *energy-exchange time* by

$$t_{ex} := \frac{\varepsilon^2}{< \Delta\varepsilon^2 / \Delta t >} \quad . \tag{2.6.21}$$

This definition is meaningful if the test particle has a velocity of the order of the thermal velocity of the field particles. The energy change then fluctuates in its sign, and Equation (2.6.21) describes the net effect. Writing the exchange of energy in terms of parallel and perpendicular velocity relative to the initial velocity $\mathbf{v}$ of the test particle,

$$\Delta\varepsilon = \frac{m}{2}\left\{[(v+\Delta v_{\text{par}})^2 + \Delta v_{\text{perp}}^2] - v^2\right\} \quad . \tag{2.6.22}$$

After averaging over the statistical ensemble we find in the lowest order of Δ-terms

$$< \Delta\varepsilon^2 > = m^2 v^2 < \Delta v_{\text{par}}^2 > \quad , \tag{2.6.23}$$

and, again using the Fokker-Planck method,

$$t_{ex} = \frac{v^2}{4 < \Delta v_{\text{par}}^2/\Delta t >} = \frac{v^3}{4A_d G(v/v_{tF})} \quad . \tag{2.6.24}$$

The energy-exchange time is for instance used to determine the time to equalize the temperatures of two plasma species (Exercise 2.5). For infinitely heavy field particles, t_{ex} approaches infinity as the field particles become stationary, the test particle would move in their fixed potential, and collisions are elastic. Equation (2.6.24) becomes unreliable for large v/v_{tF} where secondary terms contribute.

C. *Momentum Loss*

The effects of collisions on macroscopic motions such as friction, viscosity, electric resistance, and wave damping frequently require an evaluation of the average loss of forward momentum. The slowing down of the initial velocity of a test particle is a combination of the frictional forces and deflections by multi-particle interactions. Which effect dominates depends on the mass of the field particles. It is measured by the *slowing-down time*,

$$t_s := -\frac{v}{< \Delta v_{\text{par}}/\Delta t >} = \frac{2v(v_{tF})^2}{[1+m_T/m_F]A_d G(v/v_{tF})} \quad . \tag{2.6.25}$$

For $v \gg v_{tF}$ it is proportional to v^3. Contrary to t_{ex}, the slowing-down time is finite for infinitely heavy field particles and is twice the deflection time, since forward momentum loss is dominated by deflection. Nevertheless, the slowing-down time for electron motions close to (or below) the thermal speed of the field particles is different and needs careful evaluation.

D. *Energy Loss*

A very important collision time is the energy loss of a fast particle. At $v_T \gg v_{tF}$, the test particle only loses energy by interactions. The *energy loss time* is defined by

$$t_\varepsilon := -\frac{\varepsilon}{< \Delta\varepsilon/\Delta t >} \quad . \tag{2.6.26}$$

Using Equation (2.6.22), we rewrite it as

$$\frac{1}{t_\varepsilon} = -\frac{2v < \Delta v_{\rm par}/\Delta t >}{v^2} - \frac{< \Delta v^2_{\rm par}/\Delta t >}{v^2} - \frac{< \Delta v^2_{\rm perp}/\Delta t >}{v^2} \quad . \tag{2.6.27}$$

This equation can be expressed in the previously derived terms according to the definitions (2.6.25), (2.6.24), and (2.6.9).

$$\frac{1}{t_\varepsilon} = \frac{2}{t_s} - \frac{4}{t_{\rm ex}} - \frac{1}{t_d} = \frac{A_d}{v^3}\left[(1 + \frac{m_T}{m_F})(\frac{v}{v_{tF}})^2 G(v/v_{tF}) - \Psi(v/v_{tF})\right] \quad . \tag{2.6.28}$$

For $v \gg v_{tF}$ it simplifies to

$$t_\varepsilon \approx \frac{v^3 m_F}{A_d m_T} \quad . \tag{2.6.29}$$

Let an electron be the test particle for illustration. Equation (2.6.29) then yields a very long energy loss time for protons as the field particles. The physical reason is again the large inertia of protons, making interactions with them nearly loss-free. Thus the relevant collision partners and field particles are the thermal electrons. For fast electrons, the energy loss time becomes equal to the electron-electron deflection time (Eq. 2.6.14). Also for proton test particle, the much more mobile field electrons absorb most of the collision energy.

Relativistic electrons lose energy within a time of

$$t_\varepsilon^{\rm rel} \approx 1.59 \cdot 10^{12} \, \frac{\varepsilon_{MeV}}{n_e} \quad , \tag{2.6.30}$$

where ε_{MeV} is the particle energy in MeV (Benz and Gold, 1971).

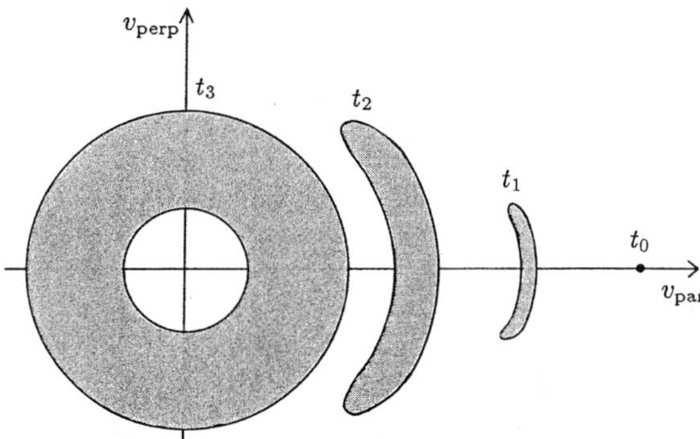

Fig. 2.12. Evolution of the probability distribution of a super-thermal electron due to small-angle scattering starting at t_0 with a velocity $\mathbf{v_0}$. Since energy exchange and friction are slow, changes in direction dominate. After $t_\varepsilon \gtrsim t_s \gtrsim t_d$ the ring distribution contracts toward zero velocity.

E. *Discussion*

How are the different collision times related to each other? As an example, let us follow in time a fast electron (test particle) interacting with field particles. The small-angle deflections by microscopic fields in a plasma scatter the test particle in a stochastic way described by a probability distribution $f_T(\mathbf{x}, \mathbf{v}, t)$ in space and velocity. The distribution is initially a delta function in six dimensions, $\prod_{i=1}^{3} \delta(x_i - x_{0i})\delta(v_i - v_{0i})$, centered at the initial position and velocity. The four collision times derived above indicate how this probability function spreads out, and they suffice for a statistical description of the behavior of a test particle (Figure 2.12). If the electron initially is super-thermal (i.e. $v_T \gg v_{tF}$), pitch-angle scattering predominates. The probability distribution grows only slowly in width parallel to $\mathbf{v}$, since energy-exchange is extremely slow (Eq. 2.6.24). After a time t_d the distribution forms a ring in the (v_{perp}-v_{par}) plane (about at t_3 in Fig. 2.12), when the forward momentum is also lost. The radius of the ring decreases by energy loss. Energy exchange widens the ring (see Eq. 2.6.23), which finally shrinks toward the compact distribution of the field particles near the origin of velocity space after about t_ε.

Table 2.1. Collision times in seconds for non-relativistic test particles with $v_T \gg v_{tF}$ in a fully ionized hydrogen plasma with solar helium abundance and using cgs units. The values in the table have to be corrected by the factor 20/lnΛ, being usually of order unity.

test →	electron	electron	proton	proton
field →	electrons	ions	electrons	ions
t_d	$3.1 \cdot 10^{-20} \frac{v_T^3}{n_e}$	$2.7 \cdot 10^{-20} \frac{v_T^3}{n_e}$	$1.0 \cdot 10^{-13} \frac{v_T^3}{n_e}$	$9.0 \cdot 10^{-14} \frac{v_T^3}{n_e}$
t_s	$3.1 \cdot 10^{-20} \frac{v_T^3}{n_e}$	$5.4 \cdot 10^{-20} \frac{v_T^3}{n_e}$	$1.1 \cdot 10^{-16} \frac{v_T^3}{n_e}$	$1.0 \cdot 10^{-13} \frac{v_T^3}{n_e}$
t_ε	$3.1 \cdot 10^{-20} \frac{v_T^3}{n_e}$	$6.2 \cdot 10^{-17} \frac{v_T^3}{n_e}$	$5.7 \cdot 10^{-17} \frac{v_t^3}{n_e}$	$1.1 \cdot 10^{-13} \frac{v_T^3}{n_e}$

It is evident from Table 2.1 that a super-thermal particle (electron or proton) is deflected by ions equally well as by electrons (except for a small difference due to field particle densities converted into n_e). This is not true for energy loss. An electron is slowed down twice as much by thermal electrons as by ions, and an ion is retarded almost exclusively by the friction of electrons. This difference between electron and ion slow-down has a simple explanation. A beam of super-thermal electrons is decelerated also by loss of directed momentum, not only by energy loss of individual particles. Thus super-thermal electrons are decelerated in forward momentum and deflected at similar rates. Ions are slowed down faster than deflected.

In conclusion, we note that the collision times of a charged particle in a plasma increase with velocity to a power of 3 or 5. It is opposite to billiard balls (or neutral atoms), in which collision time decreases with velocity. This unusual property is specific to the inverse-square law of particle interaction and has far-reaching consequences in astrophysics: Once charged particles are accelerated, they keep their energy much longer than the collision time of the thermal background particles. Energetic particles may exist long after their acceleration and can serve as indicators of violent processes in the past, like the smoke of a gun. As an example, we just mention cosmic rays in interstellar space, which – once accelerated in a supernova or a stellar flare – have t_ε easily exceeding 10^9 years. Nuclear collisions, following a different law, effectively reduce this time.

F. *Thermal Collision Times*

The different collision times become practically identical for a test particle moving at the thermal velocity and interacting with its own species. Deflection and energy exchange occur at similar rates. This *thermal collision time* – also called self-collision time – is a characteristic time of a plasma species rather than of a single particle. Its value generally is taken as t_d evaluated at the root mean square thermal velocity in three dimensions, $(3k_BT/m)^{1/2}$, and amounts to

$$t_t = \frac{v^3}{0.714 A_d} \approx \frac{0.267}{\ln \Lambda} \frac{T^{3/2}}{Z^4 n} \left(\frac{m}{m_e}\right)^{1/2} \quad [\mathrm{s}] \ , \tag{2.6.31}$$

where Z is the charge of the particles in units of e, the elementary electron charge, T is in K, n in cm^{-3}, and Λ has to be evaluated from Equation (2.6.16) or (2.6.17). The thermal collision time sets the time scale in which the bulk of a plasma regains thermal equilibrium after a disturbance. We note that (*i*) this takes $(m_p/m_e)^{1/2}$ longer for protons than for electrons, and (*ii*) that for 20 keV electrons, to give an example, t_ε is more than three orders of magnitude larger than t_t for thermal electrons at 10^6 K.

Waves that include oscillating electrons (an example was given in Section 2.5) are damped by the collisional randomization of electron momentum on ions. The process occurs within the thermal *electron-ion collision time*, $t_{e,i}$, calculated from the forward momentum loss (Eq. 2.6.25),

$$t_{e,i} = 1.907 \cdot 10^{-2} \frac{T^{3/2}}{\sum_i Z_i^2 n_i} \left(\frac{20}{\ln\Lambda}\right) \quad [\mathrm{s}] \ . \tag{2.6.32}$$

The ions are the field particles. For a fully ionized plasma with solar abundances $\sum_i Z_i^2 n_i \approx 1.16 \, n_e$.

Exercises

2.1: Particles moving along a curved magnetic field experience a drift due to their inertia. Calculate the centrifugal drift (Eq. 2.1.29) by defining a ‘centrifugal force’,

$$\mathbf{F}_c := \frac{m v_z^2}{R_c^2} \mathbf{R}_c \quad , \tag{2.6.33}$$

where R_c is the curvature radius of the magnetic field.

2.2: Calculate the bounce period (Eq. 2.2.4) of a particle orbiting in a symmetric magnetic loop having a parabolic form: $B = B_0(1 + s^2/H_B^2)$, where s is the distance from the top.

2.3: Prove that Equation (2.3.1) is consistent with the Boltzmann equation (1.4.11) in the absence of collisions and other forces.

2.4: Calculate the distance after which a beam evolving out of a hot Maxwellian distribution exceeds three times the mean thermal velocity, v_{tc}, of the cold population, the approximate threshold for instability. Assume $v_{th}/v_{tc} = 10$, $T_c = 2 \cdot 10^6$K, and $\tau = 1$s, values typical for solar type III radio bursts.

2.5: Prove that the equilibration time to equalize the temperatures of two plasma species T and F defined by

$$t_{eq} :=\mid T_F - T_T \mid (dT_T/dt)^{-1} \tag{2.6.34}$$

amounts to

$$t_{eq} = \frac{m_T m_F}{8\sqrt{2\pi} q_T^2 q_F^2 n_F \ln\Lambda} \left(\frac{k_B T_T}{m_T} + \frac{k_B T_F}{m_F} \right)^{3/2} \quad . \tag{2.6.35}$$

[*Hint:* Calculate $< \Delta\varepsilon_T >_F$ and average over velocity of species T.] As an example evaluate t_{eq} for the solar wind, where the electron temperature usually exceeds the ion temperature. How long would it take to relax the observed difference near Earth between electrons (say $T_e = 10^5$ K) and protons ($T_p = 3 \cdot 10^4$ K) in the solar wind if the density ($n_e = 5$ cm^{-3}) did not change?

Further Reading and References

Adiabatic invariants and drifts

Schmidt, G.: 1979, *Introduction to High Temperature Plasmas* (2nd edition), Chapter 2, Academic Press, New York.

Collisions

Spitzer, L.: 1962, *Physics of Fully Ionized Gases*, Interscience Publ., New York.

Schmidt, G.: 1979, *op.cit.*, Chapter 11.

Emslie, A.G.: 1978, 'The Collisional Interaction of a Beam of Charged Particles with a Hydrogen Target of Arbitrary Ionization Level', *Astrophys. J.* **224**, 241.

Petrosian, V.: 1985, 'Directivity of Bremsstrahlung Radiation from Relativistic Beams and the Gamma Rays from Solar Flares', *Astrophys. J.* **299**, 987.

References

Benz, A.O., and Gold, T.: 'Magnetically Trapped Particles in the Lower Solar Atmosphere Directivity', *Solar Phys.* **21**, 157.

Boström, R., Koskinnen, H., and Holback, B.: 1987, 'Low-Frequency Waves and Small-Scale Solitary Structures Observed by Viking', *ESA SP* **275**, 185.

CHAPTER 3

MAGNETOHYDRODYNAMICS

In this chapter we use statistical approaches to coronal plasmas, 'smearing out' individual particles to a fluid. Although their individuality is lost, the particles' collective physical properties are retained. Magnetohydrodynamics (MHD) is a fluid theory. It is appropriate for coronal phenomena that take place on a relatively large scale and are slow. The main branches of MHD are: equilibria, waves, instabilities, and reconnection, on each of which there are already excellent books. These basic processes have been applied to dynamo theory, magneto-convection, flows in the photosphere and chromosphere, coronal loops, prominences, flares, coronal heating, and stellar winds, again on each of which whole books have been written.

This chapter introduces MHD from basic principles and at an elementary level. Its aim is to prepare the ground for the kinetic plasma theory which deals with much smaller-scale and faster processes, and which is the subject of Chapters 5 – 10.

3.1. Basic Statistics

3.1.1. BOLTZMANN EQUATION

The results of Section 2.6 on collisions in plasmas now allow a deeper understanding of the Boltzmann equation introduced in Chapter 1. If the number of particles of species α is conserved, their distribution function in space and velocity must obey an equation of continuity

$$\frac{\partial f_\alpha}{\partial t} + \mathbf{v} \cdot \frac{\partial f_\alpha}{\partial \mathbf{x}} + \ddot{\mathbf{x}} \cdot \frac{\partial f_\alpha}{\partial \mathbf{v}} = 0 \quad . \tag{3.1.1}$$

This states mathematically that a decrease of the number of particles in an elementary volume of phase space, $f_\alpha(\mathbf{x}, \mathbf{v}, t) d^3x \; d^3v$, is equal to the loss of particles from the volume by particle motion in space and velocity.

We now deal specifically with electromagnetic forces and put

$$\ddot{\mathbf{x}} = \frac{q_\alpha}{m_\alpha} (\mathbf{E} + \frac{1}{c} \mathbf{v} \times \mathbf{B}) \quad . \tag{3.1.2}$$

The electric and magnetic fields can be produced collectively or externally, or originate from neighboring particles. The former two are the fields an observer would measure at low spatial or temporal resolution, and they form the macroscopic parts $\mathbf{E}_m$ and $\mathbf{B}_m$. The latter are caused by particles within a Debye radius that generate microscopic fields $\Delta\mathbf{E}$ and $\Delta\mathbf{B}$ fluctuating rapidly in space and time. Thus

$$\mathbf{E} = \mathbf{E}_m + \Delta\mathbf{E} \quad , \tag{3.1.3}$$

$$\mathbf{B} = \mathbf{B}_m + \Delta\mathbf{B} \quad . \tag{3.1.4}$$

We rewrite Equation (3.1.2), dropping the subscripts m and α wherever no confusion is possible,

$$\ddot{\mathbf{x}} = \frac{q}{m}(\mathbf{E} + \frac{1}{c}\mathbf{v} \times \mathbf{B}) + \frac{\Delta\mathbf{F}}{m} \quad , \tag{3.1.5}$$

summarizing the microscopic forces in $\Delta\mathbf{F}$. Equation (3.1.1) becomes the Boltzmann equation (1.4.11),

$$\frac{\partial f}{\partial t} + \mathbf{v} \cdot \frac{\partial f}{\partial \mathbf{x}} + \frac{q}{m}(\mathbf{E} + \frac{1}{c}\mathbf{v} \times \mathbf{B}) \cdot \frac{\partial f}{\partial \mathbf{v}} = -\frac{\Delta\mathbf{F}}{m} \cdot \frac{\partial f}{\partial \mathbf{v}} =: \left(\frac{\partial f}{\partial t}\right)_{\text{coll}} \quad . \tag{3.1.6}$$

It contains a subtlety: As the left side contains only variables averaged over the ensemble, the right side must also be evaluated to express the average effect of collisional encounters (as was done in Section 2.6).

Equation (3.1.6) is the basis for the physics of fluids as well as kinetic plasma physics. In the fluid approach, the Boltzmann equation is integrated in velocity space. This is the topic of the following two sections. In Chapter 4 we shall come back to Equation (3.1.6), but neglect the collision term. The reduced, collisionless equation is generally referred to as the Vlasov equation. The choice of approach is mainly a question of the time scale or wave frequency.

3.1.2. Velocity moments of the Boltzmann equation

In the fluid description of a gas the information on the particle velocity distribution is relinquished and replaced by values averaged over velocity space. This is obviously reasonable if the velocity distribution contains little information; in particular, if it is close to Maxwellian and remains so during the course of the process. Let us define such an average for a general variable $A(v)$ with

$$< A(v) > := \frac{\int A(v) f d^3v}{\int f d^3v} \quad . \tag{3.1.7}$$

Using this notation, we define for one species (with mass m):

particle density $$n := \int f d^3v \tag{3.1.8}$$

average velocity $$\mathbf{V} :=< \mathbf{v} > \tag{3.1.9}$$

stress tensor $$\hat{\mathbf{P}} := nm < \mathbf{v} \circ \mathbf{v} > \tag{3.1.10}$$

pressure tensor $$\hat{\mathbf{p}} := nm < (\mathbf{v} - \mathbf{V}) \circ (\mathbf{v} - \mathbf{V}) > \tag{3.1.11}$$

mean thermal velocity in i-direction $\quad v_t^i := \sqrt{< (\mathbf{v_i} - \mathbf{V_i})^2 >}$ (3.1.12)

mean thermal energy density $\quad \mathcal{E} := \frac{1}{2}mn < (\mathbf{v} - \mathbf{V})^2 >=: \frac{3}{2}nk_BT$ (3.1.13)

The definitions of stress and pressure make use of the fact that a coronal plasma is very close to an *ideal gas.* Equation (3.1.13) also defines a temperature T corresponding to Equation (1.4.10).

A. *Conservation of Particles*

Now integrate the Boltzmann equation (3.1.6) in velocity space:

$$\int \frac{\partial f}{\partial t} d^3v + \int \mathbf{v} \cdot \frac{\partial f}{\partial \mathbf{x}} d^3v + \frac{q}{m} \int (\mathbf{E} + \frac{1}{c}\mathbf{v} \times \mathbf{B}) \cdot \frac{\partial f}{\partial \mathbf{v}} d^3v = \int \left(\frac{\partial f}{\partial t}\right)_{\text{coll}} d^3v \quad . \tag{3.1.14}$$

The third term on the left side is a scalar product, thus a sum of three terms. Each can be integrated by parts and then vanishes. The collision term represents the change of the density by collisions, and is zero due to particle conservation,

$$\int \left(\frac{\partial f}{\partial t}\right)_{\text{coll}} d^3v = \left(\frac{\partial n}{\partial t}\right)_{\text{coll}} = 0 \quad . \tag{3.1.15}$$

Thus the remaining terms are

$$\frac{\partial n}{\partial t} + \frac{\partial}{\partial \mathbf{x}} \cdot (n\mathbf{V}) = 0 \quad . \tag{3.1.16}$$

The integration of the Boltzmann equation (3.1.16) has yielded the *equation of continuity* of particle density – a result of particle conservation.

B. *Conservation of Momentum*

Let us multiply the Boltzmann equation by mv_k, where k denotes here one of the three velocity coordinates, and integrate over all velocity space. The first term then becomes

$$m\frac{\partial}{\partial t} \int v_k \cdot f d^3v = \frac{\partial}{\partial t} \cdot (mnV_k) \quad , \tag{3.1.17}$$

and corresponds to the temporal change of momentum density. The second term is the force per unit volume due to a pressure gradient,

$$m \int v_k \mathbf{v} \cdot \frac{\partial f}{\partial \mathbf{x}} d^3v = \frac{\partial}{\partial \mathbf{x}} \cdot (nm < v_k \mathbf{v} >) = \sum_i \frac{\partial P_{ki}}{\partial x_i} \quad . \tag{3.1.18}$$

The subscripts k and i refer to tensor and vector elements. Replacing the electromagnetic force, $q(\mathbf{E} + (\mathbf{v} \times \mathbf{B})/c)$, and the other forces by the general symbol $\mathbf{F}$, we find for the third term

$$\int v_k \mathbf{F} \cdot \frac{\partial f}{\partial \mathbf{v}} d^3v = -\int f \frac{\partial}{\partial \mathbf{v}} \cdot (\mathbf{F} v_k) d^3v = -n < F_k > \tag{3.1.19}$$

and for the collision term

$$m \int v_k \left(\frac{\partial f}{\partial t}\right)_{\text{coll}} d^3v = (\frac{\partial}{\partial t} nmV_k)_{\text{coll}} =: S_k \quad , \tag{3.1.20}$$

representing the change of momentum by collisions. Combining Equations (3.1.17) – (3.1.20), the first moment of the Boltzmann equation becomes

$$\frac{\partial}{\partial t}(nmV_k) + \sum_i \frac{\partial P_{ki}}{\partial x_i} = n < F_k > + S_k \quad . \tag{3.1.21}$$

Equation (3.1.21) expresses the conservation of momentum and is usually referred to as the *equation of motion*. These equations hold for each particle species. The different species are coupled by the collision term, **S**, of each species.

As an immediate application of the above derivations, Exercise 3.1 considers the electric field that arises from the smaller gravitational force on the electrons compared to ions. In a stationary equilibrium this electric field reduces the weight of the protons by half, and instead the electrons are pressed into the star. The extent of electron and proton atmospheres and their scale heights become equal. Compared to the Lorentz force of coronal magnetic fields, the effect of the electric field on individual particle orbits is usually small.

C. *Conservation of Energy*

Analogously, the second moment yields energy conservation,

$$\frac{\partial \mathcal{E}}{\partial t} + \frac{\partial}{\partial \mathbf{x}} \cdot (n < \frac{1}{2} mv^2 \mathbf{v} >) = n < \mathbf{F} \cdot \mathbf{v} > + \Pi \quad . \tag{3.1.22}$$

The right side of Equation (3.1.22) consists of two terms. The first represents the work done by the force **F**. It may include, for example, acceleration by an electric field, emission of radiation, or a heat input. Π is the change in energy density due to collisions.

The second term on the left side of Equation (3.1.22), the energy flux, enters as a new variable, just as particle and momentum flux have appeared in the equations of particle and momentum conservation, respectively. The term includes changes in flow velocity, as well as changes in thermal conduction, $-\nabla \cdot (\hat{\kappa} * \nabla T)$. Classical thermal conduction is controlled by particle collision. The thermal conductivity tensor, $\hat{\kappa}$, is diagonal. Parallel to the magnetic field, thermal conduction is primarily by the more mobile electrons. Across the magnetic field, ions – having a larger gyroradius – are primarily responsible.

The thermal conductivity parallel to the magnetic field is proportional to the density multiplied by the thermal velocity and the mean free path ($l_{\text{mfp}} = v_{te} t_t^e$). In equilibrium the diffusion of electrons builds up an electric field such that the current is cancelled. This electric field reduces the heat flow by a factor $a \approx 0.5$. For solar abundances one finds from Spitzer (1962)

$$\kappa_z \approx a\, n_e v_{te}^2 t_t^e k_B \approx 1.72 \cdot 10^{-5} \frac{T_e^{5/2}}{\ln\Lambda} \quad [\text{erg s}^{-1}\ \text{K}^{-1}\ \text{cm}^{-1}] \ . \tag{3.1.23}$$

For thermal conduction perpendicular to the magnetic field, we replace the thermal velocity in Equation (3.1.23) by the ion gyroradius R_i divided by the thermal collision time of ions, t_t^i, namely

$$\kappa_\perp \approx \frac{a n_i k_B R_i^2}{t_t^i} \approx \ 8.9 \cdot 10^{-13} \frac{n_e^2 (\ln\Lambda)^2}{T^3 B^2} \kappa_z \quad [\text{erg s}^{-1}\ \text{K}^{-1}\ \text{cm}^{-1}] \ . \tag{3.1.24}$$

Equation (3.1.24) assumes $\Omega_i t_t^i \gg 1$. Solar abundances have been assumed, T is in degrees kelvin, B in gauss, and Λ has been given by Equations (2.6.16) and (2.6.17). Note that Equations (3.1.23) and (3.1.24) assume plasmas having nearly Maxwellian velocity distributions and particles having a mean free path much shorter than the temperature scale length. These approximations become questionable in the transition layer, in flares, and in solar and stellar winds.

3.1.3. Elementary Magnetohydrodynamics (MHD)

Some processes in coronal physics are slow, and if a plasma process is slow enough, the physics becomes much simpler. Let us introduce a characteristic time of the process, t_{char}, which may be a travel time, a wave period or the inverse of a growth rate. We shall assume that the process is slow enough for particle collisions to smear out deviations from a Maxwellian velocity distribution, and for differences between particle species in temperature and average velocity to become unimportant. Furthermore, we assume that spatial limitations and boundary effects are unimportant. It may then be advantageous to sum particle number, momentum and energy density over all species α, and use the equations of conservation. An important consequence is that the sum over the collision terms of all species vanishes, since the total momentum is conserved,

$$\sum_\alpha \mathbf{S}^\alpha = 0 \quad . \tag{3.1.25}$$

We shall use the following definitions:

total mass density $\qquad \rho := \sum_\alpha m_\alpha n_\alpha \qquad$ (3.1.26)

total stress tensor $\qquad \hat{\mathbf{P}} := \sum_\alpha \hat{\mathbf{P}}_\alpha \qquad$ (3.1.27)

mass flow velocity $\qquad \mathbf{V} := (\sum_\alpha m_\alpha n_\alpha \mathbf{V}_\alpha)/\rho \qquad$ (3.1.28)

barycentric pressure $\qquad \hat{\mathbf{p}} := \sum_\alpha m_\alpha n_\alpha < (\mathbf{v} - \mathbf{V}) \circ (\mathbf{v} - \mathbf{V}) >_\alpha \qquad$ (3.1.29)

$$= \hat{\mathbf{P}} - \mathbf{V} \circ \mathbf{V} \rho \tag{3.1.30}$$

The tensor $\hat{\mathbf{p}}$ is defined in relation to the motion of mass, being dominated by the ions (baryons). Furthermore, one finds from the definitions (1.4.6) and (1.4.7)

charge density $$\rho^* = \sum_\alpha q_\alpha n_\alpha \quad , \tag{3.1.31}$$

current density $$\mathbf{J} = \sum_\alpha q_\alpha n_\alpha \mathbf{V}_\alpha \quad . \tag{3.1.32}$$

A. *MHD Equations and Approximations*

We now multiply Equation (3.1.16) by m_α and sum over α to get the equation of mass conservation,

$$\frac{\partial \rho}{\partial t} + \nabla \cdot (\rho \mathbf{V}) = 0 \quad . \tag{3.1.33}$$

Summing the equation of momentum conservation (3.1.21) over the particle species yields, for electromagnetic forces and gravity,

$$\frac{\partial}{\partial t}(\rho V_k) + \frac{\partial}{\partial x_i} P_{ki} = \sum_\alpha n_\alpha [q_\alpha (\mathbf{E} + \frac{1}{c}\mathbf{V}_\alpha \times \mathbf{B})_k + m_\alpha g_k] \quad , \tag{3.1.34}$$

which we rewrite in vector form using Equations (3.1.30) and (3.1.33),

$$\rho \frac{\partial \mathbf{V}}{\partial t} + \rho (\mathbf{V} \cdot \nabla) \mathbf{V} + \nabla * \hat{\mathbf{p}} = \rho^* \mathbf{E} + \frac{1}{c} \mathbf{J} \times \mathbf{B} + \rho \mathbf{g} \quad . \tag{3.1.35}$$

The equation states that the change in momentum density (first term on the left side) is caused by the flow of momentum (second term), pressure gradient (third term) and external forces (right side). The equations for mass and momentum conservation, (3.1.33) and (3.1.35), contain the variables $\rho, \mathbf{V}, \hat{\mathbf{p}}, \rho^*, \mathbf{E}, \mathbf{J}, \mathbf{B}$. The combined system of equations including Maxwell's (Eqs. 1.4.1 – 1.4.5) comprises a smaller number of equations and initial conditions than variables. The following approximations are generally used in MHD to close the system:

(1) Charge neutrality, $\rho^* = 0$, is suggested by the Debye shielding effect for large-scale and slow processes. Ampère's equation (1.4.2) then requires $\nabla \cdot \mathbf{J} = 0$.

(2) $\mathbf{J} = \sigma \mathbf{E}'$ in the frame of reference of the system moving with $\mathbf{V}$.

$= \sigma(\mathbf{E} + (\mathbf{V} \times \mathbf{B})/c)$ in the laboratory system.

The electric conductivity σ has to be determined from electron–ion collisions (Section 9.2). Often $\sigma = \infty$ is assumed; then $\mathbf{E} = -(\mathbf{V} \times \mathbf{B})/c$. This is called *ideal MHD*.

(3) We assume that the characteristic time of the process exceeds the rate of collisions (Eq. 2.6.31) and the gyroperiods,

$$t_{\text{char}} \gg t_{\text{coll}}, \frac{2\pi}{\Omega_i}, \frac{2\pi}{\Omega_e} \quad . \tag{3.1.36}$$

Then the distributions of particle velocity, and consequently the pressure, are isotropic. The relation

$$p_{ki}^{\alpha} = \delta_{ki} p^{\alpha} \ = n_{\alpha} k_B T_{\alpha} \tag{3.1.37}$$

then follows from Equations (3.1.11) and (3.1.13) where an ideal gas has been assumed. Equation (3.1.37) is called the *equation of state.* The total pressure is the sum of the partial pressures,

$$p = \sum_{\alpha} p_{\alpha} \quad . \tag{3.1.38}$$

This is a consequence of Equation (3.1.36), requiring that the collisions have time to equalize the temperatures of all species. Thus,

$$p = \sum_{a} n_{\alpha} k_B T = \frac{\rho}{\overline{m}} k_B T \quad . \tag{3.1.39}$$

For a fully ionized plasma with solar abundances, $\sum_{\alpha} n_{\alpha} = 1.92 n_e$, and the mean mass is $\overline{m} = 0.60 m_p$.

(4) In place of the energy equation (3.1.22), a process may be approximated by one of the following special equations of state adapted to the problem:

incompressible $\Longleftrightarrow \rho(t) = \text{const.}$ (implying $\nabla \cdot \mathbf{V} = 0$, cf. Eq. 3.1.33)

isothermal $\Longleftrightarrow \rho \propto p$

adiabatic $\Longleftrightarrow p\rho^{-5/3} = \text{const}$.

Furthermore, several implicit assumptions are generally made: (*i*) In the evaluations of σ and p the plasma is assumed to be close to thermodynamic equilibrium; and (*ii*) the internal structure of the plasma is neglected. This means that the size of the plasma and the phenomena of interest (e.g. the wavelength) are assumed to exceed the ion gyroradius, the Debye length, and the mean free path. If we also neglect relativistic effects and the displacement current, the basic equations of MHD are

$$\frac{\partial \rho}{\partial t} + \nabla \cdot (\rho \mathbf{V}) = 0 \tag{3.1.40}$$

$$\rho \frac{\partial \mathbf{V}}{\partial t} + \rho(\mathbf{V} \cdot \nabla)\mathbf{V} + \nabla p = \frac{1}{c} \mathbf{J} \times \mathbf{B} + \rho \mathbf{g} \tag{3.1.41}$$

$$\nabla \times \mathbf{B} = \frac{4\pi}{c} \mathbf{J} \tag{3.1.42}$$

$$\nabla \times \mathbf{E} = -\frac{1}{c}\frac{\partial \mathbf{B}}{\partial t} \tag{3.1.43}$$

$$\nabla \cdot \mathbf{B} = 0 \tag{3.1.44}$$

$$\mathbf{J} = \sigma(\mathbf{E} + \frac{1}{c}\mathbf{V} \times \mathbf{B}) \tag{3.1.45}$$

Energy Equation or Specialized Equation of State .

Instead of the energy equation supplemented by a heat conduction equation, the system of equations may be closed by an appropriately specialized equation of state. MHD in these various forms finds many applications in astrophysics from the liquid metallic core of planets to the flows of stellar winds.

B. *Electric Fields*

The displacement current $1/4\pi\ \partial\mathbf{E}/\partial t$ (Eq. 1.4.2) has been neglected in Equation (3.1.42) for the following reason. Assume a scale length of the fields, H, and a characteristic time for the process, t_0. Faraday's equation (3.1.43) implies $E \approx -vB/c$, where $v = H/t_0$. Therefore, the displacement current is about $v^2B/(Hc)$, and is much smaller than $c\nabla \times B \approx cB/H$ for $v^2 \ll c^2$.

The electric field, **E**, can be calculated readily from Ohm's law (3.1.45), and **J** follows from Ampère's law (3.1.42). Putting them into Faraday's equation (3.1.43) yields the *induction equation*

$$\frac{\partial \mathbf{B}}{\partial t} + \nabla \times (\mathbf{B} \times \mathbf{V}) = \frac{c^2}{4\pi\sigma}\nabla^2\mathbf{B} \quad , \tag{3.1.46}$$

where we have used the vector identity (A.10) and $\nabla \cdot \mathbf{B} = 0$.

It is important to note that in ideal MHD **B** and **V** are the fundamental variables. **J** and **E** are secondary and can be calculated if required. **B** and **V** are determined by the induction equation and the equation of motion. Once they are found, **J** and **E** follow. Note in particular that **J** is not driven by **E** in ideal MHD, and so standard circuit theories are inappropriate.

C. *MHD Properties*

Equations (3.1.40) – (3.1.45) are only the starting point of MHD. They can describe an enormous number of phenomena. Over the years, much practical knowledge on MHD plasma behavior has accumulated, some key elements of which we present here.

• The strong coupling between a magnetic field and matter, a very important property of plasmas, follows immediately from the induction equation (3.1.46). This consequence of Equation (3.1.46) can best be appreciated by integrating it over a plane surface A. The left side then expresses the total change in magnetic flux through the surface, $\Phi := \int_A \mathbf{B} \cdot d\mathbf{s}$. Let the boundary A' of A be defined by some floating corks in the fluid. The boundary thus moves with the mass flow velocity defined in Equation (3.1.41). The first term of Equation (3.1.46) is the change of flux due to a variation in **B**, the second term corresponds to the change caused by the motion of the boundary of A. Thus we write the left side of the integrated Equation (3.1.46) as a total derivative,

$$\frac{d\Phi}{dt} := \int_A \frac{\partial \mathbf{B}}{\partial t} \cdot d\mathbf{s} + \oint_{A'} (\mathbf{B} \times \mathbf{V}) \cdot d\mathbf{l} \approx \operatorname{sig}(\nabla^2 \mathbf{B} \cdot d\mathbf{s}) \frac{c^2 \Phi}{4\pi H_B^2 \sigma} \quad . \tag{3.1.47}$$

The scale length $H_B^2 = B/\nabla^2 B$ has been introduced. The time

$$\tau = \frac{4\pi H_B^2 \sigma}{c^2} \tag{3.1.48}$$

is needed for diffusion of the magnetic field through the plasma. It is an upper limit of the decay time for magnetic flux concentration. For photospheric conditions ($H_B \approx 10^8$cm, $\sigma \approx 10^{12}$ Hz) the diffusion time is $\tau \approx 4$ years. Obviously, sunspots with typical lifetimes of weeks cannot build up or decay by diffusion of field lines. It will be shown in Section 9.2.1 that the conductivity σ is inversely proportional to the collision rate. Equation (3.1.48) states that the magnetic diffusion involves a drag on the particle motion due to collisions. The same drag and finite conductivity also dissipate energy (Ohmic heating). Therefore, the decay of the magnetic field by diffusion is due to dissipation of energy through Ohmic heating and *vice versa*.

If one studies processes shorter than the diffusion time, σ may be considered infinite. Equation (3.1.47) then states that the magnetic flux is conserved in a surface moving with the plasma. The *magnetic field is frozen into the matter*, meaning that a fluid element is attached to its field line like a pearl on a string. Density differences can be smoothed out easily along field lines, but not across them, which is why it has become common to think of a field line as a real object, though it is only a mathematical construct. The magnetic field line and the plasma stay together whether the matter is moving and pulling the field along or *vice versa*.

• The induction equation (3.1.46) relates the temporal change of the magnetic field to convection (second term) and diffusion (right side). An MHD process is conveniently characterized by the ratio of these terms, a dimensionless parameter known as the *magnetic Reynolds number*

$$R_m := \frac{4\pi}{c^2} V_\perp \, \sigma \, H_B \quad , \tag{3.1.49}$$

where $V_\perp$ is the velocity component perpendicular to **B**, and H_B is the scale length of the magnetic field. R_m is infinite for ideal MHD and much larger than unity for most coronal processes.

• It is convenient to eliminate **J** in the equation of momentum conservation (3.1.41) to give

$$\rho \frac{\partial \mathbf{V}}{\partial t} + \rho(\mathbf{V} \cdot \nabla)\mathbf{V} + \nabla(p + \frac{B^2}{8\pi}) = \frac{(\mathbf{B} \cdot \nabla)\mathbf{B}}{4\pi} + \rho \mathbf{g} \quad . \tag{3.1.50}$$

The right side differs from zero if the field curves or converges and so exerts an anisotropic tension. Equation (3.1.50) contains an important relation between magnetic field and plasma pressure. Let us consider a stationary plasma (i.e. $\mathbf{V} = 0, \partial/\partial t = 0$). The first two terms of Equation (3.1.50) then vanish. The rest of the equation states that p and $B^2/8\pi$, the magnetic energy density, are in equilibrium with tensions produced by magnetic field inhomogeneity. For straight magnetic field lines the right side of Equation (3.1.50) vanishes, and the combined pressures of the plasma and the magnetic field become constant in space. Localized strong field regions then must have lower plasma pressure to balance the outside pressure. As the magnetic pressure in coronae is usually higher than the plasma pressure, large variations of the latter can be accommodated by small variations of the magnetic field. This is the reason for extreme inhomogeneity of coronae.

Consider as an example an idealized 'flux tube', a plasma structure outlined by magnetic field lines. Let its density be one tenth of the outside density; assume the same temperature and no magnetic field outside. The magnetic field of the flux tube must then balance 9/10 of the outside pressure. At the bottom of the solar photosphere, where $p_{\text{out}} \approx 2 \cdot 10^5$ dyne cm^{-2}, the magnetic field would be of order 2100 G, comparable to the maximum field strength measured in sunspots.

- The ratio of the thermal pressure to the magnetic pressure,

$$\beta := \frac{8\pi p}{B^2} \approx \left(\frac{\omega_p^e}{\Omega_e} \frac{2v_{te}}{c} \right)^2 \quad , \tag{3.1.51}$$

is an important dimensionless parameter known as the *plasma beta*. The second equation uses definitions of Chapter 2 and is accurate for a hydrogen plasma with $T_e = T_i$. Coronae are usually low-beta plasmas ($\beta \ll 1$), the solar wind has a beta of order unity or higher.

- In a stationary atmosphere with uniform or negligible magnetic field, Equation (3.1.50) yields

$$\frac{\partial p}{\partial h} = -\rho g(h) \quad . \tag{3.1.52}$$

The right side is the gravitational force per unit volume. For constant g and in isothermal conditions, Equation (3.1.52) becomes the barometric equation having an exponential solution

$$n(h) = n(h=0) \exp(-h/H_n) \tag{3.1.53}$$

with a density scale height

$$H_n = \frac{p}{\rho g} = 5.00 \cdot 10^3 \frac{T}{g_\odot} \text{ cm} \quad , \tag{3.1.54}$$

where $g_\odot$ is in units of the gravity in the solar photosphere (cf. Appendix C), T is in degrees kelvin, and solar abundances have been used.

3.2. MHD Waves

Waves are an important example of collective particle motion. Some or all particles in a volume element are slightly displaced by a local disturbance, but a restoring force – due to a pressure gradient, an electric or magnetic field, gravity, etc. – drives them back to the initial position. They overshoot and oscillate collectively around the equilibrium position. It is not the individual particle that is of primary interest, but rather the collective phenomenon, the average properties of the oscillation, and its propagation as a wave.

Spacecraft in the solar wind plasma have observed a bewildering variety of oscillations. In general, the wiggling particle motions, electric and magnetic field oscillations, and the combination of all three may be exceedingly complex. We shall restrict ourselves to (*i*) small disturbances of (*ii*) homogeneous, unlimited background plasmas. Under these conditions the plasma oscillations occur in only a few basic wave modes. These types of waves have characteristic *polarization* (relations between the wave vector, the vectors of particle motion, and wave fields) and *dispersion relations* (which relate the wave frequency ω to the wave vector $\mathbf{k}$).

There are two often neglected consequences of the fact that the equations of a plasma are not linear: (*i*) The superposition of two solutions does not necessarily solve the equations any longer; and (*ii*) the possibilities for waves are not exhausted by finding all the periodic and small-amplitude (linear) solutions. Nevertheless, we shall examine small-amplitude waves, as they exhibit the basic physics and form the basis of our understanding of oscillatory plasma phenomena. They are frequently sufficient for the description of waves.

Here we only derive the basic theory of MHD waves. More complete treatments can be found in many specialized textbooks. The derivation of the wave modes in the MHD approximation is valid at the low-frequency end of the spectrum. The method, however, follows a standard pattern for waves in all regimes and will be used similarly at higher frequencies. One first considers a plasma in equilibrium and perturbs it slightly such that the deviations are much smaller than the initial values. The main idea is to approximate the system of equations by a system that is *linear* in the variables of the deviation. The resulting 'linear' disturbance is analyzed to see whether it propagates as a wave having the form $\exp[i(\mathbf{k}\cdot\mathbf{x}-\omega t)]$. The goal of the mathematical discussion is to find the dispersion relation and the polarization.

3.2.1. Linearization

The MHD equations (3.1.40) – (3.1.45) with the choice of the adiabatic equation of state appropriate for most waves,

$$p\rho^{-5/3} = \text{const} \quad , \tag{3.2.1}$$

can be linearized by

$$\mathbf{B} = \mathbf{B}_0 + \mathbf{B}_1(\mathbf{x},t), \quad p = p_0 + p_1(\mathbf{x},t), \quad \mathbf{V} = \mathbf{V}_1(\mathbf{x},t) \quad , \tag{3.2.2}$$

$$\rho = \rho_0 + \rho_1(\mathbf{x}, t) \quad , \tag{3.2.3}$$

if the variables with subscript 1 (disturbance) are much smaller than the stationary, homogeneous variables (subscript 0). These *zero-order* variables cancel out when Equations (3.2.2) and (3.2.3) are inserted into the MHD equations. Let us assume ideal MHD ($\sigma = \infty$) and combine the equations of Faraday and Ohm, (3.1.42) and (3.1.45), to eliminate $\mathbf{E_1}$. Upon neglecting the products of first-order terms, we get a homogeneous system of linear equations,

$$\frac{\partial \rho_1}{\partial t} + \rho_0 \nabla \cdot \mathbf{V}_1 = 0 \tag{3.2.4}$$

$$\rho_0 \frac{\partial \mathbf{V}_1}{\partial t} + \nabla(p_1 + \frac{2\mathbf{B}_0 \cdot \mathbf{B}_1}{8\pi}) - \frac{(\mathbf{B}_0 \cdot \nabla)\mathbf{B}_1}{4\pi} = 0 \tag{3.2.5}$$

$$\nabla \times (\mathbf{V}_1 \times \mathbf{B}_0) = \frac{\partial \mathbf{B}_1}{\partial t} \tag{3.2.6}$$

$$\frac{p_1}{p_0} = \frac{5}{3} \frac{\rho_1}{\rho_0} \quad . \tag{3.2.7}$$

Counting equations and first-order variables, we find the same number. We now show, as one would expect for linearized waves, that the equations do not determine the amplitude of the disturbance. The temporal derivative of Equation (3.2.5) is

$$\rho_0 \frac{\partial^2 \mathbf{V}_1}{\partial t^2} + \nabla(\frac{\partial p_1}{\partial t} + \frac{\mathbf{B}_0 \cdot \frac{\partial}{\partial t}\mathbf{B}_1}{4\pi}) - \frac{(\mathbf{B}_0 \cdot \nabla)\frac{\partial}{\partial t}\mathbf{B}_1}{4\pi} = 0 \quad , \tag{3.2.8}$$

in which we eliminate all first-order variables except $\mathbf{V}_1$ using Equations (3.2.4), (3.2.6), and (3.2.7),

$$\begin{aligned} \rho_0 \frac{\partial^2 \mathbf{V}_1}{\partial t^2} - p_0 \frac{5}{3} \nabla(\nabla \cdot \mathbf{V}_1) + \frac{1}{4\pi} \nabla[\mathbf{B}_0 \cdot (\nabla \times (\mathbf{V}_1 \times \mathbf{B}_0))] & \\ - (\frac{\mathbf{B}_0}{4\pi} \cdot \nabla)[\nabla \times (\mathbf{V}_1 \times \mathbf{B}_0)] = 0 & \end{aligned} \quad . \tag{3.2.9}$$

This equation describes the evolution of an arbitrary initial disturbance of the mass flow velocity, $\mathbf{V}_1$, in space and time. The amplitude of $\mathbf{V_1}$ is a constant factor in space and time, and cancels. The other first-order variables can be evaluated in a similar way. For instance, $\mathbf{B_1}$ can be found from Equation (3.2.6).

3.2.2. Dispersion relation and polarization (parallel propagation)

Most wave equations like (3.2.9) look very complicated. Usually they can be interpreted by evaluating them for simple cases, like parallel or perpendicular wave propagation to the magnetic field. These simplifications should not be misunderstood as helpless attempts in the face of sheer complexity, nor as likely cases (other propagation angles to the magnetic field may be equally frequent). The extreme cases and approximations identify important physics. Even the names of waves refer to physical properties in simple limits rather than to a particular mode. For this reason a mathematically identical wave (a particular branch of the solution) may have different names at parallel and perpendicular propagation, at low and high frequency. This may be surprising or confusing in the beginning, but it indicates important differences in the physics of the wave.

We shall first solve Equation (3.2.9) for waves propagating parallel to the external magnetic field, $\mathbf{B_0}$, assumed to be in the z-direction. For simplicity we take the disturbance as an infinitely extended plane wave with an amplitude independent of space and time. Thus we write

$$\mathbf{V}_1 = \mathbf{V}_1^{\perp} + \mathbf{V}_1^{z} = (\overline{\mathbf{V}}_1^{\perp} + \overline{\mathbf{V}}_1^{z})e^{i(kz-\omega t)} \tag{3.2.10}$$

The bar indicates that the quantity is an amplitude. It will later be omitted when no ambiguities are possible. The z coordinate in the exponent marks the only spatial variation. It causes the phase of the wave to propagate in the positive z-direction if $k > 0$, and in the negative z-direction if $k < 0$. We use here the convention that the wave frequency is always positive. (Note that the inverse convention, $\omega < 0$ and $k > 0$ for negative wave direction, is also widely used in the literature).

Since Equation (3.2.9) is linear and the zero-order terms are constant, the use of Equation (3.2.10) corresponds to a Fourier transformation. In fact, more complicated cases can only be solved by the proper Fourier method. Plugging in Equation (3.2.10) (or Fourier transformation) turns the derivatives into factors,

$$\frac{\partial}{\partial t} \rightarrow -i\omega, \ \nabla \rightarrow i\mathbf{k} \quad .$$

Equation (3.2.9) becomes

$$(\frac{5}{3}p_0 k^2 - \rho_0\omega^2)\mathbf{V}_1^z + (\frac{k^2 B_0^2}{4\pi} - \rho_0\omega^2)\mathbf{V}_1^{\perp} = 0 \quad . \tag{3.2.11}$$

This relation must hold for any wave amplitude $\overline{\mathbf{V}}_1^z$ and $\overline{\mathbf{V}}_1^{\perp}$, which are independent of each other and arbitrary. For a non-trivial solution, either of the two expressions in parentheses and the other amplitude must be equal to zero. The two possibilities correspond to two wave modes studied below in detail.

From the first parentheses in Equation (3.2.11) we get

$$\frac{\omega^2}{k^2} = \frac{5p_0}{3\rho_0} =: c_s^2 \quad . \tag{3.2.12}$$

A connection between ω and k as in Equation (3.2.12) is generally called a *dispersion relation.* The wave is named *sound wave* and the phase velocity of this wave, $\omega/k = c_s$, is the sound velocity. The restoring force, as indicated by the numerator, is a gradient in pressure. The factor 5/3 is the ratio of specific heats corresponding to the 3 degrees of freedom of each particle in a plasma. In a completely ionized plasma with solar abundances $c_s = 1.51 \cdot 10^4 \sqrt{T}$ [cm s^{-1}], where T is in degrees kelvin. The group velocity of the wave, being the speed at which the wave energy is carried, equals $\partial\omega/\partial k$ and in this case is also c_s.

The second expression in parentheses in Equation (3.2.11) describes a completely different wave,

$$\frac{\omega^2}{k^2} = \frac{B_0^2}{4\pi\rho_0} =: c_A^2 \quad . \tag{3.2.13}$$

The wave exists only if $\mathbf{B_0} \neq 0$. It is the *Alfvén wave*, named after its discoverer who received the Nobel Prize in 1969 for this and other contributions. (It is also called the shear Alfvén wave to distinguish it from the compressional Alfvén wave introduced later.) The phase velocity of the waves, c_A, is called the Alfvén velocity. We shall use the relation $c_A \approx c(m_e/m_p)^{1/2}\Omega_e/\omega_p^e$ for a hydrogen plasma, and, for solar abundances, its numerical value $2.03 \cdot 10^{11} B_0/\sqrt{n_e}$ [cm s^{-1}].

What is the difference between the two wave modes? Let us investigate their polarization (meaning the directions in which the sinusoidal disturbances oscillate). The sound wave has an amplitude $\mathbf{V}_1^z$ and the particles oscillate only in the z-direction,

$$\mathbf{V}_1 = \overline{V}_1^z \mathbf{e}_z e^{i(kz-\omega t)} \quad , \tag{3.2.14}$$

where $\mathbf{e}_z$ is the unit vector in the z-direction. Such a wave in which the particles oscillate in the direction of propagation is called *longitudinal*. Since $\mathbf{V}_1$ is parallel to $\mathbf{B_0}$, it follows from Equation (3.2.6) that

$$-i\omega\mathbf{B}_1 = ik\mathbf{e}_z \times (\mathbf{V}_1 \times \mathbf{B}_0) = 0 \quad . \tag{3.2.15}$$

Faraday's equation and Equation (3.2.15) also require that $\mathbf{E}_1 = 0$. Equations (3.2.4) and (3.2.7) give

$$\rho_1 = \rho_0 \frac{k}{\omega} V_1^z \tag{3.2.16}$$

$$p_1 = \frac{5}{3} p_0 \frac{k}{\omega} V_1^z \quad . \tag{3.2.17}$$

Thus a sound wave parallel to $\mathbf{B}_0$ does not cause any electric or magnetic disturbances in a plasma, but oscillates in density and pressure like in a neutral gas. These waves are purely hydrodynamic.

On the other hand, the particles in Alfvén waves oscillate in transverse motion to the magnetic field and the direction of propagation, since

$$\mathbf{V}_1 = \overline{\mathbf{V}}_1^{\perp} e^{i(kz-\omega t)} \quad . \tag{3.2.18}$$

The magnetic disturbance is perpendicular to $\mathbf{B}_0$ and 180° out of phase with $\mathbf{V}_1$, as it follows from Equation (3.2.6) that

$$\mathbf{B}_1 = -\frac{1}{\omega}\mathbf{k} \times (\mathbf{V}_1 \times \mathbf{B}_0) = -\frac{\text{sign}(k)kB_0}{\omega}\mathbf{V}_1 \quad , \tag{3.2.19}$$

$$\mathbf{J}_1 = \frac{c}{4\pi} ik\mathbf{e}_z \times \mathbf{B}_1 \quad . \tag{3.2.20}$$

However, from Equations (3.2.4) and (3.2.7), we find

$$\rho_1 = 0 \ \text{ and } \ p_1 = 0 \quad . \tag{3.2.21}$$

Alfvén waves are therefore purely magnetodynamic, and the perturbations do not compress the plasma. They cause an oscillating ripple on the magnetic field line and may be compared to oscillations of a violin string. As suggested by the expression (3.2.13) of the phase velocity, the restoring force is magnetic tension, $B_0^2/4\pi$, indeed analogous to strings. It is the result of the magnetic gradient term in the equation of motion (3.2.5). Alfvén waves are most unusual in that they are also solutions of the full, non-linear equations, as can easily be shown by substitution. For this reason, the dispersion relation of Alfvén waves does not change for large amplitudes, and the wave does not dissipate energy by non-linear effects. As an important consequence, Alfvén waves can transport energy over long distances, even if the plasma changes gradually, as in a corona.

3.2.3. Perpendicular propagation

Let us now look at the case where the waves propagate in the x-direction with $\mathbf{B}_0$ still in the z-direction,

$$\mathbf{V}_1 = (\overline{\mathbf{V}}_1^x + \overline{\mathbf{V}}_1^y + \overline{\mathbf{V}}_1^z)e^{i(kx-\omega t)} \quad . \tag{3.2.22}$$

This is representative of all perpendicular waves without loss of generality. Analogous to the parallel case, Equation (3.2.9) gives

$$(-\rho_0\omega^2 + \frac{5}{3}p_0k^2 + \frac{k^2B_0^2}{4\pi})\mathbf{V}_1^x + (-\rho_0\omega^2)\mathbf{V}_1^y + (-\rho_0\omega^2)\mathbf{V}_1^z = 0 \quad . \tag{3.2.23}$$

Again the zeros of the expressions in parentheses are solutions for wave modes. For the second and third expression we find $\omega = 0$; there is no MHD wave perpendicular to $\mathbf{B}_0$ with transverse ($\perp \mathbf{B}$) particle oscillation. We have a mode from the first parenthesis with

$$\frac{\omega^2}{k^2} = \frac{5}{3}\frac{p_0}{\rho_0} + \frac{B_0^2/4\pi}{\rho_0} \ := \ c_s^2 + c_A^2 \quad . \tag{3.2.24}$$

The wave is longitudinal and possesses a combination of acoustic and electromagnetic properties. It is named a *fast magnetoacoustic wave* since it is faster than both sound and Alfvén waves. It is a longitudinal wave. For $c_s \to 0$, the fast magnetoacoustic wave does not behave like the Alfvén wave derived in Equation (3.2.13), although $w/k = c_A$. It is then called the *compressional Alfvén wave*.

3.2.4. General case

Having explored the physics of the waves in the simple cases of parallel and perpendicular propagation, we now look at the intermediate angles to outline briefly how the modes change with angle. The three modes found for parallel and perpendicular directions keep their identity in the general case of skew propagation. This may have been expected from the fact that the disturbances of the three waves are mutually perpendicular in parameter space. However, the physics of the waves changes considerably.

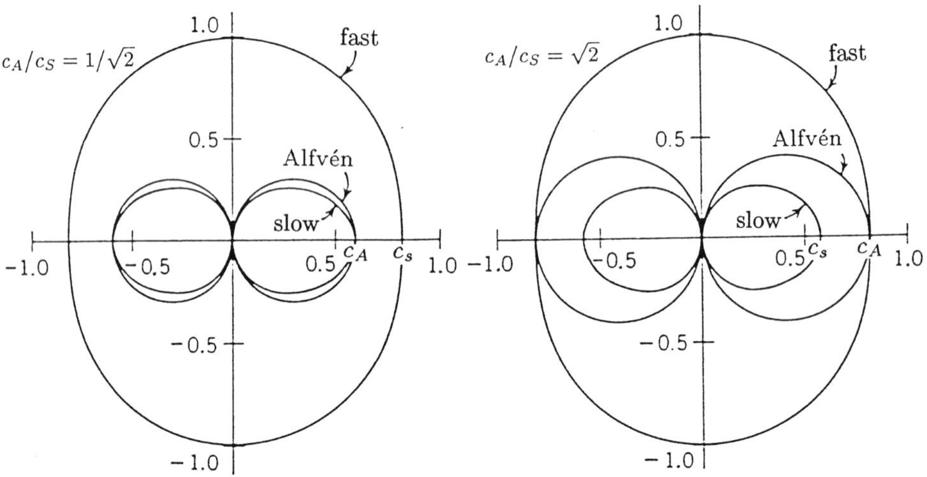

Fig. 3.1. Polar diagram of vector phase velocities. The length of the radius vector at an inclination angle θ to the equilibrium magnetic field equals the phase speed (ω/k) for waves propagating in that direction. Speeds have been normalized with respect to $(c_A^2 + c_s^2)^{1/2}$.

In non-parallel directions, sound waves are not purely longitudinal, and the Lorentz force takes part in the oscillation through magnetic tension and magnetic pressure. The pure sound wave is a singularity of the parallel and $B = 0$ cases. At a general angle, the mode has a magnetoacoustic nature. However, it does not necessarily coincide with the fast magnetoacoustic mode!

In the parallel direction and for $c_A \geq c_s$, the fast magnetoacoustic wave propagates at the Alfvén velocity and becomes a purely transverse wave driven by magnetic tension. It produces no compression and is physically identical to the Alfvén wave (right part of Figure 3.1). However, for $c_A \leq c_s$ (left part of Figure 3.1) the fast magnetoacoustic wave in the parallel direction is identical to the sound wave. Except for the singularities at parallel and perpendicular propagation, there is always a *fast and a slow magnetoacoustic mode.*

The phase velocity of the Alfvén wave approaches zero at perpendicular phase velocity, since the relevant field component of the magnetic tension decreases with the sine of the inclination angle. Its speed is *intermediate* and always faster than the slow magnetoacoustic mode. The name 'intermediate mode' is not commonly used for linear Alfvén waves, but for non-linear waves (Chapter 9). The slow mode at parallel propagation can be a sound wave or an Alfvén wave, depending on the ratio of c_A/c_s. Note that the expressions 'slow', 'intermediate' and 'fast' are mostly name tags of mathematical solutions and do not express the physics of the waves.

Figure 3.1 summarizes the phase velocities of the three MHD modes for different inclinations to the magnetic field and two values of the ratio of Alfvén speed to sound speed. In general, the group velocity, $\partial\omega/\partial k$, has different values and directions from the phase velocity. A remarkable example is formed by the Alfvén (intermediate) waves whose group velocity is always field aligned and thus carries energy only along $\mathbf{B}$ regardless of the inclination of $\mathbf{k}$.

Some physical properties persist over all angles and c_A/c_s ratios. For fast and slow waves both $\mathbf{V_1}$ and $\mathbf{B_1}$ remain in the plane defined by $\mathbf{B}$ and $\mathbf{k}$. On the other hand, $\mathbf{V_1}$ and $\mathbf{B_1}$ of the Alfvén waves are perpendicular to this plane. For the fast mode, the magnetic pressure and the density always oscillate in phase, but magnetic pressure and density oscillations are 180° out of phase in the slow mode.

Exercises

3.1: Assume a highly ionized hydrogen corona in equilibrium at rest ($V^e = V^p = 0$) with $p_e \approx p_p$. Prove that the gravity of the star creates an electric field

$$eE = \frac{1}{2}(m_p - m_e)g \approx \frac{1}{2}m_p g \tag{3.2.25}$$

in the upward direction. It prohibits the sedimentation of the heavier protons at the bottom of the corona, as one would expect for an atmosphere of neutral particles. Compare the force (3.2.25) to the Lorentz force $F_L = eBv/c$, where $v/c \approx 1/100$ for a thermal electron in the corona and B $\approx$ 1 G.

3.2: Calculate β (plasma beta) at the site of a flare before the energy is released, assuming thermal equilibrium and pressure equilibrium, an electron density of 10^{10} cm^{-3}, a magnetic field of 100 G, and a temperature of $5 \cdot 10^6$ K. Let us first assume for simplicity that the flare locally increases only the electron temperature by a factor of 100 and leaves the other plasma parameters unchanged (neglect non-thermal particles). What is now β and what is the consequence? Assume solar abundances, thus $n_i = 0.92\ n_e$. What would happen if only the local magnetic field were annihilated?

3.3: Describe the properties of the slow mode MHD waves traveling parallel to $\mathbf{B}$ for both $c_A > c_s$ and $c_A < c_s$.

3.4: Prove that in an ideal and non-relativistic MHD plasma the ratio of electric energy density to magnetic energy density is always $(V_\perp/c)^2$, and show that the kinetic and magnetic wave energy are equal for an Alfvén wave.

Further Reading and References

Magnetohydrodynamics

Cowling, T.G.: 1976, *Magnetohydrodynamics*, 2nd edition, A. Hilger, Bristol, England.

Priest, E.R.: 1982, *Solar Magnetohydrodynamics*, D. Reidel, Dordrecht, Holland.

MHD equilibrium and instability

Bateman, G.: 1978, *MHD Instability*, MIT Press, Cambridge, MA, USA.

Manheimer, W.M. and Lashmore-Davies, C.: 1984, *MHD Instabilities in Simple Plasma Configuration*, Naval Research Laboratory, Washington, DC.

Van Hoven, G.: 1981, 'Simple-Loop Flares: Magnetic Instabilities', in *Solar Flare MHD* (ed. E.R. Priest), Gordon and Breach Science Publishers, New York.

Thermal conductivity

Spitzer, L.: 1962, *Physics of Fully Ionized Gases*, Interscience, New York.

CHAPTER 4

WAVES IN A COLD, COLLISIONLESS PLASMA

In the tenuous plasmas of coronae where fluid turbulence and vortex motions are frequently inhibited by the magnetic field, waves take their place to carry away the energy and momentum of local disturbances. Many kinds of waves are possible in a plasma depending on frequency, species of oscillating particles, restoring force, boundary conditions, inhomogeneity, propagation angle to the magnetic field, etc. In this chapter we shall give an overview of the basic wave modes in homogeneous plasmas ranging from the MHD waves, at periods of the order of minutes under solar conditions, to the high-frequency electromagnetic waves escaping from the corona as radio to X-ray emissions. The common physics of waves as collective phenomena has already been emphasized in Section 3.2 on MHD waves.

What happens in a wave whose frequency exceeds the collision rate? In principle, each particle or group of particles could oscillate in its own way. The velocity distribution may oscillate and not remain Maxwellian. In a first approach that is *not kinetic*, we simply ignore thermal motions in this chapter and replace the velocity distributions by δ-functions. This is what we mean by the adjective 'cold'. The oscillations of the distributions therefore do not play a role here and will be the topic of the next chapter. As in MHD, the cold plasma is considered as a fluid, and the individuality of particles is neglected. The equations of particle and momentum conservation (the moments of Boltzmann's equation) are similar to MHD, except that there are no temperature effects and the different plasma species are not locked to each other by collisions.

4.1. Approximations and Assumptions

Can a plasma be both free of collisions (hot) and have negligible thermal motions (cold)? The adjectives cold, collisionless, non-relativistic, infinite, etc. mean different simplifications in the fundamental equations. Such simplifications are important for understanding the physics in plasma phenomena. The approximations hold as long as the neglected effects are smaller than the phenomenon under scrutiny. Note that there are simplifications that exclude certain effects altogether. The sound wave, for example, does not appear under the cold plasma assumptions, and can only be recovered by allowing for thermal motions. The danger of plasma physics is to exceed the limits of validity set by these simplifying assumptions. In practice, one often does not know these limits. To give an example, interplanetary

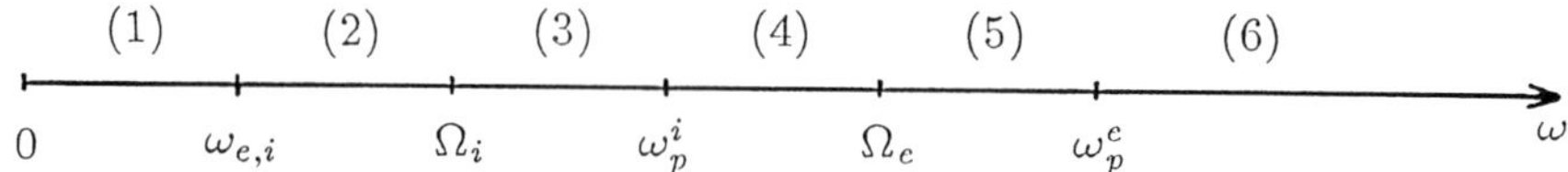

Fig. 4.1. Characteristic frequencies in the order typical for the solar corona. The numbers refer to frequency ranges discussed in the text. $\omega_{e,i}$ is the electron-ion collision frequency (Eq. 2.6.32).

density variations make it questionable to consider the interplanetary medium as an infinite, homogeneous background plasma even for localized processes.

Characteristic frequencies divide a plasma into ranges where different assumptions apply. The parameters of cosmic plasmas vary over many orders of magnitude (see Table 4.1), and this division is not fixed. Depending on the frequency of a wave, the plasma species behave differently. The frequency ranges given as an example in Figure 4.1 for the solar corona have the following properties:

(1) Collisions dominate $\rightarrow$ MHD waves.

(2) Collisionless plasma, but $\rho^* \approx 0 \rightarrow$ Chew–Goldberger–Low waves (similar to MHD, but the pressure can be different in directions parallel and perpendicular to $\mathbf{B}_0$).

(3) Ions are *unmagnetized*, meaning that the wave oscillates faster than the ion gyration time. The orbit of an ion within a wave period can be approximated by a straight line and is independent of the magnetic field. $\rightarrow$ Ions and electrons behave differently and have to be described by two fluids.

(4) Ions are practically immobile. $\rightarrow$ Only electrons are important (one fluid).

(5) Electrons are *unmagnetized*.

(6) Electrons are also immobile. $\rightarrow$ The plasma increasingly comes to resemble a vacuum.

In every range there is, in principle, a different system of equations suitably describing linear waves. The waves keep their identity, but gradually change their character from one range to the next.

Table 4.1. Typical parameters of various plasmas and their characteristic frequencies.

Plasma	n [cm^{-3}]	T [$^\circ K$]	B [G]	$\nu_{e,i}$ [Hz]	$\Omega_e/2\pi$ [Hz]	ν_p [Hz]
intergalactic	10^{-6}	10^5	10^{-8}	10^{-12}	$3 \cdot 10^{-2}$	10
interstellar	10^{-2}	10^2	10^{-5}	10^{-3}	30	10^3
interplanetary	10	10^5	10^{-4}	10^{-5}	$3 \cdot 10^2$	$3 \cdot 10^4$
solar corona	10^8	10^6	10	10	$3 \cdot 10^7$	10^8
ionosphere	10^6	10^3	0.1	$3 \cdot 10^3$	$3 \cdot 10^5$	10^7
center of star	10^{24}	10^8	10^6?	10^{14}	$3 \cdot 10^{12}$?	10^{16}
white dwarf	10^{30}	10^8	10^5	*	$3 \cdot 10^{11}$	10^{12}
tokamak	10^{16}	10^8	10^5	10^6	$3 \cdot 10^{11}$	10^{12}

4.2. Cold Plasma Modes

4.2.1. LINEARIZATION

A plasma is said to be cold if processes are investigated that do not depend on thermal motion or pressure. The behavior of waves in a cold plasma is derived from the moments of Boltzmann's equation (Eqs. 3.1.16 and 3.1.21) without the MHD approximation and summation over species. These multi-fluid equations – upon neglecting the pressure term ($p = 0$) and collisions ($S^\alpha = 0$) – contain the physics of a cold, collisionless plasma in the frequency ranges 2 – 6 of Figure 4.1. For each species α there is an equation for particle and momentum conservation,

$$\frac{\partial n^\alpha}{\partial t} + \nabla \cdot (n^\alpha \mathbf{V}^\alpha) = 0 \tag{4.2.1}$$

$$\frac{\partial \mathbf{V}^\alpha}{\partial t} + (\mathbf{V}^\alpha \cdot \nabla)\mathbf{V}^\alpha = \frac{q_\alpha}{m_\alpha}(\mathbf{E} + \frac{1}{c}\mathbf{V}^\alpha \times \mathbf{B}^\alpha) \quad . \tag{4.2.2}$$

In addition we are using the full set of Maxwell's equations in their classical form,

$$\nabla \cdot \mathbf{B} = 0 \qquad \nabla \times \mathbf{E} = -\frac{1}{c}\frac{\partial \mathbf{B}}{\partial t} \qquad \rho^* := \sum_\alpha q_\alpha n_\alpha \tag{4.2.3}$$

$$\nabla \cdot \mathbf{E} = 4\pi\rho^* \qquad \nabla \times \mathbf{B} = \frac{4\pi}{c}\mathbf{J} + \frac{1}{c}\frac{\partial \mathbf{E}}{\partial t} \qquad \mathbf{J} := \sum_\alpha q_\alpha n_\alpha \mathbf{V}^\alpha \quad . \tag{4.2.4}$$

The method to extract the linear modes (periodic solutions to small disturbances of an equilibrium) is analogous to the MHD waves (Section 3.2). We study the temporal evolution of the variables of the system having the form $A = A_0 + A_1(\mathbf{x}, t)$, where $A_0 = const$ is the value of the homogeneous, stationary plasma. A_1 is a small deviation compared to the stationary value, satisfying $|A_1| \ll |A_0|$. We shall furthermore assume that $\mathbf{E}_0 = 0$, $\mathbf{J}_0 = 0$, and $\rho_0^* = 0$. The zero-order terms satisfy Equations (4.2.1) – (4.2.4). When the variables of the disturbed plasma are put in, the zero-order terms cancel. Products of first-order variables can be considered of higher order and are neglected. The remaining system of equations is linear,

$$\frac{\partial n_1^\alpha}{\partial t} + \mathbf{V}_0^\alpha \cdot \nabla n_1^\alpha + n_0^\alpha \nabla \cdot \mathbf{V}_1^\alpha = 0 \tag{4.2.5}$$

$$\frac{\partial \mathbf{V}_1^\alpha}{\partial t} + (\mathbf{V}_0^\alpha \cdot \nabla)\mathbf{V}_1^\alpha = \frac{q_\alpha}{m_\alpha}(\mathbf{E}_1 + \frac{1}{c}\mathbf{V}_1^\alpha \times \mathbf{B}_0 + \frac{1}{c}\mathbf{V}_0^\alpha \times \mathbf{B}_1) \tag{4.2.6}$$

$$\nabla \times \mathbf{E}_1 = -\frac{1}{c}\frac{\partial \mathbf{B}_1}{\partial t} \tag{4.2.7}$$

$$\nabla \cdot \mathbf{E}_1 = 4\pi\rho_1^* \tag{4.2.8}$$

$$\rho_1^* = \sum_\alpha q_\alpha n_1^\alpha \tag{4.2.9}$$

$$\nabla \times \mathbf{B}_1 = \frac{4\pi}{c}\mathbf{J}_1 + \frac{1}{c}\frac{\partial \mathbf{E}_1}{\partial t} \tag{4.2.10}$$

$$\nabla \cdot \mathbf{B}_1 = 0 \tag{4.2.11}$$

$$\mathbf{J}_1 = \sum_\alpha q_\alpha (n_1^\alpha \mathbf{V}_0^\alpha + n_0^\alpha \mathbf{V}_1^\alpha) \quad . \tag{4.2.12}$$

This system of equations is very general and includes a large variety of waves. It also contains the displacement current in Equation (4.2.10) which will be important for high-frequency waves. We shall restrict the discussion to two special cases revealing the most important wave modes and their properties:

(1) $\mathbf{B}_0 \neq 0, \mathbf{V}_0^\alpha = 0$ for all particle species α.

(2) $\mathbf{B}_0 = 0, \mathbf{V}_0^\alpha \neq 0$ with at least two species moving in relation to each other. This case will be studied in Section 4.6.

For a homogeneous stationary background, the Fourier transformation can be carried out by the simple form for plane waves, assuming that all first-order terms have the form

$$A_1(\mathbf{x}, t) = \bar{A}_1 \exp[i(\mathbf{k} \cdot \mathbf{x} - \omega t)] \quad . \tag{4.2.13}$$

All derivatives then become factors,

$$\frac{\partial}{\partial t} \to -i\omega, \ \nabla \to i\mathbf{k} \quad .$$

The dispersion relation can be calculated from Faraday's law (Eq. 4.2.7) if the magnetic field, $\mathbf{B_1}$, is expressed by the electric field. For this purpose we first have to calculate the electric conductivity and the dielectric tensor, and then use Ampère's equation (4.2.10).

4.2.2. OHM'S LAW

Electric conductivity is the relation between electric field and current density. Thus we search for an equation of the form of Ohm's law. It follows from the linearized momentum equation (4.2.6) and the assumptions $\mathbf{B}_0 \neq 0$ and $\mathbf{V}_0^\alpha = 0$ that

$$-i\omega \mathbf{V}_1^\alpha = \frac{q_\alpha}{m_\alpha}(\mathbf{E}_1 + \frac{1}{c}\mathbf{V}_1^\alpha \times \mathbf{B}_0) \quad . \tag{4.2.14}$$

Using the gyrofrequency in vector form, $\mathbf{\Omega}^\alpha := q_\alpha \mathbf{B}_0/(m_\alpha c)$ (Eq. 2.1.4), this transforms into

$$-i\omega \mathbf{V}_1^\alpha + \mathbf{\Omega}^\alpha \times \mathbf{V}_1^\alpha = \frac{q_\alpha}{m_\alpha}\mathbf{E}_1 \quad . \tag{4.2.15}$$

This linear, homogeneous system of three equations can easily be inverted. Let us use a frame of reference in which the z-axis coincides with the direction of $\mathbf{B_0}$. The gyrofrequency vector, $\mathbf{\Omega}^\alpha$, has a non-vanishing component in the z-direction only. We shall use the z-component of the gyrofrequency defined by $\Omega_z^\alpha := q/|q| \ \Omega_\alpha$. The value of Ω_z^α is positive or negative depending on the sign of the charge. Then

$$\mathbf{V}_1^\alpha = \frac{q_\alpha}{m_\alpha} \begin{pmatrix} \frac{i\omega}{\omega^2-(\Omega_z^\alpha)^2} & \frac{-\Omega_z^\alpha}{\omega^2-(\Omega_z^\alpha)^2} & 0 \\ \frac{\Omega_z^\alpha}{\omega^2-(\Omega_z^\alpha)^2} & \frac{i\omega}{\omega^2-(\Omega_z^\alpha)^2} & 0 \\ 0 & 0 & \frac{i}{\omega} \end{pmatrix} * \mathbf{E}_1 \quad , \tag{4.2.16}$$

$$\mathbf{V}_1^\alpha = \widehat{\mathbf{M}}_\alpha * \mathbf{E}_1 \quad . \tag{4.2.17}$$

Equations (4.2.16) and (4.2.17) define a mobility tensor, $\widehat{\mathbf{M}}_\alpha$, describing the effect of the wave electric field on the mean velocity of particle species α. We immediately get Ohm's law and the conductivity tensor $\hat{\sigma}$,

$$\mathbf{J}_1 = \sum_\alpha q_\alpha n_\alpha \mathbf{V}_1^\alpha = (\sum_\alpha q_\alpha n_\alpha \widehat{\mathbf{M}}_\alpha) * \mathbf{E}_1 =: \hat{\sigma} * \mathbf{E}_1 \quad . \tag{4.2.18}$$

It is remarkable that there is a finite conductivity, even without collisions. Inspecting Equation (4.2.16) we find, however, that the conductivity relates to the wave character of the disturbance, and it becomes infinite in the z-direction for $\omega = 0$ in agreement with our assumption of ideal MHD. The conductivity appearing in waves is the result of particle inertia hindering free mobility in the wave fields.

4.2.3. Dielectric tensor

As a next step we calculate a formal dielectric tensor. So far we have considered the plasma as a set of particles in empty space, and the dielectric properties of the background medium were those of a vacuum. The physical reason for the appearance of dielectric properties is simply that the electromagnetic field and the current of a wave depend on each other. The mobility of the free charges tends to weaken the electric field and thereby induces a magnetic field. One defines a **D** vector in Ampère's equation (4.2.10) through

$$i\mathbf{k} \times \mathbf{B}_1 = \frac{4\pi}{c}\mathbf{J}_1 - \frac{i\omega}{c}\mathbf{E}_1 =: -\frac{i\omega}{c}\mathbf{D}_1 \quad . \tag{4.2.19}$$

Using Ohm's law (Eq. 4.2.18) we eliminate $\mathbf{J}_1$ and put $\mathbf{D}_1 := \hat{\epsilon} * \mathbf{E}_1$, where we have defined the dielectric tensor

$$\hat{\epsilon} := \hat{\mathbf{1}} - \frac{4\pi}{i\omega}\hat{\sigma} = \hat{\mathbf{1}} - \sum_\alpha \frac{4\pi q_\alpha n_\alpha}{i\omega}\widehat{\mathbf{M}}_\alpha \tag{4.2.20}$$

$$= \begin{pmatrix} 1 - \sum_\alpha \frac{(\omega_p^\alpha)^2}{\omega^2 - (\Omega_z^\alpha)^2} & -i\sum_\alpha \frac{\Omega_z^\alpha (\omega_p^\alpha)^2}{\omega(\omega^2 - (\Omega_z^\alpha)^2)} & 0 \\ i\sum_\alpha \frac{\Omega_z^\alpha (\omega_p^\alpha)^2}{\omega(\omega^2 - (\Omega_z^\alpha)^2)} & 1 - \sum_\alpha \frac{(\omega_p^\alpha)^2}{\omega^2 - (\Omega_z^\alpha)^2} & 0 \\ 0 & 0 & 1 - \frac{(\omega_p)^2}{\omega^2} \end{pmatrix} \tag{4.2.21}$$

$$=: \begin{pmatrix} \epsilon_0 & i\epsilon_1 & 0 \\ -i\epsilon_1 & \epsilon_0 & 0 \\ 0 & 0 & \epsilon_\parallel \end{pmatrix} \quad . \tag{4.2.22}$$

We have defined the plasma frequencies for each species as well as for the whole plasma by

$$(\omega_p^\alpha)^2 := \frac{4\pi q_\alpha^2 n_\alpha}{m_\alpha} \quad , \tag{4.2.23}$$

$$\omega_p^2 := \sum_\alpha (\omega_p^\alpha)^2 \quad . \tag{4.2.24}$$

The reason for the appearance of the plasma frequency is obvious; ω_p is proportional to the ratio of the electric force to inertia (mass), which also controls the oscillation investigated in Section 2.5. The dielectric properties are therefore usually dominated by the lightest particles – the electrons. The dielectric tensor reduces to unity for very high frequencies or effectively infinite inertia and becomes equal to the value in a vacuum.

4.2.4. DISPERSION RELATION

Let us now Fourier transform the linearized Faraday equation (4.2.7)

$$i\mathbf{k} \times \mathbf{E}_1 = \frac{i\omega}{c}\mathbf{B}_1 \quad . \tag{4.2.25}$$

$\mathbf{B}_1$ can be eliminated using Equation (4.2.19). The result is a system of three linear, homogeneous equations,

$$(k^2\hat{\mathbf{1}} - \mathbf{k} \circ \mathbf{k} - \frac{\omega^2}{c^2}\hat{\epsilon}) * \mathbf{E}_1 = 0 \quad . \tag{4.2.26}$$

The tensor product $\circ$ is defined in Appendix A. Equation (4.2.26) only has a non-trivial solution if the determinant of the system equals zero. We thus require

$$\det[(\frac{ck}{\omega})^2\hat{\mathbf{1}} - (\frac{c}{\omega})^2\mathbf{k} \circ \mathbf{k} - \hat{\epsilon}] = 0 \quad . \tag{4.2.27}$$

It is a general dispersion relation for waves in a cold, stationary plasma. The combination $ck/\omega =: \mathcal{N}$ is called the *refractive index*. It controls the refraction (bending) of propagation in inhomogeneous media.

There are four basic roots for linear plasma modes: two *electromagnetic* and two *electrostatic* modes. The electromagnetic waves are related to the two polarization modes of radiation in a vacuum. The charged particles in a plasma participate in the wave oscillations and modify the character of the waves. The electrostatic modes are related to the different oscillation properties of ions and electrons. We have encountered an electrostatic mode in Section 2.5 as the eigenmode of electron oscillations around essentially inert ions. The mode is named *electron plasma wave*. The other electrostatic mode, called *ion wave*, will appear as lower hybrid wave (Section 4.4.1) and as ion acoustic wave (Section 5.2.6). The name 'electrostatic' expresses the fact that there is no magnetic force involved. In the more realistic case, however, when an external magnetic field is present and the wave is not parallel to this field, these waves in general also have magnetic components.

The four basic wave modes do not exist at all frequencies and at all angles to the magnetic field. Furthermore, they behave differently at different frequencies and angles. For this reason, the same mode may carry different names relating to essential physical properties under different approximations. The name of the wave generally characterizes the basic physical principle, not the mode. As for the MHD waves, the physics is best studied in the limiting cases. For an overview we shall ultimately connect the waves at different angles in Section 4.5 to recover the four basic modes.

4.3. Parallel Waves

For a first orientation we have a look at waves that propagate parallel to the magnetic field. We write $\mathbf{k} = (0, 0, k)$, and the determinant (4.2.27) is easily calculated,

$$\epsilon_{\|}[(\frac{c^2k^2}{\omega^2} - \epsilon_0)^2 - \epsilon_1^2] = 0 \quad . \tag{4.3.1}$$

Equation (4.3.1) has three solutions: one electrostatic and two electromagnetic. The parallel electrostatic ion wave does not appear in cold plasma and will be discussed in Section 5.2.6.

4.3.1. Electrostatic waves

A first solution to Equation (4.3.1) is readily extracted by putting

$$\epsilon_{\|} = 0 \quad . \tag{4.3.2}$$

According to the definition of $\epsilon_{\|}$ in Equation (4.2.22),

$$\omega = \omega_p \quad . \tag{4.3.3}$$

We shall show that we have recovered the plasma eigenmode of Section 2.5, except that now all plasma species are allowed to oscillate freely. Thus, the plasma frequency in (4.3.3) is the root mean square over the plasma frequencies of all species. Nevertheless, we shall refer to these waves as *electron plasma waves*, since the oscillation energy of the electrons exceeds that of the ions by the mass ratio $m_i/(m_e Z_i^2)$ (Exercise 4.1). Equation (4.2.26) can only be satisfied (for $\mathbf{E}_1 \neq 0$) if $\mathbf{E}_1$ is parallel to $\mathbf{B}_0$. The mobility equation (4.2.16) then requires that $\mathbf{V}_1$ is also parallel. In other words, the particles oscillate in the same direction as the wave vector. Such a wave is called *longitudinal.*

Furthermore, Ampère's equation (4.2.19) becomes

$$i\mathbf{k} \times \mathbf{B}_1 = -\frac{i\omega}{c}\hat{\epsilon} * \mathbf{E}_1 = 0 \quad . \tag{4.3.4}$$

On the other hand, Equation (4.2.11) requires

$$i\mathbf{k} \cdot \mathbf{B}_1 = 0 \quad . \tag{4.3.5}$$

Both can only be satisfied if $\mathbf{B}_1 = 0$. A wave with this property is generally called *electrostatic.* This special case of a parallel propagating electron plasma wave, where no magnetic effects appear – neither from the background plasma nor in the wave – is named the *Langmuir wave* after one of the pioneers of plasma physics in the 1920s.

The dispersion relation (4.3.3) already indicates how different electron plasma waves are from MHD waves derived in Section 3.2. Only one frequency is possible, but the wavelength is arbitrary. The phase velocity, ω/k, varies with wavelength; a wave with this property is called *dispersive*. The group velocity vanishes as $\partial\omega/\partial k = 0$. A wave packet thus does not propagate, nor does the wave transport energy. It corresponds to the plasma eigenmode introduced in Section 2.5. We note that these extreme properties of electron plasma waves are considerably moderated when we shall allow temperatures $T \neq 0$ in the following chapter.

Electron plasma waves are excellent density diagnostics in astrophysics. Even allowing for thermal effects, their frequency is closely related to the electron density in the source region (Eqs. 4.2.23 and 4.3.3). But how can these non-propagating waves be observed? The conversion of plasma waves into observable electromagnetic waves is discussed in Section 6.3. Such *plasma wave emission* is the generally accepted process for solar type III radio bursts. An alternative way to make electron plasma waves observable is their investigation by radar, which is practicable in the Earth's magnetosphere.

4.3.2. Electromagnetic Waves

Two more solutions of Equation (4.3.1) can be found by putting the term in brackets equal to zero. First, let

$$\frac{c^2k^2}{\omega^2} = \epsilon_0 + \epsilon_1 := \epsilon_L \quad . \tag{4.3.6}$$

The left side of Equation (4.3.6) is the square of the refractive index, and the equation, combined with the definitions (4.2.22) of ϵ_0 and ϵ_1, is the dispersion relation of a further mode supported by a cold, collisionless plasma. To determine the polarization of this wave, we put the dispersion relation (4.3.6) into Equation (4.2.26) and get

$$\begin{pmatrix} \epsilon_1 & -i\epsilon_1 & 0 \\ i\epsilon_1 & \epsilon_1 & 0 \\ 0 & 0 & -\epsilon_{\parallel} \end{pmatrix} * \mathbf{E}_1 = 0 \quad . \tag{4.3.7}$$

This requires

$$E_{1x} - iE_{1y} = 0 \quad , \tag{4.3.8}$$

$$E_{1z} = 0 \quad . \tag{4.3.9}$$

Equation (4.3.9) states that the wave is *transverse*, meaning that $\mathbf{E}_1$ is perpendicular to $\mathbf{k}$. To comprehend Equation (4.3.8) one has to remember that the first-order variables contain a complex exponential, $\exp[i(k_z z - \omega t)]$, but only the real part is observable. It is straightforward to show that the wave is *left circularly polarized.* (Left here means that the $\mathbf{E}_1$ vector for an observer at a given location rotates counterclockwise when looking along the vector $\mathbf{B}_0$.) Note that neither the sense of rotation (defined by $\mathbf{B}_0$) nor any other property changes for a wave with negative k propagating in the negative z-direction. We may in general call the wave defined by Equation (4.3.6) an *L-wave.*

It follows immediately from the Faraday equation (4.2.7) that $\mathbf{B}_1$ is perpendicular to both $\mathbf{k}$ and $\mathbf{E}_1$. The L-mode thus is *electromagnetic.* It differs from the well-known electromagnetic wave in a vacuum by its inclusion of a current ($\mathbf{J}_1 \perp \mathbf{k}$). This has important consequences if a wave is near resonance to one of the characteristic frequencies of the plasma.

The third solution is found in the analogous way,

$$\frac{c^2k^2}{\omega^2} = \epsilon_0 - \epsilon_1 := \epsilon_R \quad , \tag{4.3.10}$$

being the dispersion relation of the *R-wave.* Again one gets the polarization from Equation (4.2.26). The E_{1z} component vanishes and

$$E_{1x} + iE_{1y} = 0 \quad . \tag{4.3.11}$$

The wave is transverse and *right circularly polarized.* It is not just the mirror image of the L-wave, but differs from it by having other resonances. This we are now going to investigate.

4.3.3. Dispersion relations of the L and R waves

L-waves and R-waves are the realizations of the two electromagnetic modes for cold plasma and parallel propagation. They are circularly polarized and can resonate with gyrating particles. In this section we study the effects of resonances on the waves from their dispersion relations. For simplicity, we assume only one species of ions. One derives easily

$$\epsilon_{\substack{L\\R}} = \epsilon_0 \pm \epsilon_1 = 1 - \frac{(\omega_p/\omega)^2}{(1 \mp \frac{\Omega_i}{\omega})(1 \pm \frac{\Omega_e}{\omega})} \quad , \tag{4.3.12}$$

using the modulus of the gyrofrequencies ($\Omega_i := |\mathbf{\Omega}_i|$ and $\Omega_e := |\mathbf{\Omega}_e|$). The upper sign stands for the L-mode, the lower sign for the R-mode. In the following we study this dispersion relation for waves at different frequencies.

- For $\omega \ll \omega_p, \Omega_i, \Omega_e$, both L and R waves have

$$\frac{k^2c^2}{\omega^2} = 1 + \frac{\omega_p^2}{\Omega_i\Omega_e} = 1 + \frac{c^2}{c_A^2} \quad . \tag{4.3.13}$$

This is the dispersion relation for *Alfvén waves* (Eq. 3.2.13),

$$\omega^2 = \frac{k^2c_A^2}{1 + (c_A/c)^2} \quad , \tag{4.3.14}$$

with an additional term stemming from the displacement current, being neglected in MHD. In Section 3.2 we have assumed the form of a linearly polarized Alfvén wave, which can be done in two independent perpendicular directions. Since L and R waves have the same dispersion relations, they can be superposed to a linearly polarized Alfvén wave. At low frequencies, L and R waves are thus identical to Alfvén waves.

- For $\omega \gg \omega_p, \Omega_i, \Omega_e$, one finds from the definitions (4.2.22) and (4.3.12)

$$\epsilon_L = \epsilon_R = 1 \quad . \tag{4.3.15}$$

The dispersion relation then follows immediately as

$$\omega^2 = k^2 c^2 \quad . \tag{4.3.16}$$

The equation is identical to the dispersion relation of electromagnetic waves in a vacuum; so the waves are the same as ordinary radiation.

4.3.4. Resonances at the Gyrofrequencies

It is not surprising that the dispersion relation of the L-wave (Eq. 4.3.12) has a singularity at Ω_i, the gyrofrequency of the left circling ions. At this frequency the ions rotate in phase with the wave, feel a constant electric field $\mathbf{E_1}$, and quickly exchange energy with the wave. L-waves below Ω_i are called *ion cyclotron waves.* If propagating into a region where it is in resonance, a wave can be reflected or absorbed, depending on the damping processes. We note that an L-wave propagating into the negative z-direction is also in resonance with the ions.

The analogous process occurs for R-waves at Ω_e. R-waves between the electron and ion gyrofrequencies have peculiar properties deserving special attention. In the range

$$\Omega_i \ll \omega < \Omega_e \ll \omega_p \quad , \tag{4.3.17}$$

Equations (4.3.10) and (4.3.12) can be approximated by

$$\frac{k^2c^2}{\omega^2} \approx \frac{\omega_p^2}{\omega(\Omega_e - \omega)} \quad . \tag{4.3.18}$$

In the form of phase velocity, the dispersion relation (4.3.18) becomes

$$v_{ph} = \frac{\omega}{k} \approx \frac{c\Omega_e}{\omega_p}\sqrt{\frac{\omega}{\Omega_e}(1 - \frac{\omega}{\Omega_e})} = c_A\sqrt{\frac{m_i}{m_e}}\sqrt{\frac{\omega}{\Omega_e}(1 - \frac{\omega}{\Omega_e})} \quad . \tag{4.3.19}$$

These R-waves are called *whistlers*, since their group velocity depends on frequency,

$$v_{gr} = \frac{\partial\omega}{\partial k} \approx 2c_A\sqrt{\frac{m_i\omega}{m_e\Omega_e}}\left(1 - \frac{\omega}{\Omega_e}\right)^{3/2} = 2v_{ph}(1 - \frac{\omega}{\Omega_e}) \quad . \tag{4.3.20}$$

When pulses of whistler waves are excited, the high-frequency waves will arrive first. This property gives the waves their peculiar name. With ordinary long-wave radio receivers one occasionally hears a whistling sound at a decreasing pitch originating from whistler waves excited by terrestrial lightning and propagated through the magnetosphere.

4.3.5. CUTOFFS NEAR ω_p

It is important to discuss carefully the behavior of observable waves near the plasma frequency. For $\omega \gg \Omega_i$ one can approximate Equation (4.3.12) by

$$\left(\mathcal{N}_{\substack{L\\R}}\right)^2 := \epsilon_{\substack{L\\R}} \approx 1 - \frac{(\omega_p/\omega)^2}{1 \pm \Omega_e/\omega} \quad . \tag{4.3.21}$$

The right side – being equal to the square of the refractive index $\mathcal{N}$ – must be positive for the modes to exist. This condition creates a *cutoff* in frequency below which the waves are evanescent (they do not propagate since their refractive index is purely imaginary). The refractive index equals zero at the cutoff. A propagating wave that meets a cutoff is usually reflected. For R-waves (minus sign in Eq. 4.3.21) the cutoff is at

$$\omega_x \approx \sqrt{\omega_p^2 + \frac{1}{4}\Omega_e^2} + \frac{1}{2}\Omega_e \quad . \tag{4.3.22}$$

For $\omega_p^2 \gg \Omega_e^2$, the cutoff frequency is about $\omega_p + \frac{1}{2}\Omega_e$. In the other extreme, for $\omega_p^2 \ll \Omega_e^2$, the cutoff frequency $\omega_x \approx \Omega_e(1 + \omega_p^2/\Omega_e^2)$. It is easily shown that for R-waves the cutoff is always above $\max(\omega_p, \Omega_e)$, above the two elementary frequencies. The cutoff determines the lowest frequency of waves that can escape from a stellar atmosphere and that can be observed (see Fig. 4.2).

According to Equation (4.3.21) L-waves have a cutoff at $\omega_z = (\omega_p^2 + \frac{1}{4}\Omega_e^2)^{1/2} - \frac{1}{2}\Omega_e$. However, this is a singularity of strictly parallel propagation and is of no practical importance. For all other propagation angles there is a resonance as well as a cutoff at a higher frequency (as will become clear in Section 4.4), namely the plasma frequency. Thus in realistic circumstances, L-waves escape for

$$\omega > \omega_p \quad . \tag{4.3.23}$$

The local plasma frequency is therefore the lowest frequency of electromagnetic radiation that can leave a corona from a given height.

The L and R waves are evanescent between the resonance at Ω_i and Ω_e, and their cutoffs at ω_p and ω_x, respectively. This part of the spectrum is called the *stop* region.

Cutoffs and resonances are important for the understanding of electromagnetic radiation from stars. Broadband emission processes cannot excite waves in the stop region between resonance and cutoff in the source. Secondly, a propagating wave may be absorbed. While propagating in an inhomogeneous region, its

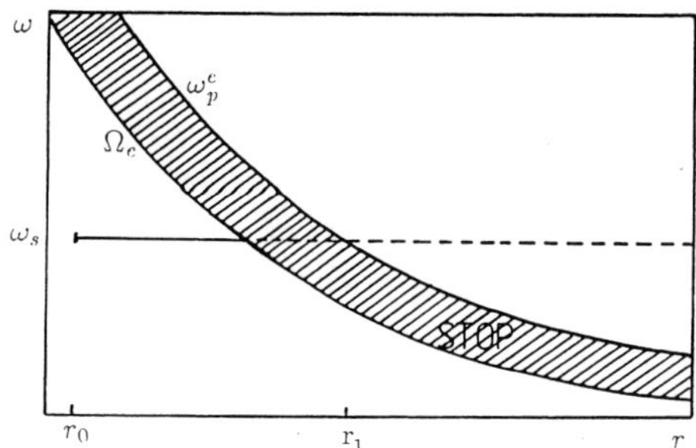

Fig. 4.2. Schematic drawing of whistler wave propagation in a corona. Plasma frequency and electron gyrofrequency vs. radial distance r from the center of the star. No electromagnetic waves can propagate in the 'stop' region.

frequency remains constant. It may reach a stop region if the local characteristic frequencies decrease along the path. Whistler waves, as an example, cannot leave a corona, since their frequency at some higher altitude will exceed the local electron gyrofrequency. In Figure 4.2 a whistler of frequency ω_s originates at r_0. It propagates only until ω_s is about the local gyrofrequency Ω_e, where it is absorbed and/or reflected. Waves observable at ω_s from the outside must originate at $r > r_1$. In extreme cases, tunneling through the stop region is possible. It has been observed in regions with a steep density gradient in the Earth's ionosphere.

4.4. Perpendicular Propagation

4.4.1. Electrostatic waves

The electrostatic modes can easily be extracted from Equation (4.2.26) by scalar multiplication by $\mathbf{k}$ from the left, and the result is

$$\frac{\omega^2}{c^2}\mathbf{k} \cdot (\hat{\epsilon} * \mathbf{E}_1) = 0 \quad . \tag{4.4.1}$$

Electrostatic waves are longitudinal, i.e. $\mathbf{E}_1 \parallel \mathbf{k}$, as required by the Faraday equation (4.2.25) and $\mathbf{B}_1 = 0$. Equation (4.4.1) leads immediately to the general dispersion relation for longitudinal, electrostatic waves,

$$\mathbf{k} \cdot (\hat{\epsilon} * \mathbf{k}) = 0 \quad . \tag{4.4.2}$$

It includes the parallel case (Eq. 4.3.2). In the perpendicular case, $\mathbf{k} = (k, 0, 0)$, Equation (4.4.2) gives

$$\epsilon_0 = 0 \quad . \tag{4.4.3}$$

The definition of ϵ_0, Equation (4.2.22), then yields for $\omega \gtrsim \omega_p \gg \Omega_e$

$$\omega^2 \approx \omega_p^2 + \Omega_e^2 \quad . \tag{4.4.4}$$

This mode is called an *upper hybrid wave*, since it is a combination of space charge oscillations with the gyration of electrons. Its frequency, given by Equation (4.4.4), is the upper hybrid frequency. The wave is the analog of the parallel electron plasma waves and may be a cause of radio emission, as will be discussed in Chapters 8 – 10. In the general case with angle θ between $\mathbf{k}$ and B_0, Equation (4.4.4) is replaced by

$$\omega^2 \approx \omega_p^2 + \Omega_e^2 \sin^2\theta \tag{4.4.5}$$

(Exercise 4.2). It also includes the parallel case ($\epsilon_{\|} = 0$), the electron plasma wave of the previous section.

The conditions $\Omega_i \ll \omega \ll \Omega_e$ lead to a completely different perpendicular wave. We find

$$\epsilon_0 \approx 1 + \frac{(\omega_p^e)^2}{\Omega_e^2}(1 - \frac{\Omega_e \Omega_i}{\omega^2}) = 0 \quad , \tag{4.4.6}$$

and by simple manipulation

$$\omega^2 \approx \frac{(\omega_p^i)^2}{1 + (\omega_p^e/\Omega_e)^2} \quad . \tag{4.4.7}$$

This electrostatic wave is known as the *lower hybrid wave*. Its frequency is called lower hybrid frequency, amounting to $\omega^2 \approx \Omega_i \Omega_e$ for $\omega_p \gg \Omega_e$.

For $\omega_p \ll \Omega_e$ Equation (4.4.7) yields a wave frequency of about ω_p^i. Why does the plasma frequency of the ions appear? This is an interesting piece of wave physics. The wave frequency is small compared to the gyrofrequency of electrons; thus they circle many times per wave period. Electrons therefore remain closely attached to their magnetic field line. The ions, however, need much longer to gyrate than a wave period and move a practically linear orbit during this time. As a result, they appear to be not bound to the magnetic field and can freely move within a wave period. The lower hybrid wave is an oscillation of space charge of the ions. This is in contrast to electron plasma waves, where electrons oscillate around the inert ions (Section 4.3.1). Electrons and ions have changed their roles! For example, lower hybrid waves can be excited by perpendicular ion currents (Chapter 9) and can accelerate electrons parallel to the magnetic field. They are a manifestation of the second electrostatic mode, the ion plasma waves.

4.4.2. ELECTROMAGNETIC WAVES

One may expect that the general dispersion relation, Equation (4.2.27), of waves in a cold, collisionless plasma has the same number of solutions in the cases of parallel and perpendicular propagation. Thus we search for the electromagnetic modes putting $\mathbf{k} = (k, 0, 0)$ into Equation (4.2.27) and find two more modes:

(1) *Ordinary mode*

$$\omega^2 = k^2c^2 + \omega_p^2 \tag{4.4.8}$$

The frequency of these waves always exceeds ω_p, where they have a cutoff. They are linearly polarized with $\mathbf{E}_1 \parallel \mathbf{B}_0$ and $\mathbf{B}_1 \perp \mathbf{B}_0$. As $\mathbf{E}_1 \perp \mathbf{k}$, they are electromagnetic. Since the oscillation of the particles in transverse waves is always parallel to $\mathbf{E}_1$, and in this case also to $\mathbf{B}_0$, the magnetic field does not influence the wave. Therefore B_0 does not appear in Equation (4.4.8). The wave and particles behave as in a non-magnetic plasma. This characteristic property has led to the name *ordinary wave* or *o-mode.* For large **k**-vectors it is mathematically the same branch of solution as the *L-mode* for $\mathbf{k} \parallel \mathbf{B}_0$. A complication arises at small **k**, which will be discussed in Section 4.5.

(2) *Extraordinary mode*

There is another mode called *extraordinary* or *x-mode.* The dispersion relation for $\omega \gg \Omega_i$ is

$$\frac{k^2c^2}{\omega^2} \cong 1 - \frac{\omega_p^2(1-\omega_p^2/\omega^2)}{\omega^2 - (\omega_p^2 + \Omega_e^2)} \quad . \tag{4.4.9}$$

The wave is again electromagnetic and linearly polarized, but has $\mathbf{E}_1 \perp \mathbf{B}_0$. As $\mathbf{V}_1 \parallel \mathbf{E}_1$, the particle oscillations (mostly electrons are involved) are perpendicular to the magnetic field. The higher the electron gyrofrequency, the greater the influence of the magnetic field. The wave differs from the non-magnetic mode, thus it is named *extraordinary.* As one may expect, this oscillation corresponds to the *R-mode* at parallel direction. The refractive index vanishes and causes a cutoff at

$$\omega_x = \sqrt{\omega_p^2 + \frac{1}{4}\Omega_e^2} + \frac{1}{2}\Omega_e \quad . \tag{4.4.10}$$

The resonance (singularity of Eq. 4.4.9) is at the upper hybrid frequency being, however, always below the cutoff frequency.

We note that for both electromagnetic modes the frequencies of wave-particle resonance change with propagation angle between **k** and $\mathbf{B}_0$. The cutoff frequencies are the same for parallel and perpendicular propagation, since the lowest frequency is at $k \to 0$ in both modes. In the next section we shall connect the two regimes through intermediate angles.

4.5. Oblique Propagation and Overview

The dispersion relations of the high-frequency waves are shown without derivation in Figure 4.3 as surfaces in $(\omega, k_z, k_\perp)$-space. These waves, also known as *magnetoionic modes*, are the modes supported by the electron gas (Appleton–Hartree approximation). Waves due to the motion of ions (such as the lower hybrid mode) have lower frequency and have been omitted. Four surfaces are clearly visible in Figure 4.3. They are called *branches*. The modes we have previously discussed correspond to cuts along the parallel or perpendicular k-axis.

For later use, thermal effects (assuming T=25 000 K) are included. They limit the $\mathbf{k}$-vector to values smaller than the inverse Debye length (or $kR_e \leq \omega_p/\Omega_e$, Chapter 5). Figure 4.3 presents the case of $\omega_p = 3.22\ \Omega_e$.

In the front edge of Figure 4.3 ($k_z = k_\perp = 0$), the three high frequency modes have their cutoffs at about $\omega_p \pm \frac{1}{2}\Omega_e$, and ω_p, respectively, as expected from $\omega_p^2 \gg \Omega_e^2$. At large $\mathbf{k}$, the correspondence between parallel and perpendicular modes is simple: The parallel R and L modes connect to the x and o modes, respectively, at perpendicular propagation angle. The modes remain transverse, changing gradually from circular polarization into elliptic and finally linear polarization. For intermediate $k_\perp$ ($10^{-2} \ll k_\perp R_e \ll 1$) the Langmuir mode changes into the perpendicular, electrostatic upper hybrid wave. It has cyclotron harmonics at even larger $k_\perp$ called Bernstein waves, a kinetic plasma mode to be discussed in Chapter 8 (for further information see Melrose, 1980). For clarity, the Bernstein modes in the second and fourth bands above Ω_e are indicated in Figure 4.3 only for perpendicular propagation.

At small $\mathbf{k}$, only the R-mode and the whistler branch transform simply. Note the remarkable property of the parallel electron plasma oscillations (Langmuir waves) connecting to the perpendicular o-mode at small k through a region of predominantly electrostatic plasma waves! The L-mode at small k, on the other hand, connects to a predominantly electromagnetic wave (called *z-mode*), becoming gradually electrostatic as $k_\perp$ increases. The transverse character of the z-mode is a thermal effect. For this reason the wave has not appeared in our analysis of cold plasma modes. It exists in the range $\omega_z < \omega < \omega_{uh}$ near perpendicular direction.

Only o(L) and x(R) mode waves can escape from an atmosphere. The general dispersion relation of these waves is

$$\left(\mathcal{N}_{\substack{o\\x}}\right)^2 := \frac{k^2c^2}{\omega^2} = 1 - \frac{X}{1 - \frac{1}{2}Y_T^2(1-X)^{-1} \pm [\frac{1}{4}Y_T^4(1-X)^{-2} + Y_L^2]^{1/2}} \quad , \tag{4.5.1}$$

where

$$X := (\omega_p/\omega)^2 \tag{4.5.2}$$

$$Y := \Omega_e/\omega \ , \quad Y_T := Y\sin\theta \ , \quad Y_L := Y\cos\theta \quad . \tag{4.5.3}$$

The modes discussed so far in this chapter are referred to as *normal*. If a plasma does not comply with our assumptions (being, for example, inhomogeneous or moving), new modes can appear. One such case will be discussed in the following section.

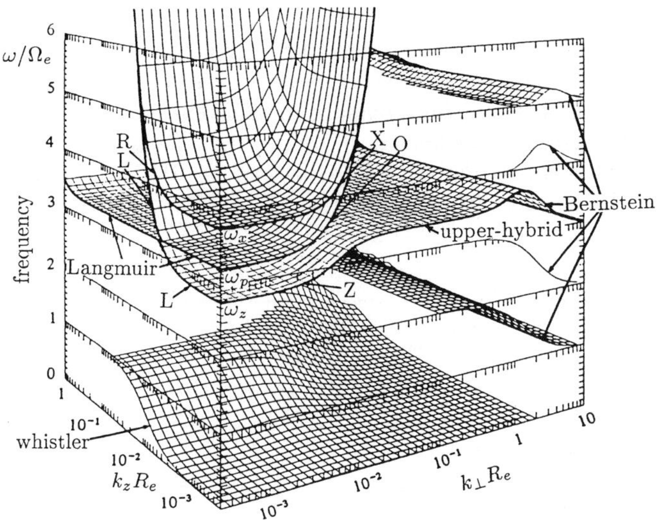

Fig. 4.3. Dispersion surfaces (branches) of high-frequency, collisionless modes in a homogeneous, dense plasma. Regions of strong damping are omitted (after André, 1985).

4.6. Beam Mode

Let us now consider a moving particle species in a plasma that is again cold and collisionless. Moving particle species or beams are ubiquitous in the universe. It is the second case of Section 4.2 with $\mathbf{B}_0 = 0$ and $\mathbf{V}_0^\alpha \neq 0$. We simplify to motions only in the z-direction and consider longitudinal waves. Thus $\mathbf{E}_1 \parallel \mathbf{k} \parallel \mathbf{V}_1^\alpha$. Substituting into Equations (4.2.5) – (4.2.9) yields

$$-i\omega n_1^\alpha + V_0^\alpha ikn_1^\alpha + n_0^\alpha ikV_1^\alpha = 0 \tag{4.6.1}$$

$$-i\omega \mathbf{V}_1^\alpha + \mathbf{V}_0^\alpha ik\mathbf{V}_1^\alpha = \frac{q_\alpha}{m_\alpha}(\mathbf{E}_1 + \frac{1}{c}\mathbf{V}_0^\alpha \times \mathbf{B}_1) \tag{4.6.2}$$

$$ikE_1 = 4\pi \sum_\alpha q_\alpha n_1^\alpha \quad . \tag{4.6.3}$$

We restrict ourselves to electrostatic waves assuming $B_1 = 0$. The equation of momentum conservation, (4.6.2), yields $\mathbf{V}_1^\alpha$ in relation to $\mathbf{E}_1$, and from the continuity Equation (4.6.1) one extracts n_1^α. Equation (4.6.3) can be rewritten as

$$ikE_1 = 4\pi \sum_\alpha \frac{ikq_\alpha^2 n_0^\alpha}{m_\alpha} \frac{E_1}{(\omega - kV_0^\alpha)^2} \quad , \tag{4.6.4}$$

$$1 - \sum_\alpha \frac{(\omega_p^\alpha)^2}{(\omega - kV_0^\alpha)^2} = 0 \quad . \tag{4.6.5}$$

The dispersion relation (4.6.5) corresponds to the electron plasma wave (Eq. 4.3.3) if one puts $V_0^\alpha = 0$. The Doppler shift, kV_0^α, however produces an oscillation at a frequency lower than ω_p. The waves are referred to as the *beam mode*. They have an important property that we shall illustrate in the following example.

Let us assume that the electrons are in motion, and the ions at rest ($\mathbf{V}_0^i = 0, \mathbf{V}_0^e \neq 0$). Equation (4.6.5) becomes

$$1 = \frac{(\omega_p^i)^2}{\omega^2} + \frac{(\omega_p^e)^2}{(\omega - kV_0^e)^2} =: H(\omega, k) \quad . \tag{4.6.6}$$

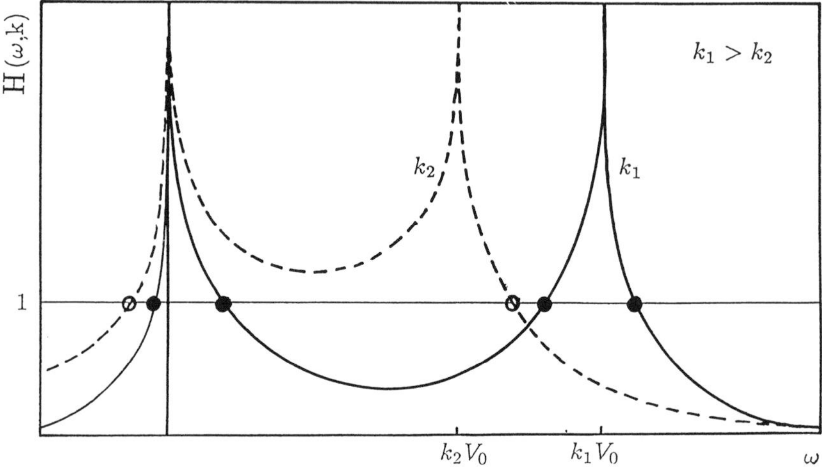

Fig. 4.4. The function $H(\omega, k)$ defined in Equation (4.6.6) vs. ω for two different values of k.

The function $H(\omega, k)$ is shown in Figure 4.4 for two particular waves with different wave vectors. The solution of Equation (4.6.6) are the points of intersection of $H(\omega, k)$ with the horizontal line at $H = 1$. A critical wave vector, k_c, exists below which there are only two real solutions. The two others are complex conjugated,

$$\omega_3 = \omega_r - i\gamma_k \quad , \tag{4.6.7}$$

$$\omega_4 = \omega_r + i\gamma_k \quad . \tag{4.6.8}$$

The solution (4.6.7) is a damped oscillation and is irrelevant. Equation (4.6.8) is of great interest, since it describes a wave with exponentially growing amplitude. Since there is always some small disturbance at the thermal level, this means that the plasma is not stable. The kinetic energy of the moving particles is transformed into wave energy at an increasing rate. Wave amplitude and growth form a feedback cycle; it is an exponential instability. Plasma physics is rich in such phenomena. This particular example is called the *two-stream instability*. The growth of the waves may be extremely fast (Exercise 4.3).

If the growth rate, γ, of Equation (4.6.8) exceeds the damping rate due to collisions in the background plasma, the neglect of collisions is justified. If not, the waves do not grow. The relevant rate is the thermal collision time between the oscillating electrons (test particles) and ions (field particles). It has been evaluated in Equation (2.6.32). The collisional interactions exert a frictional force on the particle motion. The details depend on the excited wave mode and the corresponding fraction of wave energy residing in kinetic motion. As an example, this ratio can be calculated (starting at the equation of motion 4.6.2) to be one half for electron plasma waves at $\omega = \omega_p$. From the work of Comisar (1963), we quote the collisional damping rate of electron plasma waves,

$$\gamma_{\rm coll} = \frac{87 n_e}{T_e^{3/2}} \left(\frac{\ln\Lambda}{20} \right) \quad , \tag{4.6.9}$$

close to the thermal electron-ion collision rate found in Equation (2.6.32). Solar abundances and fully ionized ions have been used to transform n_i into n_e.

The two-stream instability is in fact an extreme case that rarely (if ever) occurs in astrophysical plasmas, since it assumes monoenergetic particles. In the following chapter we shall study the instability in the presence of a finite spread of the velocity distribution. This will take into account that only a fraction of the particles with particular velocities may participate. Such an approach is termed *kinetic* as opposed to the *hydrodynamic* (or 'reactive') instability considered here. As a rule, the instability threshold and growth rate of collisionless waves must be evaluated from a kinetic investigation. Nevertheless, a hydromagnetic treatment, as in this section, may quickly indicate the type of waves to be expected.

Exercises

4.1: Prove that in electron plasma waves the oscillation energy of electrons exceeds that of the ions by the mass ratio $m_i/(Z_i^2 m_e)$ (assuming one ion species only).

4.2: Derive the dispersion relation of electron plasma waves at arbitrary angle α between $\mathbf{k}$ and the magnetic field for $\omega \gg \Omega_e$ (generalization of Eq. 4.4.4).

4.3: Evaluate the dispersion relation of electrostatic waves in cold plasma consisting of two electron beams at velocities $+v$ and $-v$ in opposite directions (neglect ions). At which wave number does the instability grow fastest? Show that the highest growth rate is $\omega_p/2^{3/2}$.

Further Reading and References

Cold plasma waves

André, M.: 1985, 'Dispersion Surfaces', *J. Plasma Phys.* **33**, 1.

Godfrey, B.B., Shananhan, W.R., and Thode, L.E.: 1975, 'Linear Theory of Cold Relativistic Beam Propagating along an External Magnetic Field', *Phys. Fluids* **18**, 346.

Wave damping by collisions

Comisar, G.G.: 1963, 'Collisional Damping of Plasma Oscillations', *Phys. Fluids* **6**, 76.

Wood, W.P. and Ninham, B.W.: 1969, 'Collisional Damping of Plasma Oscillations', *Aust. J. Phys.* **22**, 453.

Reference

Melrose, D.B.: 1986, *Instabilities in Space and Laboratory Plasmas*, Cambridge University Press, Cambridge, England.

CHAPTER 5

KINETIC PLASMA AND PARTICLE BEAMS

In the previous chapter, the collisionless waves have been reviewed using the cold plasma approximation. The two-stream instability of cold beams demonstrates that the situation changes drastically when particles have motion. Now let us study the conditions for the appearance of waves in *kinetic plasma processes* including thermal and non-thermal particle motions. The goal of this chapter is to understand instability, saturation, and (collisionless!) damping of waves in kinetic plasma theory. For this purpose we start with electron beams and study the sources of a particular kind of solar radio emission (so called *type III* or fast-drift bursts). They are the best known major kinetic plasma process in astrophysics and the showpiece of solar radio astronomy. Yet they are still not fully understood nor exhaustively observed (in particular near the Sun). Other cases of waves and instabilities in kinetic plasma are presented in the following chapters.

Beam-plasma processes and instabilities are important examples of kinetic plasma phenomena. Instabilities are not just by-products of beams, but generally control the beam's evolution. The basic principles are derived in this chapter and will be applied to interpret cosmic beams in Chapters 6 and 7. Other aspects of the phenomenon, such as particle acceleration or wave propagation, are more generally treated later (Chapters 9 – 11). The mathematical part in Section 5.2 may be considered as a prototype for treating singularities in the kinetic response of a plasma to oscillations, and is presented comprehensively. The final section presents some simple applications. Fundamental type III observations and basic interpretations are summarized as typical examples of the extremely difficult but necessary interplay of observations and theory in plasma astrophysics. The reader should be warned that the observations are often minimal, in view of the large number of free parameters, and that the selection of the relevant physics is already a major step in the interpretation. It is all the more remarkable that some aspects of the phenomena – occurring far away under exotic conditions – are reliably understood today.

5.1. Radio Observations of Solar Electron Beams

Soon after the serendipitous discovery of solar radio emission by its jamming Second World War radar, multi-frequency observations led to the important insight that coronal radio emissions have various and physically different origins. In addition to the steady *quiet* radiation (thermal bremsstrahlung) and the slowly vary-

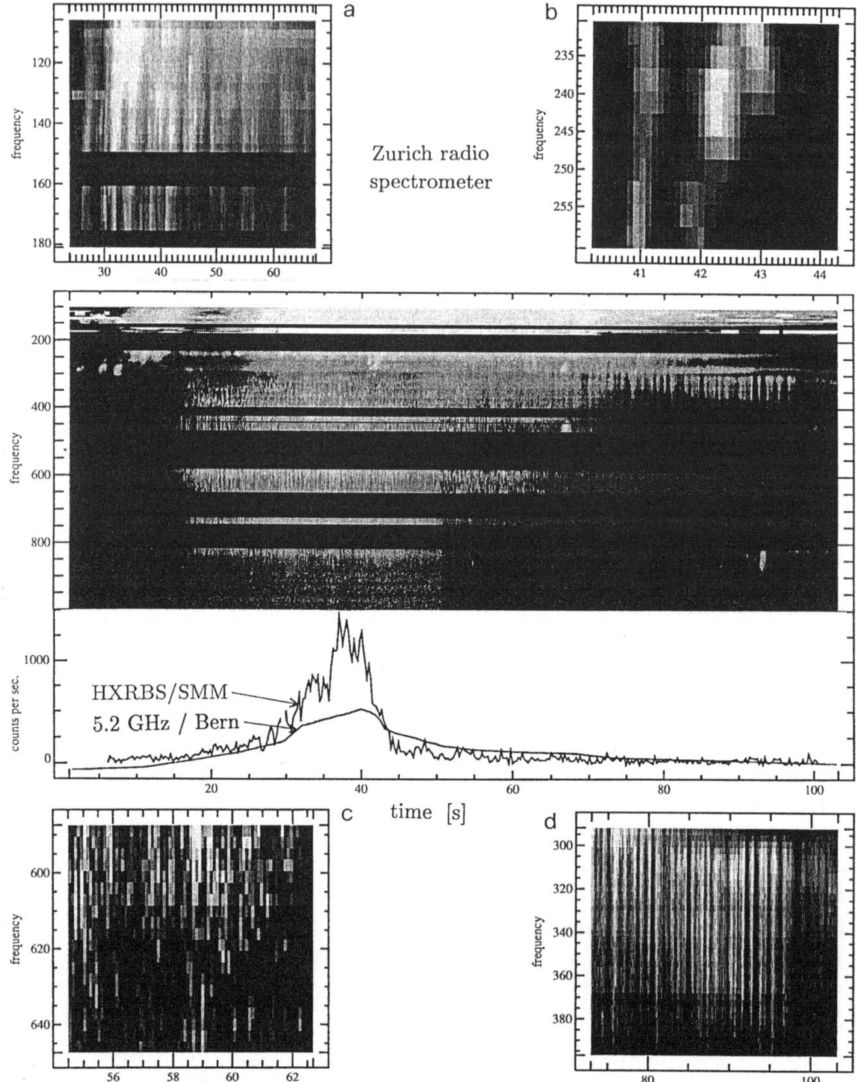

Fig. 5.1. Radio emission of a solar flare as observed by a Zurich spectrometer. The digitally recorded data are presented as a *spectrogram*, showing consecutive spectra in the vertical direction. The spectra were taken every 0.1 seconds. Frequency/time pixels of enhanced radio flux are bright. Frequencies disturbed by terrestrial interference are omitted. The time scale is in seconds, the frequency scale is in MHz. Parts of the spectrogram showing different types of emission are enlarged (squares) and discussed in the text. *a:* Metric type III bursts, *b:* U-burst, *c:* narrowband spikes, *d:* decimetric pulsations. The simultaneous hard X-ray emission ($\varepsilon > 25$ keV) and broadband centimetric emission at 5.2 GHz (lower curve, peak flux 265 10^{-19} erg s^{-1} cm^{-2} Hz^{-1}), are shown for comparison in timing.

ing component (thermal bremsstrahlung and, partially, gyroresonance emission), highly variable *bursts* have been classified into five morphological types designated in the 1950s by the roman numerals I – V. This classification has since been the basis for interpretations and models in the range of meter wavelengths. Observations at frequencies $\gtrsim$ 300 MHz ($\lambda \lesssim$ 1 m) meet the difficulties of an increasing contribution of the quiet Sun and decreasing burst brightness. This range has only recently been surveyed with sufficient sensitivity. The metric classification has generally not been found useful in the decimeter (0.3 – 3 GHz) and centimeter (3 – 30 GHz) regions, except for type III bursts (the focus of this chapter) and type IV events (Chapter 8).

An example of radio and hard X-ray observations of kinetic plasma processes during a solar flare is shown in Figure 5.1. Neither the ground-based radio telescopes, nor the satellite-borne hard X-ray detector have spatial resolution. What matters are time coincidences. Several different structures in the radio spectrogram (or 'dynamic spectrum') have been enlarged. They are typical of the decimeter range: fast drifting type III bursts, pulsations of quasi-periodic, broadband continua, narrowband spikes of a few tens of a millisecond duration (not resolved in time) and U-bursts (a variety of type III bursts). Note the different scales in time and frequency of the enlargements and use the middle panel for orientation. All the radio bursts in Figure 5.1 except the centimetric emission are produced by coherent mechanisms in which groups of particles (organized by waves) emit in phase. Spikes and pulsations will be discussed in Chapter 8.

Hard X-rays result primarily from collisional interactions of individual, energetic electrons with background electrons and ions. The process is called *bremsstrahlung* and will be introduced in Section 6.4. As the emission process is incoherent, it is less efficient, but linear to the number of energetic electrons. The hard X-ray count rate therefore indicates the evolution of the number of fast particles in time. The broadband centimetric radiation is emitted by mildly relativistic electrons spiraling in the magnetic field. Their gyrosynchrotron emission is also incoherent and will be discussed in Section 8.1.2.

The example of Figure 5.1 indicates that the information content of coronal observations is rarely sufficient to clearly interpret plasma processes. Spatial resolution would certainly help, but would not solve the problem, since important parameters like the magnetic field are hardly measurable. What can we do without *in situ* observations by spacecraft? Historically, the first step was always the proper classification of the observed phenomena. Careful judgement is needed to select the most relevant characteristics of observations and the matching physics. Comprehensive models or simulations usually involve more parameters than can be derived from any set of observations. Useful interpretations emphasize the most basic properties of the plasma process and test them against the observations.

5.1.1. RADIO INSTRUMENTS

Solar radio astronomy has two major means of observations: single telescope *spectrometers*, usually viewing the full solar disk, and *interferometers* making use of interference between several antennas to spatially resolve and locate the sources at one or a few frequencies:

Spectrometers measure the *flux density*, F, of the electromagnetic radiation in [erg s^{-1} Hz^{-1} cm^{-2}] in many frequency channels. Spectral observations on hundreds of frequencies are essential for the identification of the type of the bursts. Since radio bursts easily outshine the rest of the Sun at a given frequency, a full-disk spectrometer easily recognizes the bursts. The appearance of a burst changes from frequency to frequency, so spectrometers are widely used to survey the Sun in the frequency and time dimensions. A reason for this effectiveness is the cutoff and resonance of the o and x modes at ω_p and $\omega_p + \frac{1}{2}\Omega_e$, respectively (Section 4.5). As the density, and therefore the plasma frequency, decreases with altitude, radiation at frequency ω can only originate from heights where $\omega_p < \omega$. The spectrum combined with a density model gives a lower limit of source height and an indication of vertical motion. For this reason, the spectrum is usually shown with the highest frequency at the bottom of a spectrogram (as in Fig. 5.1). The abscissa (inverted frequency) represents altitude (but not linearly). Frequently, a perturbation caused by a solar flare travels through the corona, producing radiation at decreasing frequency.

Interferometers observing the Sun with many antennas are called *radioheliographs*. Of course, they are essential in relating radio sources to spatial features such as flares, sunspots or filaments, observed also at other wavelengths. Furthermore, they can observe source dimensions and thus determine the *intensity* or equivalently the *brightness temperature*, T_b, which is often a decisive parameter in identifying the emission process.

The *brightness temperature*, T_b, is defined by the equivalent temperature of a black body in thermal equilibrium, emitting the same intensity I or flux density F.

$$I := \frac{4FD^2}{\pi\ell^2} =: \frac{2h\nu^3}{c^2(e^{h\nu/k_BT_b} - 1)} \approx \frac{2\nu^2 k_B T_b}{c^2} \quad [\text{erg s}^{-1}\ \text{Hz}^{-1}\ \text{cm}^{-2}\ \text{sterad}^{-1}]\ , \tag{5.1.1}$$

where D is the source distance, and ℓ the source diameter (assumed spherical and homogeneous). The second equation defines T_b. For a black body it corresponds to Planck's equation. The factor 2 stands for the two modes of polarization. The third equation, an approximation for $h\nu/k_BT_b \ll 1$, due to Rayleigh and Jeans, applies throughout the radio domain. Note that T_b may depend on frequency as well as on viewing angle! We shall distinguish between T_b as measured by a remote observer and the *photon temperature*, T_t, in the source also defined by the Rayleigh – Jeans approximation (Eq. 6.3.3). They may differ due to propagation effects (as will be discussed in Chapter 11). Since I and T_b are proportional, the brightness temperature is frequently used in radio astronomy as a measure of intensity also for non-thermal sources. We distinguish three types of emission:

(1) For *thermal* radiation $T_b(\nu) \approx T_t(\nu) \leq T_e$, the electron temperature of the source.

(2) The emission is said to be *non-thermal* if $T_t(\nu) > T_e$. $k_B T_t$ may then reach the mean energy, $< \varepsilon >$, of the radiating population of super-thermal particles.

(3) Processes organizing particles to radiate in phase are called *coherent*. They can yield a brightness temperature with $k_B T_b$ far exceeding any single particle energy.

5.1.2. Type III radio bursts

As an example of beams interacting with kinetic plasma, we have a first look at solar type III radio bursts. They have been recognized very early as a uniform phenomenon in spectrograms. The decisive work was by J.P. Wild and collaborators in 1950. In the meantime they have been observed in the frequency range from 10 kHz to 9 GHz. Their eye-catching characteristic is that the emission drifts rapidly in frequency: in meter waves predominantly from high to low (Figure 5.2), and in microwaves ($\nu > 1$ GHz) more often from low to high frequency.

The drift rate, $d\nu/dt$, is defined as the displacement of the peak in frequency per unit time. The absolute value of the rate decreases with decreasing frequency. Drift rates reported by various authors for the frequency range from 74 kHz to 550 MHz have been fitted by the relation

$$\frac{d\nu}{dt} \approx -0.01 \ \nu^{1.84} \quad [\mathrm{MHz\ s^{-1}}] \quad , \tag{5.1.2}$$

where ν is in MHz (Alvarez and Haddock, 1973a). The negative sign indicates that only bursts drifting from high to low frequency have been selected. In the microwaves too, the drift rate has been found to continue to increase (in absolute value) with frequency, but with a smaller exponent. Individual values may differ by a factor of two or more from this average, and the drift rate of the leading edge (defined by some flux threshold) is usually higher.

Wild already made the suggestion that the fast drifting bursts are caused by beams of collisionless, energetic electrons. These streams move out through the corona along field lines and excite electron plasma oscillations at the local plasma frequency, $\omega \approx \omega_p$. The plasma waves emit electromagnetic (radio) waves at approximately the plasma frequency. Because the electron density (and consequently the plasma frequency) decreases outward from the Sun, the emission drifts to lower frequency.

Directional observations have confirmed this *plasma emission* scenario. Simultaneous interferometer measurements at a number of frequencies show that type

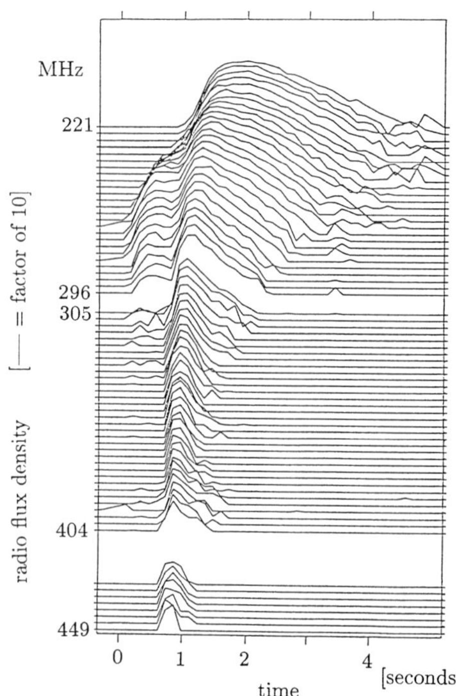

Fig. 5.2. A solar radio burst caused by an electron beam in the corona (type III), observed by a Zurich spectrometer. Flux density in logarithmic scale is presented versus time in many channels of metric frequencies (courtesy Dr. M.R. Perrenoud).

III burst sources appear at higher altitude (i.e. lower source density), the lower the frequency. They travel through the corona with velocities between about $0.1c$ and $0.6c$. This is consistent with the combination of Equation (5.1.2) and a coronal density model.

However, the two-stream instability of cold plasma theory (Section 4.6) would predict that the frequency of the unstable beam mode, $\omega \lesssim kV_0$, covers a wide range of frequencies. Why is it restricted to the narrow band around ω_p? This is an essential kinetic plasma property; it makes type III sources useful probes of the corona. We study it in the following section.

5.2. Waves and Instability in Kinetic Plasmas

Two-stream instability and cold plasma modes may provide a general scenario for the emission of electromagnetic waves by electron beams, but they are inadequate to explain even qualitatively solar type III radio bursts. Furthermore, the large growth rates derived in Section 4.6 have the awkward property that the beam would lose all its energy to the waves after a distance of a few hundred meters.

Sturrock's dilemma, as this problem has become known, will be studied in the remainder of this chapter and in Section 6.2.

The basic difference in kinetic plasma theory is that individual particles are allowed to have different velocities. A velocity distribution can be considered as the sum of cold, monoenergetic particle beams. Every beam then has two solutions of the longitudinal, electrostatic dispersion relation (4.6.6) of cold plasma. We must expect that some of them even are complex and unstable. A large number of ω and $\mathbf{k}$ seem possible, and the plasma can support many frequencies and wavelengths. Just take the collisionless particles with velocity $\mathbf{v}$, group them at regular intervals of $2\pi/k$, and you get a wave with $\omega = \mathbf{v} \cdot \mathbf{k}$. In fact, this 'thought experiment' is not entirely unrealistic, and these waves are found in the thermal field fluctuations caused by particle charges and motions. However, most of the oscillations are transitory, and only appear as waves for artificial initial conditions. We shall assume 'natural' initial conditions and clearly specify what we mean by them.

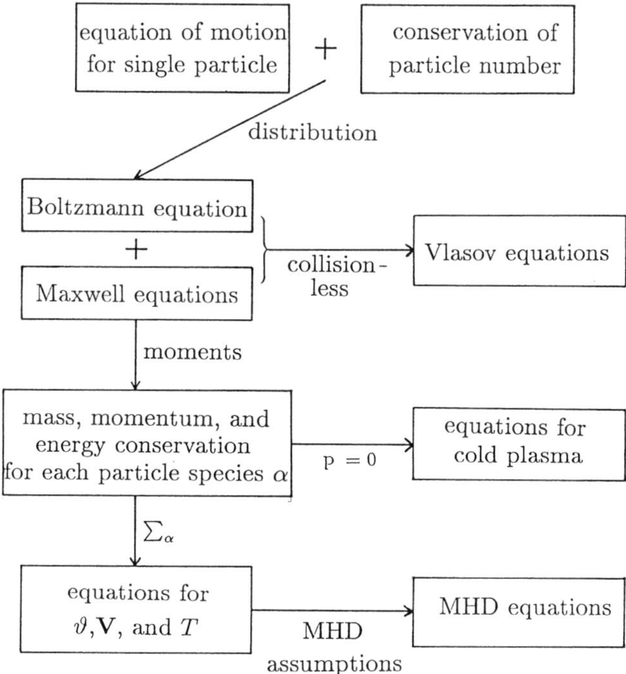

Fig. 5.3. Overview of different approaches to plasma physics.

To include individual particle velocities we have to go back to the Boltzmann equation (1.4.11). We neglect collisions (Figure 5.3), thus

$$\frac{\partial f}{\partial t} + \mathbf{v} \cdot \frac{\partial f}{\partial \mathbf{x}} + \frac{q}{m}(\mathbf{E} + \frac{1}{c}\mathbf{v} \times \mathbf{B}) \cdot \frac{\partial f}{\partial \mathbf{v}} = 0 \quad . \tag{5.2.1}$$

The closed system of equations, including (5.2.1) and Maxwell's equations (1.4.2) – (1.4.7), referred to as the *Vlasov or Maxwell-Vlasov equations*, contain the changes in the distribution of one particle species in space and velocity due to classical, macroscopic or external electric and magnetic fields. Figure 5.3 continues to MHD for completeness, but we stop now at the Vlasov equations. We shall assume that the undisturbed distribution is homogeneous and stationary, and linearize the Vlasov equations by introducing a small disturbance

$$f(\mathbf{x}, \mathbf{v}, t) = f_0(\mathbf{v}) + f_1(\mathbf{x}, \mathbf{v}, t) \quad , \tag{5.2.2}$$

with

$$\mid f_1 \mid \ll \mid f_0 \mid \text{ for all } \quad \mathbf{x}, \mathbf{v} \text{ and } t. \tag{5.2.3}$$

We restrict the discussion to longitudinal, electrostatic oscillations in the direction of the background magnetic field (Langmuir waves). Without loss of generality we choose the field direction to be parallel to the z-axis. Therefore, $\mathbf{k} \parallel \mathbf{B}_0$ and

$$f_1 = \bar{f}_1(\mathbf{v}, t)e^{ikz}, \quad E_1 = \bar{E}_1(t)e^{ikz}, \quad \mathbf{E}_1 \parallel \mathbf{e}_z \ , \quad \mathbf{B}_1 = 0 \quad . \tag{5.2.4}$$

The exponential of the disturbance acts like a Fourier transformation in space when plugged into Equation (5.2.1). The result is an equation for the oscillation amplitudes only,

$$\frac{\partial \bar{f}_1}{\partial t} + ikv_z\bar{f}_1 + \frac{q}{m}\bar{E}_1\frac{\partial f_0}{\partial v_z} = 0 \quad . \tag{5.2.5}$$

The notation '$^-$' for amplitudes (or Fourier transformed quantities) will be omitted where no confusion should be possible. Since we have already noted that the initial conditions may be important, we have not yet Fourier transformed in time (or assumed temporal oscillations in the form of 5.2.4). Instead, we transform according to Laplace, following the standard method originally developed by L.D. Landau in 1946. This is a beautiful example of complex analysis leading to an important physical result. The Laplace transform is defined by

$$f_1(\mathbf{v}, p) = \int_0^\infty f_1(\mathbf{v}, t)e^{-pt}dt \quad , \tag{5.2.6}$$

with the usual inverse transform

$$f_1(\mathbf{v}, t) = \frac{1}{2\pi i}\int_{-i\infty+p_0}^{+i\infty+p_0} f_1(\mathbf{v}, p)e^{pt}dp \quad . \tag{5.2.7}$$

The integration path of (5.2.7) (see Fig. 5.4) is defined parallel to and to the right of the imaginary p-axis and to the right of all singularities of the integrand. It is usually closed at infinity and evaluated using the techniques for complex integration developed by A.L. Cauchy. The singularities, therefore, are the critical

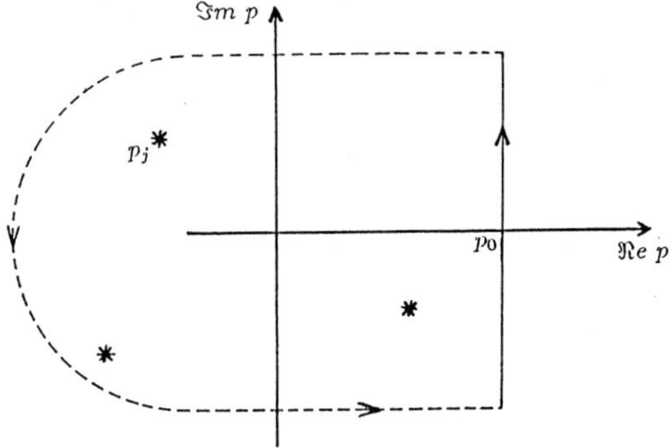

Fig. 5.4. Contour integration to carry out the inverse Laplace transformation (5.2.7). Its closure at infinity is dashed. Singularities are indicated by an asterisk.

and essential points in the complex p-plane. We shall show in the following that each represents a wave mode.

The derivative in time is Laplace transformed in the following way:

$$\frac{\partial f_1}{\partial t} \to \int \frac{\partial f_1}{\partial t} e^{-pt} dt = f_1 e^{-pt}\Big|_0^\infty + p \int f_1 e^{-pt} dt$$
$$= -f_1(\mathbf{v}, t=0) + p f_1(\mathbf{v}, p) \quad . \tag{5.2.8}$$

The initial disturbance, $f_1(\mathbf{v}, t = 0)$, thus enters the equation explicitly. The transformed Vlasov equation becomes

$$(p + ikv_z) f_1(\mathbf{v}, p) + \frac{q}{m} E_1(p) \frac{\partial f_0}{\partial v_z} = f_1(\mathbf{v}, 0) \quad . \tag{5.2.9}$$

We also transform the Poisson equation – $\nabla \cdot \mathbf{E}_1 = 4\pi\rho_1^*$ – in the same way, according to Fourier in space and to Laplace in time,

$$ikE_1(p) = 4\pi q \int f_1(\mathbf{v}, p) d^3v \quad . \tag{5.2.10}$$

The Vlasov equation (5.2.9) is used to derive

$$E_1(p) = \frac{4\pi q}{ik} \int_{-\infty}^{+\infty} \frac{g_1(v_z) dv_z}{p + ikv_z} \left(1 + \frac{4\pi q^2}{ikm} \int_{-\infty}^{+\infty} \frac{\partial f_0(v_z)/\partial v_z}{p + ikv_z} dv_z \right)^{-1} \quad , \tag{5.2.11}$$

where we have integrated in v_x and v_y, used (5.2.10) to eliminate $f_1(\mathbf{v}, p)$ in favor of E_1, and introduced the following definitions,

$$g_1(v_z) := \int f_1(\mathbf{v}, 0) dv_x dv_y \quad , \tag{5.2.12}$$

$$f_0(v_z) := \int f_0(\mathbf{v}) dv_x dv_y \quad . \tag{5.2.13}$$

Equation (5.2.11) gives the wave spectrum, $E_1(p)$. Individual oscillations can be evaluated with the inverse transformation (5.2.7),

$$E_1(t) = \frac{1}{2\pi i} \int_{-i\infty+p_0}^{i\infty+p_0} E_1(p) e^{pt} dp \quad . \tag{5.2.14}$$

Since lim $e^{pt} = 0$ for $\Re e\ p \to -\infty$, we can close the integration path through infinity (Figure 5.4). Using the Cauchy theorem, the integral can be expressed by the residues of the poles p_j,

$$E_1(t) = \sum_{j=1}^{N} e^{p_j t} \, \text{Res}[E_1(p_j)] \quad . \tag{5.2.15}$$

The poles are singularities of the integrand; they are the points in the p-plane where $E_1(p)$ is infinite. With $\omega := ip$ Equation (5.2.15) has the form of a sum of oscillations with amplitudes $\text{Res}[E_1(p_j)]$. Each oscillation represents a linear wave mode. As usual, the wave electric field is defined as the real part of $E_1(t)$.

Complex analysis has reduced the problem to that of finding the poles. There are two kinds of poles: (i) either the numerator of $E_1(p)$ becomes infinite (Eq. 5.2.11), or (ii) one of the denominators in (5.2.11) vanishes.

5.2.1. Singularities

The numerator can only be infinite for particular initial conditions. It is, for example, the case when particles with equal velocity, v_z^0, are grouped in spacings of λ, creating a wave with arbitrary frequency ($\omega = v_z^0 2\pi/\lambda$). The distribution, $g_1(v_z)$, would be proportional to $\delta(v_z - v_z^0)$. We shall exclude such artificially prepared initial conditions in the following.

The denominator of the first integral in Equation (5.2.11), $p+ikv_z$, also appears in the second integral and for this reason does not produce a pole in (5.2.14). Only zeros of the main denominator in parentheses are of interest,

$$H(k, \omega/k) := \frac{4\pi q^2}{mk^2} \int_{-\infty}^{+\infty} \frac{\partial f_0/\partial v_z}{v_z - \omega/k} dv_z = 1 \quad , \tag{5.2.16}$$

where we have replaced p by $-i\omega$. The solutions of Equation (5.2.16) only depend on f_0 and not on the initial conditions. They are therefore inherent to every plasma and independent of the initial disturbance. This is what we mean by 'natural' or normal modes, to be studied now in detail.

$H = 1$ relates ω and k, and can be considered a dispersion relation. H and the dispersion relation are equivalent to the corresponding parameters of cold plasma (Eq. 4.6.6), as can be proved by putting $f_0 = n\delta(v_z - V_0)$ and, if necessary, a term for ions (see below).

There is a subtle mathematical problem of great importance. The Cauchy theorem used in Equation (5.2.15) is valid only for *analytic* integrands. Analytic at a point p is defined by the property that the function can be differentiated in every point within a circle with center p. It means that the function can be expanded into power series. However, $H(k, \omega/k)$ is discontinuous at the $\Re e\ \omega/k$-axis, since for small γ

$$\Delta H := H(\omega/k + i\gamma/k) - H(\omega/k - i\gamma/k) \approx \left(\frac{2\pi i}{nk^2}\omega_p^2 \frac{\partial f_0}{\partial v_z}\right)\Bigg|_{\omega/k} . \qquad (5.2.17)$$

Thus $H(k, \omega/k)$ cannot be analytic, and so is $E_1(p)$ that contains H in the denominator (Eq. 5.2.11). To carry out the inverse Laplace transformation (5.2.14) by closing the integration path, $E_1(p)$ must be continued analytically from $\Re e\ p > p_0$ into the $\Re e\ p < p_0$ half-plane (Fig. 5.4, note that $\Re e\ p \equiv \Im m\ \omega =: \gamma_k$). This is achieved with the definition

$$\tilde{H} := \begin{cases} H & \text{for } \Re e\ p = \Im m\ \omega > 0 \\ H + \Delta H & \text{for } \Re e\ p = \Im m\ \omega < 0 \end{cases} . \qquad (5.2.18)$$

At the imaginary axis ($\Im m\ \omega = 0$) $\tilde{H}$ becomes analytic by the Plemelj formula,

$$\lim_{\gamma \to 0^+} \tilde{H}(\omega/k + i\gamma/k) = \frac{4\pi q^2}{mk^2}\left(\mathcal{P}\int_{-\infty}^{+\infty} \frac{\partial f_0/\partial v_z}{v_z - \omega/k} dv_z + \pi i \ \mathrm{sign}(k) \left.\frac{\partial f_0}{\partial v_z}\right|_{v_z = \omega/k}\right) , \qquad (5.2.19)$$

where the principal part is defined by

$$\mathcal{P} := \lim_{\gamma \to 0}\left(\int_{-\infty}^{\omega/k - \gamma/k} + \int_{\omega/k + \gamma/k}^{\infty}\right) . \qquad (5.2.20)$$

With these definitions $\tilde{H}$ is analytic in the whole p plane. The supplementary definitions (5.2.17) – (5.2.19) can be interpreted as a change of the integration path in Equation (5.2.16) from integration along the real v_z-axis into a contour integral (see Fig. 5.5). The integration must be carried through below the singular point $v_z - \omega/k$. This is called the *Landau prescription* after the theorist who first solved this problem.

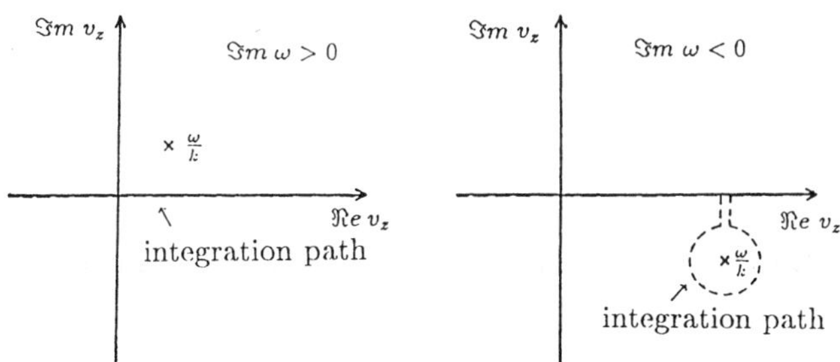

Fig. 5.5. Integration path for the integral in $\tilde{H}(k, \omega/k)$ defined in Equation (5.2.18): the 'Landau prescription'.

5.2.2. Dispersion relation

To solve the dispersion relation $\tilde{H} = 1$ for $\omega(k)$, the function $\tilde{H}$ has to be evaluated. Its integral consists of the principal part (integration along the $\Re e\ v_z$-axis, Figure 5.5) and, possibly, a contribution of the pole at ω/k.

A. *Principal Part*

Integrating by parts, we get

$$\mathcal{P}\int_{-\infty}^{+\infty} \frac{\partial f_0/\partial v_z}{v_z - \omega/k} dv_z = \left.\frac{f_0}{v_z - \omega/k}\right|_{-\infty}^{+\infty} + \mathcal{P}\int_{-\infty}^{+\infty} \frac{f_0 dv_z}{(v_z - \omega/k)^2}$$

$$= \mathcal{P}\int \frac{f_0 dv_z}{(1 - kv_z/\omega)^2} \frac{k^2}{\omega^2} \quad . \tag{5.2.21}$$

Let us assume that the majority of the particles are much slower than the phase velocity of the wave, and the denominator can be expanded for $\Re e\ (kv_z/\omega) \ll 1$. The validity of this approximation has to be checked in actual cases. As a rule, the deviations from the above result are small, if $f_0(\omega/k)$ is small and if the imaginary part of ω is small compared to the real part. Then the right side of Equation (5.2.21) becomes

$$= \frac{k^2}{\omega^2} \int f_0[1 - 2\frac{kv_z}{\omega} + 3(\frac{kv_z}{\omega})^2 - \ldots]dv_z \quad . \tag{5.2.22}$$

The terms of the series can easily be evaluated.

(1) The first term yields $k^2 n_0/\omega^2$, the same as for cold plasma.

(2) The second term vanishes, since we have assumed the average velocity V=0.

(3) The third term becomes $3k^4 k_B T n_0/(\omega^4 m)$, where $k_B T := m < v_z^2 >$ has been used (Eq. 3.1.13). It is a new correction term for kinetic plasma ($T \neq 0$).

B. *Singular Point*

For $\Im m\ \omega = 0$ the contribution of the pole is given by Equation (5.2.19). Since $\tilde{H}$ is analytic, this is also a good approximation for $\mid \Im m\ \omega \mid \ll \Re e\ \omega$, which we shall use in the following. Summing over the terms of the principal part and of the singular point, we arrive at

$$\tilde{H} = \frac{\omega_p^2}{\omega^2}\left(1 + \frac{3k^2 k_B T}{\omega^2 m} + i\frac{\pi\omega^2}{k^2 n_0}\left.\frac{\partial f_0}{\partial v_z}\right|_{\omega/k}\right) = 1 \tag{5.2.23}$$

for the dispersion relation. As usual we define $\omega := \omega_r + i\gamma_k$, and again assume that $(\omega/k)^2 \gg \langle v_t \rangle^2$, the mean thermal velocity of the particles. Then the solutions, ω_r for the real part of Equation (5.2.23) and γ_k (imaginary part), are

$$\omega_r^2 \approx \omega_p^2(1 + 3\frac{k_B T\ k^2}{m\ \omega_p^2}) = \omega_p^2(1 + 3k^2\lambda_D^2) \quad , \tag{5.2.24}$$

$$\gamma_k \approx \frac{\pi}{2}\frac{\omega_p^2}{k^2}\frac{\omega_r}{n_0}\frac{\partial f_0(v_z)}{\partial v_z}\bigg|_{\omega_r/k} \quad . \tag{5.2.25}$$

We have used $\lambda_D = v_{te}/\omega_p$. So far we have included the dynamics of one particle species only. It is straightforward to extend the work to any number of species of charged particles. The plasma frequency in Equation (5.2.24) then is replaced by the sum over all species (Eq. 4.2.24). Equation (5.2.25) must be summed over all derivatives, each weighted by the plasma frequency of its species. In both the real and imaginary part, electrons are strongly favored over ions by the mass ratio; the more mobile electrons lose and gain energy to and from the waves much faster. The first term in the parentheses of Equation (5.2.24) is the solution for cold plasma (Eq. 4.3.3). The second term is the correction for $T \neq 0$, making the wave frequency slightly dependent on k. The phase velocity $v_{ph} = \omega/k \gg v_{te}$, as assumed in (5.2.22). The wave has a finite group velocity,

$$v_{gr} \;=\; \frac{\partial \omega}{\partial k} \;\approx\; \frac{3v_{te}^2}{v_{ph}} \quad , \tag{5.2.26}$$

a value generally much lower than the mean thermal electron speed. Furthermore, the frequency (5.2.24) is shifted to $\omega > \omega_p$ and into a region where o-mode waves with equal frequency can propagate. This will obviously be important for the emission of observable radiation of beams.

In the cold plasma approximation we have found plasma oscillations to be unattenuated. Depending on the sign of $\partial f_0/\partial v_z$, Equation (5.2.25) for kinetic plasma predicts growth or damping.

5.2.3. Landau Damping

As an important example we calculate the case where electrons are distributed according to a *Maxwellian distribution*

$$f_0(v_z) = \frac{n_0}{\sqrt{2\pi}v_{te}}\ \exp\left(-\frac{v_z^2}{2v_{te}^2}\right) \quad . \tag{5.2.27}$$

For $\omega_r \approx \omega_p$, implying $k \ll (\lambda_D)^{-1}$, the imaginary part of the frequency, Equation (5.2.25), becomes

$$\gamma_k \approx -\sqrt{\frac{\pi}{8}}\frac{\omega_p}{(k\lambda_D)^3}\ \exp\left(-\frac{1}{2(k\lambda_D)^2}\right) \quad . \tag{5.2.28}$$

The negative sign indicates wave damping, named after L.D. Landau to honor the discoverer. Where is the energy going? The damping is not by randomizing collisions (the entropy does not increase), but by a transfer of wave field energy into oscillations of resonant particles.

Note that for $k\lambda_D \gtrsim 1$ the damping rate would exceed the wave frequency. Our approximations (Eq. 5.2.22) are not valid in this regime. Nevertheless, the oscillations are strongly damped when the wavelength is smaller than the Debye length, λ_D, since the thermal motions within a plasma period wash out the spatial order of the wave (Sections 2.4 and 2.5).

For $k\lambda_D \lesssim 1/5$, Landau damping on background electrons is very inefficient and is usually below collisional damping.

5.2.4. Bump-on-tail instability

The second example to be evaluated is directly relevant to beams. Let us assume an additional population of particles moving in relation to the background plasma. The velocity derivative can then be positive for $\omega > 0$, $k > 0$, and $\omega/k \gg v_{te}$ (Fig. 5.6). In the increasing part of the distribution we find $\gamma_k > 0$ according to Equation (5.2.25). The Langmuir waves are unstable and grow exponentially in amplitude. The instability is called *bump-on-tail* or *double hump* instability. It corresponds directly to the two-stream instability in the cold plasma, but includes only a limited number of particles in resonance (having $v_z \approx \omega/k$) that actually drive the wave. Of course, all other charged particles also feel the wave's electric field and participate in the oscillation; therefore they determine the (real) frequency of the wave (Eq. 5.2.24). The bump-on-tail instability is an example of a class of instabilities called *kinetic* since it involves a characteristic of the velocity distribution and kinetic theory. The two-stream instability belongs to the class of *hydrodynamic* (or reactive) instabilities that can be derived from fluid equations integrated in velocity space.

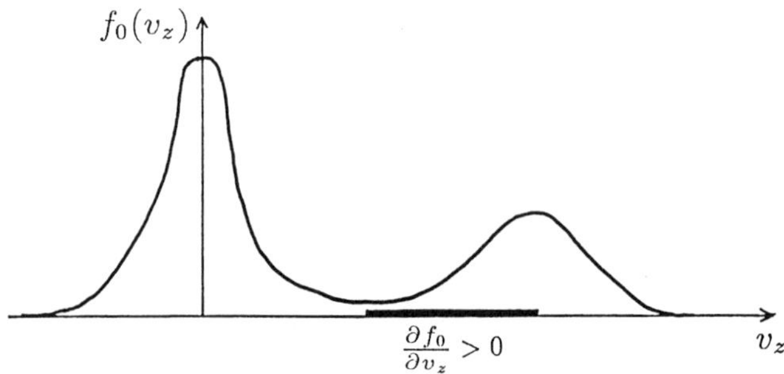

Fig. 5.6. Velocity distribution of a particle beam leading to bump-on-tail instability of growing plasma waves.

The growth rate can be approximated for $k\lambda_D \ll 1$, equivalent to $\omega \approx \omega_p^e$, with

$$\gamma_k \approx \frac{\pi}{2}\left(\frac{v_{ph}}{\Delta v}\right)^2 \frac{n_b}{n_0}\frac{m_e}{m_\alpha}\omega \quad , \tag{5.2.29}$$

where $v_{ph} = \omega/k$ is smaller but comparable to the beam velocity, Δv is the half-width of the beam distribution, and n_b is the beam density. We have assumed an electron beam. If the beam particles are of species α, an analogous derivation adds the factor (m_e/m_i) to Equation (5.2.29).

The approximation (5.2.29) does not hold for slow beams. Accurate evaluations yield growing waves for $\omega_r/k \gtrsim 3v_{te}$. Therefore, the spectrum of significant Langmuir waves is practically limited to the range from about 1.1 ω_p to ω_p (infinite phase velocity). The limited range of Langmuir waves is the reason for the limited instantaneous spectrum of type III radio bursts. This will be important in Section 6.3.4 for the polarization and escape of fundamental plasma wave emission.

5.2.5. Čerenkov Resonance

A particle moving with a velocity $\mathbf{v}$ in the oscillating field of a longitudinal, electrostatic wave, satisfying

$$\mathbf{k} \cdot \mathbf{v} = \omega \quad , \tag{5.2.30}$$

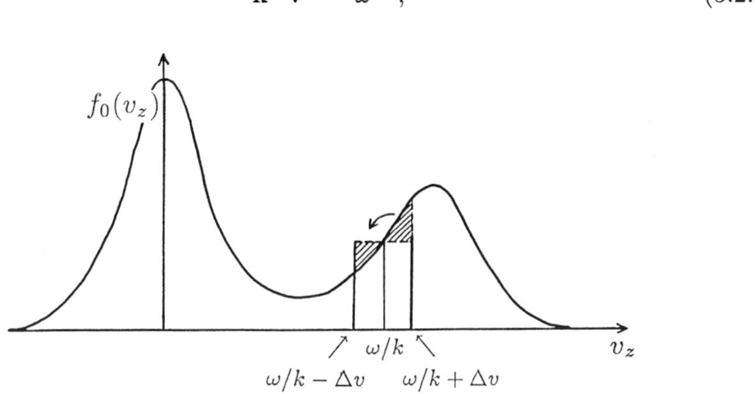

Fig. 5.7. Resonance between particles in the region $\omega/k \pm \Delta v$ and a wave having a phase velocity ω/k.

feels a constant electric field, since the wave has zero frequency in the rest frame of the particle. This is called *Čerenkov resonance*. Depending on the phase, the wave field transfers energy to the particle (like the ocean wave to a surfer), or the particle can lose energy. In both cases it starts to oscillate in a potential well of the wave. It is said to be *trapped*. Particles are trapped if they have an initial velocity between $\omega/k \pm \Delta v$, where Δv is determined by the depth of the potential well produced by the wave electric field. If a particle is initially somewhat faster than the wave but gets trapped, it effectively feeds energy to the wave. Both initially faster or slower trapped particles end up having a mean energy corresponding to the wave phase velocity, ω/k, plus oscillation energy. Instability means that the majority of the trapped particles lose energy, so the wave energy grows by the difference of particle energy before and during trapping. Figure 5.7 shows how the wave-particle interaction affects the velocity distribution. In the resonance region the particles oscillate in velocity around ω/k. Therefore, the average distribution develops a plateau, which eventually erodes the beam so that it becomes an extended tail. This evolution will be discussed in Section 6.2.1 on quasi-linear diffusion.

5.2.6. Ion acoustic waves

Electron plasma oscillations are important waves in plasmas carrying particle beams, but not the only possible mode. Also of great interest and directly observed in interplanetary space are collisionless acoustic waves. Acoustic waves are an interplay between pressure as the restoring force and inertia (Section 3.2). At frequencies too high for MHD, pressure disturbances are not propagated by collisions, but by electric fields. In the cold plasma, which has zero pressure by definition, we obviously have not encountered acoustic waves. In fact, the hot plasma correction term of the frequency of Langmuir waves (Eq. 5.2.24) has an acoustic nature. This can easily be seen if the cold plasma term (first term) is neglected (being, of course, unrealistic). The rest then reads much like the dispersion relation of sound waves (Eq. 3.2.12).

Ion acoustic waves are another electrostatic mode, based on collisionless ion oscillations. They are strongly damped in equilibrium plasma, but are of special interest for plasmas with $T_e \gg T_i$, far away from equilibrium. Important examples are the solar wind and processes preferentially heating electrons. The derivation is similar to that of the electron plasma waves. We assume electrons and ions to have Maxwellian distributions in velocity. It is instructive to allow the electrons to have, in addition, a drift velocity, V_d, relative to the ions. Let

$$f_0^i(v_z) = \frac{n_i}{\sqrt{2\pi} v_{ti}} \exp[-\frac{v_z^2}{2(v_{ti})^2}] \quad , \tag{5.2.31}$$

$$f_0^e(v_z) = \frac{n_e}{\sqrt{2\pi} v_{te}} \exp[-\frac{(v_z - V_d)^2}{2(v_{te})^2}] \quad . \tag{5.2.32}$$

The dispersion relation is $\tilde{H} = 1$, similar to Eq. (5.2.16), but $\tilde{H}$ now includes an ion term. It can be evaluated in analogy to the kinetic electron plasma (Eq. 5.2.23). It will become clear that the approximation $v_{ti} < \omega/k$ can be used. For the electrons one gets

$$\mathcal{P} \int \frac{\partial f_0^e / \partial v_z}{v_z - \omega/k} + i\pi \mathrm{Res}(\omega/k)$$

$$\approx \frac{n_e}{\sqrt{2\pi} v_{te}} \int_{-\infty}^{\infty} \frac{1}{v_z} \left(- \frac{v_z}{(v_{te})^2} \exp[-\frac{v_z^2}{2(v_{te})^2}] \right) dv_z + i\pi \left. \frac{\partial f_0^e}{\partial v_z} \right|_{v_z = \omega/k} \quad , \tag{5.2.33}$$

where $V_d, \omega/k \ll v_{te}$ have been assumed in the principle part. The effect of the electron drift is hidden in the second term, causing a positive derivative for $k > 0$. Upon combining electron and ion contributions, the dispersion relation becomes

$$-\frac{(\omega_p^e)^2}{k^2(v_{te})^2}+\left(i\pi\frac{(\omega_p^e)^2}{k^2n_e}\frac{\partial f_0^e}{\partial v_z}+i\pi\frac{(\omega_p^i)^2}{k^2n_i}\frac{\partial f_0^i}{\partial v_z}\right)\Bigg|_{v_z=\omega/k}+\frac{(\omega_p^i)^2}{\omega^2}+\frac{3(\omega_p^i)^2k^2k_BT_i}{\omega^4m_i}\approx 1 \quad . \tag{5.2.34}$$

For $(k\lambda_D)^2 \ll 1$, and $\omega_r^2 \gg \gamma_k^2$ the real part of Equation (5.2.34) has the solution

$$\frac{\omega_r}{k}\approx\left(\frac{k_BT_e}{m_i}\right)^{1/2}\left(\frac{1}{2}+\frac{1}{2}\sqrt{1+\frac{12T_i}{T_e}}\right)^{1/2} \quad . \tag{5.2.35}$$

The wave is called *ion acoustic* (or ion sound). Its phase velocity is equal to the group velocity and, for $T_e \gg T_i$, amounts to about

$$\frac{\omega_r}{k}\approx c_{is} := \sqrt{\frac{k_BT_e}{m_i}} \quad . \tag{5.2.36}$$

The velocity, c_{is}, is known as the *ion sound speed* reflecting the characteristic properties of the ion acoustic wave where the mobile electrons propagate the pressure disturbance (thus T_e) and the inert ions counteract (hence m_i). When $T_e \gg T_i$, the mean thermal speed of the electrons is so high that they immediately shield the ion density fluctuation and maintain approximate neutrality as in the Debye-Hückel theory (Section 2.4). The electric field required to hold the electrons in place – against their own pressure – also acts back on the ions. This provides the necessary restoring force and explains the speed of propagation. Density disturbances are propagated like a sound wave, except that electric effects take the role of collisions. As long as $(k\lambda_D)^2 \ll 1^2$, the frequency of ion acoustic waves $\omega_r < \omega_p^i$ is well below the ion plasma frequency, ω_p^i.

The growth rate is given by the imaginary part of Equation (5.2.34); for $T_e \gg T_i$ we get

$$\gamma_k\approx\frac{\pi}{2}\frac{v_{te}^2\omega_r}{n_e}\left(\frac{\partial f_0^e}{\partial v_z}+\frac{m_e}{m_i}\frac{\partial f_0^i}{\partial v_z}\right)\Bigg|_{v_z=\omega_r/k} \quad . \tag{5.2.37}$$

Let us look at cases with $\partial f_0^e/\partial v_z\,|_{\omega_r/k} > 0$. For $c_{is}, V_d \ll v_{te}$, $v_z = \omega_r/k$ is near the peak of the electron distribution, and

$$\frac{\partial f_0^e}{\partial v_z}\Bigg|_{\omega_r/k}\approx -\frac{(\omega_r/k-V_d)}{\sqrt{2\pi}v_{te}^3}n_e \quad . \tag{5.2.38}$$

Therefore,

$$\gamma_k \approx \sqrt{\frac{\pi}{8}}\left(\frac{V_d-\omega_r/k}{v_{te}}-\frac{\omega_r/k}{v_{te}}\left(\frac{T_e^3m_i}{T_i^3m_e}\right)^{1/2}\exp\left[-\frac{1}{2}(\frac{\omega_r}{kv_{ti}})^2\right]\right)\omega_r \quad . \tag{5.2.39}$$

For small V_d or in the absence of a current, Equation (5.2.39) yields $\gamma < 0$. The wave is Landau damped by both electrons and ions. The second term, due to the ions, dominates for $T_e \lesssim 15\ T_i$ and causes rapid damping at equal temperatures.

Ion acoustic waves appear only in plasmas with $T_e \gg T_i$, explaining the definition (5.2.36). The threshold for growing waves and instability becomes approximately

$$V_d \gtrsim \frac{\omega_r}{k}\left(1+\left(\frac{T_e^3 m_i}{T_i^3 m_e}\right)^{1/2} \exp\left[-\frac{1}{2}(\frac{\omega_r}{k v_{ti}})^2\right]\right) \quad . \tag{5.2.40}$$

The second term is small when $T_e \gtrsim 15\ T_i$. Then the waves grow if $V_d \gtrsim c_{is}$, i.e. as soon as the maximum of the electron distribution at V_d exceeds c_{is}. This is important for strong electric currents (Chapter 9). Instability feeds energy of the current into growing waves and increases the resistivity far above its collisional value. The proof of these derivations is left as an exercise (cf. also Exercise 5.2 for the case $T_e = T_i$). The ion acoustic current instability will be discussed further in Section 9.3.3.

For $k\lambda_D \gtrsim 1$ the frequency derived from Equation (5.2.34) is $\omega_r \approx \omega_p^i$, and the waves are called *ion plasma waves.* Why does the ion plasma frequency appear? The electrons cannot shield charge fluctuations smaller than the Debye length. They simply act as a smooth background. The ions oscillate like ordinary plasma oscillations (Section 2.5), but with the role of ions and electrons reversed – hence the waves oscillate at the ion plasma frequency. They are in general even more strongly damped than ion acoustic waves and are of little significance.

5.2.7. THERMAL LEVEL OF WAVES

Instabilities grow from small fluctuations. Even in a stable plasma there is always a finite level of waves. Waves are emitted by particles as they move about in the plasma and are absorbed by the other particles of the plasma. The balance between emission and absorption yields a thermal level of field fluctuations. In thermodynamic equilibrium, the situation is analogous to the problem of the black-body radiation in a cavity. Normal modes are excited with an energy density of $k_B T_e$ per degree of freedom (assuming $T_e = T_i$ and the Rayleigh-Jeans limit $\hbar\omega \ll k_B T$). This is a result of the fact that both electromagnetic and longitudinal waves obey the Bose-Einstein statistics. Each normal mode of plasma oscillation (eigenmode) represents a degree of freedom. The thermal wave energy density per mode $\mathbf{k}$ in equilibrium is

$$W_{\mathbf{k}} = k_B T \quad . \tag{5.2.41}$$

The total wave energy density becomes

$$W_{\text{tot}} = \int W_{\mathbf{k}} \frac{d^3k}{(2\pi)^3} \quad . \tag{5.2.42}$$

The denominator comes from the number of waves in d^3k, being $d^3k\mathcal{V}/(2\pi)^3$, where $\mathcal{V}$ is the volume of the plasma. Strongly damped waves (such as $k\lambda_D \gg 1$) are only weakly excited, far below the level of Equation (5.2.41). This is where a plasma deviates from a vacuum. For example, Langmuir waves in a Maxwellian plasma are effectively excited only for $k \lesssim 1/\lambda_D$. Therefore, the total energy density of the waves, averaged in time and space, is

$$W_{\text{tot}} \approx \frac{nk_BT}{n\lambda_D^3} \quad , \tag{5.2.43}$$

or about the thermal kinetic energy density divided by the number of particles in the Debye sphere.

5.3. Plasma Waves in the Solar Corona

In this section we make a first step in the interpretation of the type III radio bursts by unstable electron beams. When a beam propagates through a corona, the plasma resonates at $\omega \approx \omega_p$ according to the dispersion relation (5.2.24) of kinetic plasma. The electrostatic waves are a source of electromagnetic emission near the plasma frequency and produce the characteristic signature in the radio spectrum. Electron beams and Langmuir waves have been detected directly by spacecraft in interplanetary space during low-frequency type III bursts. Here we discuss only the observed frequency and source location of the radio emission. These two observable parameters are useful information on the solar corona and interplanetary space. The second step, the conversion of the electrostatic waves into observable radio emission, its intensity and polarization, will be approached in the next chapter.

5.3.1. Plasma density

Historically, the radio emission of beams has been used to derive the first density models of the upper corona and the near-Sun interplanetary medium. Since the observed frequency depends primarily on density (Eq. 5.2.24), the emission frequency is an excellent diagnostic of the source electron density. The major uncertainty is frequently whether the radiation is emitted at $\omega \approx \omega_p$ (fundamental) or $\omega \approx 2\omega_p$ (harmonic), making a factor of 4 difference in the derived density. If this question can be decided by the spectrum or the polarization (Section 6.3), there remain two small deviations from the simple relation between observed frequency and density:

(1) The kinetic plasma correction moves the frequency of electron plasma waves slightly above ω_p. The effect is enhanced for fundamental emission where lower frequencies are strongly absorbed (Section 11.2) and the density is overestimated by about 10% .

(2) Electron plasma waves at an angle θ to $\mathbf{B}_0$ have a higher frequency than ω_p by about $\frac{1}{2}\omega_p \sin^2\theta/(\Omega_e/\omega_p)^2$ (Eq. 4.4.5, assuming $\omega_p^2 \gg \Omega_e^2$). The effect is small in the upper corona of the Sun. The unmagnetized version of electron plasma waves, the Langmuir waves, are often used for simplicity.

Problems arise when positional measurements are used to relate directly frequency (or density) to height. Where the methods can be compared, the radio measurements yield densities which are higher by an order of magnitude than the values derived from optical Thomson scattering. The discrepancy is caused by propagation effects and inhomogeneity: (*i*) The observed radio emission originates predominantly from dense regions, whereas the optical measurement is an average along the line of sight. Moreover, it appears that type III sources tend to follow high density streamers (e.g. Gopalswamy *et al.*, 1987). (*ii*) The radio waves are strongly refracted near the plasma frequency and do not travel along straight paths. The effect of *ducting* (Section 11.5.1) initially guides the radiation in the radial direction. An Earth-bound observer sees the radiation from an apparent position at higher altitude than the actual source.

Type III observations yield density profiles along field lines. This may be useful in coronal active regions where inhomogeneity makes average densities meaningless. The positional errors are smallest for harmonic emission near the center of the disk. Also useful is a method not relying on apparent position, following from the next subsection.

5.3.2. Drift

Type III bursts can sometimes be followed through the upper corona in the meter range (300 – 30 MHz) down to the lower frequency limit of ground-based observations (10 – 20 MHz) and into the range of spacecraft (1 MHz – 10 kHz) in interplanetary space at a distance of several astronomical units.

The well accepted plasma emission scenario relates observed frequency, ν, to density and, for a given density model, $n_e(h)$, to altitude, h. The drift rate can be written in partial differentials

$$\frac{d\nu}{dt} = \frac{\partial \nu}{\partial n_e} \frac{\partial n_e}{\partial h} \cos\phi \, \frac{\partial s}{\partial t} \quad , \qquad (5.3.1)$$

where ϕ is the angle between the beam direction and the vertical ($\partial h = \partial s \cos\phi$). Using a constant ratio between emission frequency and plasma frequency (Eq. 5.2.24), $\partial \nu / \partial n_e = \nu / 2n_e$. For the electron density we adopt an exponential model as a first approximation, thus $\partial n_e / \partial h = n_e / H_n$. H_n may be derived from white-light coronal observations or from the barometric equation (3.1.52). Finally, Equation (5.3.1) has to be corrected for the finite speed of light. The differential in observing time, ∂t, is shortened by the relativistic motion of the source along its path ∂s,

$$\partial t = \frac{\partial s}{v_s} - \frac{\partial s \cos\theta}{v_{gr}} \quad . \qquad (5.3.2)$$

θ is the angle between the beam direction and the radiation path to the observer, v_s is the source velocity, and v_{gr} is the group velocity of the radiation. Note that the electron oscillations of the Langmuir waves relate to the frame of the background plasma, and are not Doppler shifted by the beam motion. Putting $v_{gr} \approx c$, Equation (5.3.1) can be written as

$$\frac{d\nu}{dt} \approx -\frac{\nu v_s \cos\phi}{2H_n(1-\beta\cos\theta)} \quad . \tag{5.3.3}$$

The obvious discrepancy of this linear dependence on frequency with the empirical relation (5.1.2) can have two reasons: (*i*) The beam velocity may decrease with altitude. In addition, the source velocity introduced in Equation (5.3.2) is not identical to the average beam particle velocity, but is the speed of the region of fastest growth of the instability. Evidence for deceleration exists only for kilometer wavelengths, and it cannot account for the major part of the observed deviation from Equation (5.3.3). (*ii*) The scale height may increase with altitude. (If the change is slow, i.e. $\partial H_n/\partial h \ll H_n/h$, approximation 5.3.3 is still valid.) The prevailing view is that the coronal density decrease in the direction of beam propagation (i.e. along an open magnetic field line) is not barometric, since the temperature increases at the base of the corona and since the solar wind accelerates at high altitude. Making use of the simple relation between density and frequency, a density model along the beam path can be derived from spectral observations without spatial resolution. The geometry of the path may be estimated from the flare location (Hα position), and the beam velocity must be known independently and be constant. Equation (5.3.3) then gives the density scale height in relation to the observed drift rate and frequency (Exercise 5.4).

5.3.3. Field Geometry

Electron beams propagate along magnetic field lines. Their sources trace out the field geometry. Beam paths have been inferred directly from radio-heliographic observations of type III burst positions. Some deviate considerably from the radial direction.

A. *U-Bursts*

In some cases radio bursts start with decreasing frequency, but the drift rate changes gradually from negative (as in ordinary metric type III bursts) to positive, indicating a motion changing from upward to downward. Such bursts are called *U-bursts*, because of their shape in the frequency-time plane of spectrometer records. Examples are shown in Figure 5.1b (spectrum) and Figure 5.8 (map). The latter shows the case of an electron beam travelling on a large magnetic loop. The beam first appears as a small source at the height where $\nu_p \approx 160$ MHz, it expands while it rises to the height of 43 MHz, and then returns to the solar surface.

U-bursts are interpreted as the signature of electron beams following closed magnetic field lines. On the ascending leg, the electron stream travels into regions of lower density, hence decreasing frequency, and *vice versa* on the descending leg. The total duration (separation between the two legs) ranges from 5–40 s in the

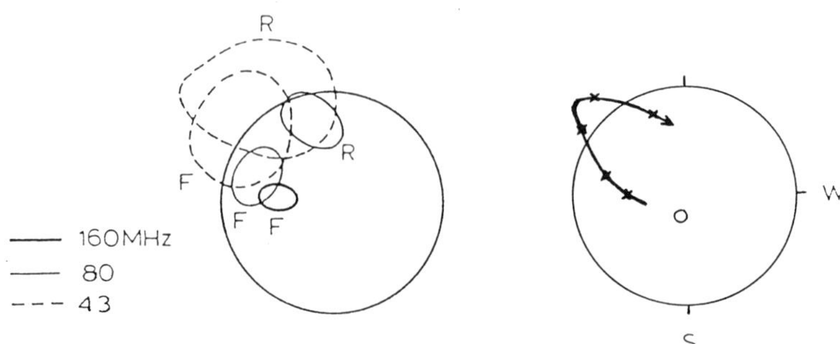

Fig. 5.8. *Left:* Half-brightness contours at three frequencies and at peak times outlining the path of an electron beam in the solar corona. It followed a large magnetic loop and has been observed with the Culgoora radioheliograph by Suzuki (1978). The first source at a given frequency (upgoing) is labelled F, the second source is R (downgoing). *Right:* The observed centroids of each frequency ($\times$) are connected to outline the propagation path of the exciter. The photospheric disk is indicated for reference, and the site of the flare in Hα is shown by an open circle.

meter waves. Therefore, U-bursts outline gigantic magnetic loops. Historically, they were the first indication of magnetic connections between different active regions (sometimes even in different hemispheres) through the corona. Similar loops can now also be observed in soft X-rays (Fig. 1.2). As expected, the total duration is correlated with turning frequency; the bigger the loop, the lower the density at the top (and the turning frequency). At frequencies between 300 MHz and 1000 MHz the typical duration is only a few seconds, indicating that loops with higher density (presumably at lower altitude) are smaller.

It has long been a riddle why metric type III sources travel primarily along open magnetic field lines (less than one percent are U-bursts or fragments thereof). Why should electrons – presumably accelerated by flares in active regions with primarily low lying, small magnetic loops (Fig. 1.2) – escape on one of the few lines leading from the active region to interplanetary space? Only recently, many weak U-bursts have been found shifted into the 1-3 GHz range due to the higher density in active-region loops. They tend to have a small total bandwidth (often below 10% of the center frequency) and are extremely short (< 0.2 s), probably reflecting the limited size of the field lines. Others may be absorbed or do not live long enough to become unstable (Section 2.3). Metric type III bursts, on the other hand, may be the result of many injections into different flux tubes, but the radio emission is observable only under favorable conditions, such as a rapidly decreasing background density in open field lines.

B. *Magnetic Field Configuration Near Acceleration*

A surprising result derived both from radio source diameters as well as interplanetary particles is the extremely wide spread of field lines from the acceleration region. Some diverge into a large fraction of interplanetary space, others turn back to the chromosphere. Flare generated electrons have been observed in the near-Earth solar wind from flares occurring at solar longitudes from about 30° E to 90° W. It is consistent with the rotation of the Sun and the Archimedean spiral of the interplanetary field to find its footpoint in the Western hemisphere, but the observed range of longitudes is surprising.

Similarly, the source size increases with decreasing frequency roughly confined to the interior of a cone of opening angle 40° – 80°, having its apex in the active region. These and other data are best interpreted by strongly diverging magnetic field lines with similar opening angles. The acceleration region may thus be characterized by a complex geometry of sheared field lines and high diversity.

C. *Interplanetary Space*

If one observes the apparent position of type III bursts from high to low frequencies, the trajectory of the electron stream from the Sun into the interplanetary medium can be estimated. Figure 5.9 shows the apparent traces of about twenty interplanetary type III bursts. It is surprising that almost all trajectories end up in a direction parallel to the ecliptic, regardless of their origin and initial direction. They do not represent a random choice of interplanetary field lines, but selected lines originating from active regions. A word of caution: the apparent source diameter at 60 kHz exceeds the frame of Figure 5.9, and the position of the source centroid may be a poor approximation of the trajectory of the beam.

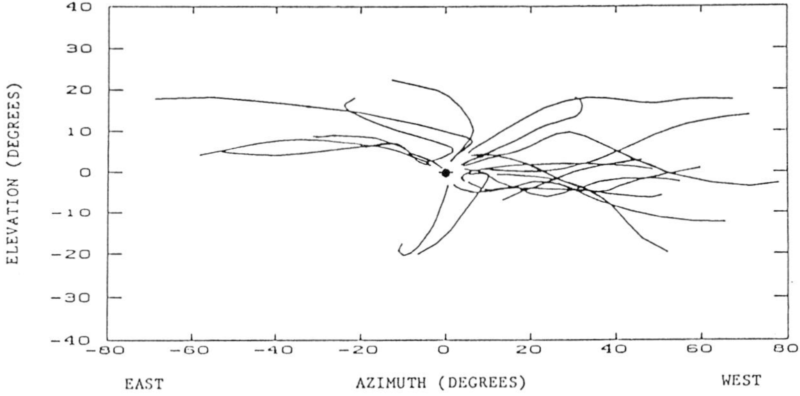

Fig. 5.9. Trajectories of apparent type III sources as measured by the radio experiment on ISEE-3 (from Dulk, 1990). The position of the burst centroids are plotted at frequencies decreasing from 1.95 MHz to 60 kHz. The coordinate system is ecliptic and centered at the Sun.

Figure 5.10 displays the source positions of an interplanetary type III burst in the ecliptic plane. The radio emission makes it possible to track the electron beam and to conjecture the geometry of the interplanetary magnetic field. Furthermore, the observation allows to measure the exciter velocity. It was found to decrease from 0.4 ± 0.2 c near the Sun to about one half of this value at 1 AU.

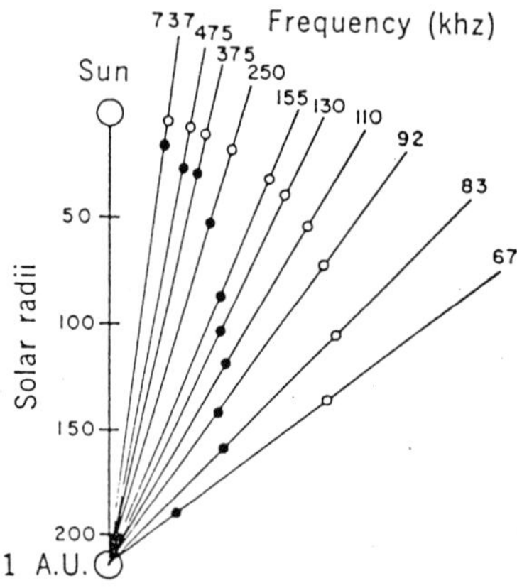

Fig. 5.10. Interplanetary electron beams follow the solar wind magnetic field shaped as an Archimedean spiral by the rotation of the Sun. The direction of a type III radio source relative to the Sun is shown as observed from an interplanetary spacecraft (IMP-6) for various frequencies given in kHz. The distance of the source is estimated in two ways. *Full circles*: from a density model of the solar wind. *Open circles*: at closest distance to the Sun along the ray (from Fainberg *et al.*, 1972).

5.3.4. Decay Time

Observing a type III burst at different frequencies yields information on the evolution of the beam in space and time. Also, the beam traverses a changing background plasma. The two effects need to be disentangled.

Type III bursts change considerably over their five decades of occurrence in frequency. Most notable is the increase in duration from a few tenth of seconds around 4 GHz to hours at 10 kHz. The average e-folding decay time in the meter and kilometer range can be fitted by

$$\tau_d \approx 10^{7.71} \nu^{-0.95} \quad [\mathrm{s}] , \tag{5.3.4}$$

where ν is in Hz (Alvarez and Haddock, 1973b). The empiric relation has been found to extend into the microwave region up to 8 GHz. The rise time tends to be shorter than the decay time.

The scaling law (5.3.4) of the decay is not compatible with the frequency dependence, ν_p^{-2}, of the thermal electron-ion collision time at constant temperature

(Eq. 2.6.32). Thus it is no surprise that early interpretations of the decay by collisional damping of the plasma waves have failed. The time profile of type III bursts appears to be a convolution of plasma wave excitation and damping. Its change with frequency is therefore the result of both beam evolution (expansion) and altering background (damping).

5.3.5. Other Radio Wave Emitting Beams

In the decameter range (3 – 30 MHz) a relatively slowly drifting variant of type III bursts also occurs outside flares in *type III storms*, a phenomenon associated with noise storms at meter waves (to be discussed in Chapter 9). The free energy appears to originate from the gradual rearrangement of new magnetic flux emerging from the photosphere. Type III-like processes are also observed in the *herringbone* feature of shockwaves (Chapter 10). These beams however exist in low density background plasma and seem to involve an insufficient number of particles to yield observable hard X-ray emission. We conclude that a great variety of beams exist in the solar atmosphere. The number of electrons per beam reported for solar flares varies by 15 orders of magnitude (!) from the hard X-ray producing downward flows to escaping metric microbursts sources.

The flux density of an intense solar type III burst seen at the distance of the nearest star would be a few μJy and is well below the sensitivity of current telescopes. Nevertheless, some more powerful variants cannot be excluded. In fact, drifting radio bursts have been observed by Jackson *et al.* (1986) and Bastian and Bookbinder (1987) from main-sequence M-stars (dMe) and interpreted as possible signatures of beams.

Radio emission is generally produced when ionized material is ejected into a stationary plasma of sufficient density to emit at an observable frequency, $\omega \approx \omega_p$. If Coulomb collisions are not frequent enough to couple the two plasmas, they act like penetrating beams. An example thereof may be the prompt radio emission of Supernova 1987A. It has tentatively been interpreted as originating from the interaction of the ejecta with a pre-existing stellar wind.

Exercises

5.1: Prove that $\epsilon := 1 - H$ is the dielectric function defined in Equation (4.2.20). H is given by Equation (5.2.16)

5.2: Derive the growth rate of the ion acoustic instability for hot ions with $T_i = T_e$. Show that in this case the threshold for instability, Equation (5.2.40), becomes

$$V_d \gtrsim 22 v_{ti} \quad . \tag{5.3.5}$$

(If V_d is driven by an electric field, the velocity distributions deviate from Maxwellian, and the threshold is slightly higher).

5.3: Prove that the electric wave energy is half of the total wave energy for *(a)* Langmuir waves and *(b)* transverse waves having $\omega > \omega_p$ (neglect B_0).

5.4: Electron beams propagating through a corona excite electron plasma oscillations near ω_p. These waves are a source of radio emission (type III burst) at the same frequency. Derive the density scale height of the corona from the observed drift rate and frequency of the emission. Let $d\nu/dt \approx 50$ MHz s^{-1}, $\nu \approx 50$ MHz, and assume a beam velocity of 10^{10}cm s^{-1}. Take a vertically propagating beam in the center of the disk.

Further Reading and References

General physics of kinetic plasma and particle beams

Gary, S.P.: 1993, *Theory of Space Plasma Instabilities*, Cambridge University Press, Cambridge, UK.

Melrose, D.B. and McPhedran, R.C.: 1991, *Electromagnetic Processes in Dispersive Media*, Cambridge University Press, Cambridge, UK.

Nicholson, D.R.: 1983, *Introduction to Plasma Theory*, J. Wiley and Sons, Inc., New York.

Mikhailovskii, A.B.: 1974, *Theory of Plasma Instabilities*, Vols. I and II, Consultants Bureau, New York.

Schmidt, G.: 1979, *Introduction to High Temperature Plasmas* (2nd edition), Academic Press, New York.

Reviews of solar electron beam radio observations

Krüger, A.: 1979, *Introduction to Solar Radio Astronomy and Radio Physics*, D. Reidel, Dordrecht.

Suzuki, S. and Dulk, G.: 1985, in *Solar Radiophysics* (eds. D.J.McLean and N.R.Labrum), Cambridge Univ.Press, Cambridge, UK.

References

Abrami, A., Messerotti, M., Zlobec, P., and Karlický, M.: 1990, 'Analysis of the Time Profile of Type III Bursts at Meter Wavelengths', *Solar Phys.* **130**, 131.

Alvarez, H. and Haddock, F.T.: 1973a, 'Solar Wind Density Model from km-Wave Type III Bursts', *Solar Phys.* **29**, 197.

Alvarez, H. and Haddock, F.T.: 1973b, 'Decay Time of Type III Solar Bursts Observed at Kilometric Wavelengths', *Solar Phys.* **30**, 175.

Bastian, T.S. and Bookbinder, J.: 1987, 'First Dynamic Spectra of Stellar Microwave Flares', *Nature* **326**, 678.

Dulk G.A.: 1990, 'Interplanetary Particle Beams', *Solar Phys.* **130**, 139.

Fainberg, J., Evans, L.G., and Stone, R.G.: 1972, 'Radio Tracking of Solar Energetic Particles through Interplanetary Space', *Science* **178**, 743.

Gopalswamy, N., Kundu, M.R., and Szabo, A.: 1987, 'Propagation of Electrons Emitting Weak Type III Bursts in Coronal Streamers', *Solar Phys.* **108**, 333.

Jackson, P.D., Kundu, M.R., White, S.M.: 1987, 'Dynamic Spectrum of a Radio Flare on UV Ceti', *Astrophys. J.* **316**, L85.

Pick, M. and van den Oort, G.H.J.: 1990, 'Observations of Beam Propagation', *Solar Phys.* **130**, 83.

Suzuki, S: 1978, 'On the Coronal Source Regions of U-Bursts', *Solar Phys.* **57**, 415.

CHAPTER 6

ASTROPHYSICAL ELECTRON BEAMS

The first astronomical elementary particles discovered were cosmic rays and their collision products penetrating to ground level on Earth. All cosmic rays are beams at some time, but only the ones of solar origin still have some directivity when arriving at Earth. We shall call a population of particles a *beam* if their mean velocity in the rest frame of the background plasma is different from zero. This includes both an enhanced tail of the velocity distribution and bumps on the tail. Other, less energetic, but clearly developed beams propagating through the magnetospheric and solar wind background plasmas have since been detected by spacecraft. Many more beams exist in the universe that cannot be studied *in situ*. In Chapter 5 we have already met solar type III bursts caused by coronal electron beams. Beams have been proposed, for example, in connection with hard X-ray flares, quasar jets, supernovae, and pulsar ejections. Signatures of beams include many kinds of radiation: bremsstrahlung of collisional interactions of beam particles with the background, atomic and nuclear excitation and subsequent emission of deexcitation line radiation, synchrotron radiation, and various coherent emissions. This chapter is limited to electron beams. The problem of return currents of intensive beams, a topic of general interest in cosmic plasmas, is outlined in Section 6.1. The non-linear evolution of the electrostatic beam instability and its saturation are presented in brief in Section 6.2. The plasma radiation emissivity is calculated in Section 6.3 and used to interpret radio observations. Hard X-ray observations are introduced in Section 6.4 and combined with radio observations to use electron beams for diagnostic purposes.

6.1. The Beam-Plasma System

Up to now it has been assumed that the particle beams are homogeneous, current-free, and of infinite length. We shall see in this section that coronal and interplanetary beams in reality are an initial value problem. They do not persist long enough to reach an equilibrium with the ambient plasma. The problem has become evident from the analysis of hard X-ray emission during solar flares. Typically, 10^{38} fast electrons ($\gtrsim$ 10 keV) accelerated by the primary energy release in the corona propagate along closed magnetic field lines and lose their energy primarily by Coulomb collisions in the denser layers of the atmosphere below. If all accelerated electrons were moving in one direction, the charges would constitute an electron flux of about 10^{36} electrons s^{-1} and an electric current of the order of

10^{26} statamps (10^{17} ampères). The magnetic field implied by Ampère's equation (3.1.42) would be overwhelming; for a beam radius of 10^9 cm one finds $B_{\rm ind} \approx 10^7$ G. This does not make sense when compared to the pre-flare value of some 100 G, thought to be the cause of the flare and thus of the electron beam. Apart from the energy problem, the beam with a flux of about 10^{36} electrons s^{-1} would imply that an enormous charge difference was built up between the corona and the chromosphere. It would evacuate all electrons of a coronal condensation above an active region in the course of one flare. What is wrong with this simple scenario?

The two dilemmas mentioned above do not occur, since the beam propagates in a background plasma. The plasma responds in driving a *return current* of roughly the strength of the beam current and in the opposite direction. It can react to the injection by two rather independent effects – both, however, originating from the full set of Maxwell's equations and closely related – loosely referred to as 'magnetic' and 'electrostatic'. The net total current densities are many orders of magnitude below the above value, and the beam current is practically *neutralized.* In this section we concentrate on such current-neutralized electron beams, isolate some essential physics, and ignore many complicated details. Current-carrying beams will be discussed in Chapter 9.

6.1.1. Magnetically driven return current

The reason why the magnetic field remains relatively small in reality is that the magnetic diffusion time in the beam-carrying plasma is much longer than the impulsive phase of flares (Eq. 3.1.48). The ring-shaped field cannot expand fast enough into the region outside the current. The magnetic field there cannot change significantly. If $\nabla \times \mathbf{B}$ vanishes initially, the full Ampère law (Eq. 1.4.2) requires that an induced electric field, $\mathbf{E}_{\rm ind}$, builds up with

$$\frac{\partial \mathbf{E}_{\rm ind}}{\partial t} \approx -4\pi \mathbf{J}_b \quad , \tag{6.1.1}$$

where $\mathbf{J}_b = -en_b\mathbf{V}_b$ is the original beam current density. $\mathbf{E}_{\rm ind}$ drives a current of background electrons in the opposite direction to prevent magnetic field growth on a timescale shorter than the magnetic diffusion time. The resulting return current, $\mathbf{J}_r$, can be calculated from the (fluid) equation of motion for the ambient electrons (Eq. 3.1.21),

$$\frac{\partial \mathbf{V}}{\partial t} + (\mathbf{V}\cdot\nabla)\mathbf{V} = -\frac{e}{m}\mathbf{E}_{\rm ind} - \frac{e}{mc}(\mathbf{V}\times\mathbf{B}) - \nu_{e,i}\mathbf{V} \quad , \tag{6.1.2}$$

where $\mathbf{V}$ is the mean velocity, and $\nu_{e,i}$ is the electron-ion collision frequency (Eq. 2.6.32). Equation (6.1.2) is related to Ohm's law, the simplest form of which one can derive from the first and the third term on the right side ($J = \sigma E$). In this section it is assumed that no charge density builds up. Then Ampère's law implies $\nabla \cdot \mathbf{J} = 0$. Putting $\mathbf{J}_r = -en\mathbf{V}$, we find

$$\left(\frac{\partial}{\partial t} + \nu_{e,i}\right) \mathbf{J}_r = \frac{(\omega_p^e)^2}{4\pi} \mathbf{E}_{\text{ind}} + \frac{e}{mc}[\mathbf{J}_r \times (\mathbf{B_0} + \mathbf{B}_{\text{ind}})] \quad . \tag{6.1.3}$$

Subscript r denotes parameters associated with the return current. As a first approximation we also neglect the Hall term (in brackets on the right side of Eq. 6.1.3). We shall discuss its effects later. Operating on Equation (6.1.3) with $[\nabla \times (\nabla \times]$ and using Ampère's and Faraday's laws and a vector identity (A.10), we find

$$\lambda_{sd}^2 \left(\frac{\partial}{\partial t} + \nu_{e,i}\right) \nabla^2 \mathbf{J}_r = -\frac{\partial}{\partial t}(\mathbf{J}_r + \mathbf{J}_b) \quad , \tag{6.1.4}$$

where we have introduced the electrodynamic skin depth $\lambda_{sd} := c/\omega_p^e$. Let the scale length of the return current be some beam radius R, i.e. $\nabla^2 \mathbf{J_r} = \mathbf{J_r}/R^2$, and $(\lambda_{sd}/R)^2 \ll 1$. Integrating Equation (6.1.4) in time demonstrates that $\mathbf{J}_b \approx -\mathbf{J}_r$ for times smaller than $1/\nu_{e,i}$. The return current flows cospatially within the beam channel and neutralizes the beam current, as one may have expected from Lenz's law.

For times much beyond $1/\nu_{e,i}$, the first term of Equation (6.1.4) can be neglected and the equation can be written in the form

$$\left(\frac{\partial}{\partial t} - \lambda_{sd}^2 \nu_{e,i} \nabla^2\right) \mathbf{J}_r = -\frac{\partial}{\partial t} \mathbf{J}_b \quad . \tag{6.1.5}$$

It becomes a diffusion equation after the beam front has passed and $\partial \mathbf{J}_b/\partial \mathrm{t} = 0$. The diffusion time is the decay time of the current and amounts to

$$\tau = \frac{1}{\nu_{e,i}} \left(\frac{R}{\lambda_{sd}}\right)^2 \quad , \tag{6.1.6}$$

equivalent to the magnetic diffusion time (Eq. 3.1.48 using the conductivity $\sigma = \omega_p^2/(4\pi\nu_{e,i})$, Eq. 9.2.5).

Hence, the current remains neutralized for roughly the time τ, after which the self-field of the beam will start to grow. The classical collision rate (as calculated for particle collisions in Eq. 2.6.32) yields diffusion times of millions of years. However, as will be discussed in Chapter 9, wave-particle interactions can also exert a drag on electrons. In fact, the ion acoustic waves growing in strong return currents (Eq. 5.2.39) or electron plasma waves driven to instability by a bump-on-tail distribution of the beam (Section 5.2.4) may provide *anomalous* collisions and enhance $\nu_{e,i}$ many orders of magnitude. Even if $\nu_{e,i} \approx 0.2\ \omega_p^i$ (a maximum value generally ascribed to Sagdeev) and currents were highly fragmented ($R = 10^6$ cm, $n_e = 10^{10}$ cm), the diffusion time is of the order of 1300 s. This is still longer than the likely duration of the initial beam in the corona.

In Figure 6.1 the basic evolution of the return current responding to a given model of the beam is plotted for times much larger than $1/\nu_{e,i}$. The beam total duration at a given location was assumed 20 s, the magnetic diffusion time $\tau = 200$ s. The return current first evolves nearly anti-symmetrically according to Equation

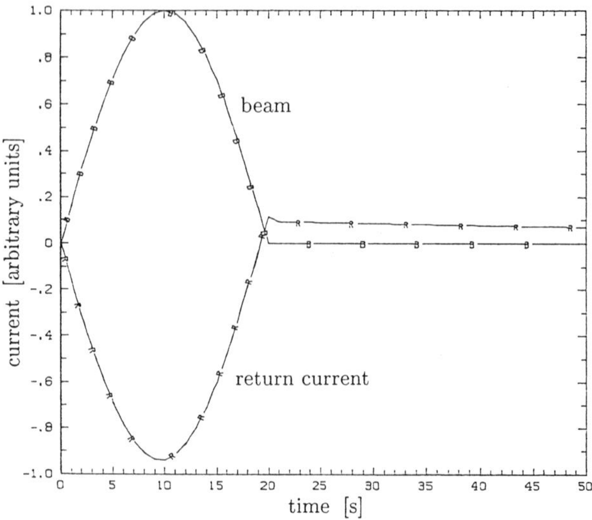

Fig. 6.1. The evolution of a normalized model beam and its return current is shown in time. $I_b < I_A$ (Eq. 6.1.7) is assumed (after Spicer and Sudan, 1984).

(6.1.5). It becomes positive (as suggested by Lenz's law) when I_b vanishes and decays with an e-folding time τ.

As the beam produced magnetic field begins to rise, beam propagation becomes more and more difficult. The beam particles acquire transverse momentum. Once their gyroradii become comparable to the beam radius, the particles may be reflected by their own magnetic self-field. This is an effect of the Hall term neglected in Equation (6.1.4). For an electron beam propagating in a weak ambient field, reflection occurs at the so-called Alfvén - Lawson limit (e.g. Hammer and Rostoker, 1970),

$$I_b \gtrsim I_A := \int J_A d^2x = \frac{mc^3\beta^3\gamma}{e[\beta^2(1-f_m)-(1-f_e)]} \quad , \tag{6.1.7}$$

where $\beta = V_b/c$, $\gamma = (1-\beta^2)^{-1/2}$, e is the elementary charge, and I_b is the total current of the beam. The initial fractional current neutralization, $f_m :=\mid I_r/I_b \mid$, is assumed to be unity ($f_m = 1$, Fig. 6.1), and f_e denotes the fractional charge neutralization (on the average $f_e = 1$, cf. next section). Since f_m decreases as $e^{-t/\tau}$, I_A diminishes and I_b may eventually exceed the limiting value. At this point, the lifetime of the beam is drastically reduced.

Below the Alfvén - Lawson limit, instability of the return current may extract considerable energy from the primary beam by *anomalous Ohmic heating*. This accelerates the diffusion of the magnetic field (Eq. 6.1.5) and reduces the life time typically by four orders of magnitude compared to the Coulomb collisions.

6.1.2. Electrostatic return current

When a beam enters a charge-neutral plasma, it creates a situation similar to the displacement experiment of Section 2.5. The charge separation sets up an electric field, $\mathbf{E}_{es}$, which slows down the beam particles and drives a return current of background electrons in the opposite direction. The situation is controlled by the electron fluid equation (6.1.2). The left side and the first term on the right yield the electron plasma waves (Section 4.1). The $\nu_{e,i}$ term introduces a damping with an e-folding time of about $1/\nu_{e,i}$. In Figure 6.2 a beam is injected at time zero. Damped oscillations develop at ω_p.

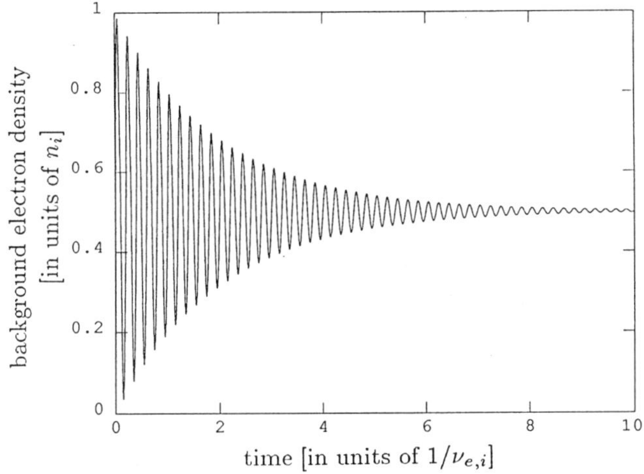

Fig. 6.2. An electron beam with a density of half the background ions is injected at $t = 0$ into a plasma with $\nu_p/\nu_{e,i} = 5$. The background electron density oscillates at the plasma frequency and reaches an average of $0.5n_i$ within the first period. The average charge density including the beam remains zero (after van den Oort, 1991).

Using the previous values characteristic for solar coronal condensations ($n_e = 10^{10}$ cm, $T = 3 \cdot 10^6$ K), the collision time $1/\nu_{e,i}$ is about 0.3 s (Eq. 2.6.32). This value suggests that the front of a beam moving at 10^{10} cm s^{-1} oscillates across a thickness of the order of 10^{10} cm. However, in the case of anomalous collisions, the thickness is reduced by many orders of magnitude. If the beam front is not as sharp as is assumed in Figure 6.2, the oscillations develop less vigorously. These may be the reasons why the oscillations have not yet been observed by their radio emission.

The lifetime of the magnetically driven return current (Section 6.1.1) is generally much longer than $1/\nu_{e,i}$. Therefore, the electrostatic field is relevant only at the front of the beam. Its main effect is to produce strong electrostatic oscillations in the background plasma and a deceleration of the beam electrons. Its role in the evolution of the beam is not yet clear, and depends in part on the details of beam formation in the energy release region.

To summarize this section, it has become clear that flare generated beams cannot exist in steady state for several reasons.

- They would spend most of their kinetic energy to induce a strong magnetic field after a magnetic diffusion time.
- The beam produced magnetic field would deflect most of the beam and destroy it even before that time.
- Magnetic induction drives a return current, which, if unstable, rapidly converts the beam energy into background heat.
- An oscillating return current builds up at the beam front from charge separation.

6.2. Non-Linear Evolution and Saturation

The electron beam in the previous section was treated in the fluid description. Now we use kinetic theory and apply the results of Chapter 5 to real beams. Interplanetary electron streams and low-frequency (kilometric) type III bursts have been observed with sufficient detail to test and further develop theoretical ideas on kinetic beam-plasma processes. Spacecraft observations have established beyond doubt the connection between type III radio bursts and electrons streaming at super-thermal velocities from the Sun. In particular, the distribution function of the electrons has been measured, and a positive slope, $\partial f/\partial v_z$, at $v_z > 0$ of the one-dimensional distribution has frequently been observed – a necessary requirement for the bump-on-tail instability (Eq. 5.2.25). The positive slope builds up by impulsively accelerated electrons escaping from the Sun with little or no collisions, the faster ones running ahead of the slower ones (Section 2.3). Figure 6.3 shows how small the region of positive slope is in reality (best visible at 0705 UT). It is sufficient to drive strong electron plasma waves (Figure 6.6).

It is also established that electron plasma waves with phase velocities corresponding to the positive-slope portion of the bump grow at the expense of the free energy in the electron stream. The back-reaction of the waves on the electron distribution is the first topic of this section. If the wave energy level is low enough, the flattening of the bump can be described by a diffusion process in velocity space. This assumption does not hold universally, but seems to apply for most of the interplanetary electron streams.

A most remarkable observation, however, that does not agree with linear theory is the level of electron plasma waves. The slope of the distribution and the travel time of the interplanetary electron stream should, according to Section 5.2, yield a substantially higher wave energy density than observed. Long before these observations, P. Sturrock noted in 1964 that even in kinetic plasma theory, the beam

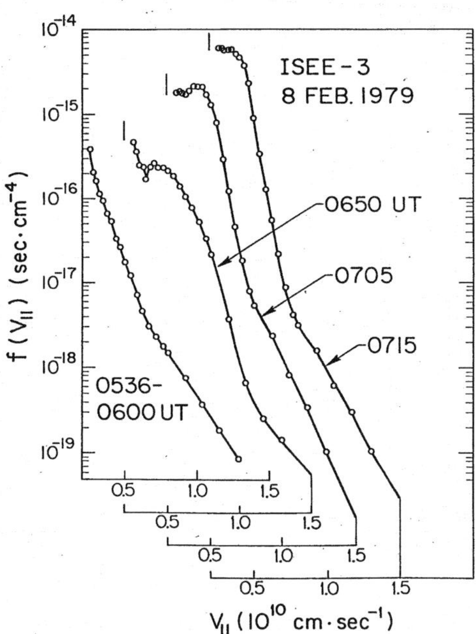

Fig. 6.3. Observation of an interplanetary electron beam by the ISEE-3 spacecraft. The velocity distribution is integrated over velocities perpendicular to the magnetic field and is shown for consecutive time intervals. The first (0536 – 0600 UT) is the distribution before the event. Each succeeding distribution is offset to the right by $5 \cdot 10^9$ cm s^{-1} (from Lin *et al.*, 1986).

would rapidly lose all its energy to the waves for any parameters he thought to be plausible. Sturrock's dilemma has found three basically different approaches for resolution: (*i*) An observer at a given location sees a beam of diminishing velocity as the fastest electrons arrive first and the slower ones later (Fig. 6.3). The early plasma waves with high phase velocity are Landau damped by the slower beam electrons at a later time. This part of the wave energy is not lost, but fed back into the stream where it accelerates slow beam electrons. (*ii*) Various non-linear processes have been suggested for removing the plasma waves from the resonance region and explaining the low wave level. (*iii*) Alternatively, density irregularities may shift continuously the plasma frequency of the background plasma and stop the linear growth of a wave having constant ω_r. Fluctuations may already be present in the solar wind due to ion acoustic waves. Which of the three processes limits the growth depends on the particular circumstances.

6.2.1. QUASI-LINEAR DIFFUSION

The linear theory of the previous section assumes a constant background distribution, f_0. Therefore it cannot be used to study the changes of f_0 or other average plasma properties caused by the growth of unstable waves. In the quasi-linear theory these changes are studied to second order in the fields, thus to first order in wave energy (weak-turbulence approximation). We write the velocity distribution $f =< f > +f_1$, where $< f >$ is the spatial average of the actual distribution f. It contains no fast local fluctuations, but the global, gradual evolution in time. The spatial changes are represented in f_1 having $< f_1 >= 0$. The fluctuations in f_1 are assumed to be caused by many waves varying rapidly in time and having $|f_1| \ll< f >$. The average distribution, $< f >$, now is allowed to change slowly under the influence of the waves. We shall prove that this change can be described by diffusion in velocity space in the form

$$\frac{\partial < f >}{\partial t} = \nabla_{\mathbf{v}} \cdot (\widehat{\mathcal{D}} * \nabla_{\mathbf{v}} < f >) \quad , \tag{6.2.1}$$

and calculate the diffusion tensor $\widehat{\mathcal{D}}$. We shall prove this general law for the special case of Langmuir waves (assuming $< \mathbf{E} >=< \mathbf{B} >= \mathbf{B}_1 = 0$) and in one dimension ($\mathbf{k}\|\mathbf{e_z}$). For collisionless plasma the Vlasov equations can be used, thus

$$\frac{\partial f}{\partial t} + v\nabla_z f + \frac{q}{m} E_1 \nabla_v f = 0 \quad . \tag{6.2.2}$$

We now average Equation (6.2.2) in space. The second term obviously cancels. Since $< E_1 < f \gg = 0$, we get

$$\frac{\partial < f >}{\partial t} = -\frac{q}{m} \nabla_v < E_1 f_1 > \quad . \tag{6.2.3}$$

The evolution of the average particle distribution is the result of a non-vanishing spatial average of $E_1 f_1$. This term is a second-order quantity.

The fast fluctuations causing f_1 are assumed to be linear waves. Thus, the linear theory developed in Section 5.2 can be applied. In particular, we shall make use of the normal modes given by the dispersion relation (5.2.24) as a function $\omega(k)$. As all waves act together on the particle distribution, we shall have to sum over the waves. For this purpose, it is natural to assume a continuous wave spectrum in $\mathbf{k}$. Thus we write

$$E_1(z,t) = \int_{-\infty}^{+\infty} E_1(k,t) \exp[i(kz - \omega t)] \frac{dk}{2\pi} \quad , \tag{6.2.4}$$

$$f_1(z, v_z, t) = \int_{-\infty}^{+\infty} f_1(k, v_z, t) \exp[i(kz - \omega t)] \frac{dk}{2\pi} \quad . \tag{6.2.5}$$

The wave amplitudes are allowed to change slowly. A relation between f_1 and E_1 is obtained from the Vlasov equation (6.2.2) assuming a dependence $\exp[i(kz-\omega t)]$ and linearizing to first order,

$$f_1(k, v_z, t) = \frac{-i}{\omega_k - kv_z} \frac{q}{m} E_1(k, t) \; \nabla_v < f > \quad . \tag{6.2.6}$$

This relation between the fluctuations of the distribution and the wave electric fields will be used to evaluate $< E_1 f_1 >$, which becomes

$$\begin{aligned} < E_1 f_1 > = \frac{1}{L} \int_{-L/2}^{+L/2} dz \int E_1(k, t) \exp[i(kz - \omega_k t)] \frac{dk}{2\pi} \\ \times \int f_1(k', v_z, t) \exp[i(k'z - \omega_{k'} t)] \frac{dk'}{2\pi} \quad , \end{aligned} \tag{6.2.7}$$

where L is the length of the region of enhanced wave level. Equation (6.2.7) can be simplified by using the standard formula

$$\lim_{L \to \infty} \int_{-L/2}^{L/2} e^{i(k+k')z} dz = 2\pi\delta(k + k') \quad . \tag{6.2.8}$$

On taking the limit $L \to \infty$ in Equation (6.2.7) and inserting (6.2.8), one obtains

$$< E_1 f_1 >= \frac{1}{L} \int E_1(k, t) f_1(-k, v, t) \exp[-i(\omega_k + \omega_{-k})] dk \quad . \tag{6.2.9}$$

For $-k$ the wave phase velocity is in the negative z-direction. It is the same wave as the one with $+k$ and negative frequency. Putting $\omega = \omega_r + i\gamma_k$, this means that $\Re e(\omega_{-k}) := \omega^r_{-k} = -\omega^r_k$ and $\gamma_{-k} = \gamma_k$. Therefore,

$$\omega_k + \omega_{-k} = 2i\gamma_k \quad . \tag{6.2.10}$$

Using Equations (6.2.6) and (6.2.10), we arrive at

$$< E_1 f_1 >= \frac{1}{L} \int E_1(k, t) E_1(-k, t) \frac{-i \exp[2\gamma_k t]}{\omega_{-k} + kv_z} \frac{q}{m} \; \nabla_v < f > \frac{dk}{2\pi} \quad . \tag{6.2.11}$$

Together with Equation (6.2.3) it proves that the quasi-linear evolution is a diffusion process (Eq. 6.2.1). The singularity at $\omega_{-k} = -kv_z$ shows that this diffusion is a resonance phenomenon. The mathematics is similar to the linear case (Section 5.2): The solution of the Vlasov equation is an initial value problem requiring a Laplace transformation in time. We have cut this short by assuming normal modes (Eqs. 6.2.4 and 6.2.5). The initial value can be implemented into Equation (6.2.11) by the Landau prescription (see Fig. 5.5), changing the integration path below the singularity in the complex k-space. The integral can then be evaluated by the Plemelj formula (5.2.19), using $\delta(\omega + kv_z) = \delta(\omega/v_z + k)/v_z$.

The imaginary part of Equation (6.2.11) vanishes, and the real part yields the diffusion coefficient

$$D = \frac{\omega_p^2}{mnv} W(k,t)\bigg|_{k=\omega_k/v} \quad . \tag{6.2.12}$$

We have introduced the spectral wave energy density

$$\overline{W}_k = \frac{E_1(\omega_k/v,t)E_1(-\omega_k/v,t)}{8\pi}\frac{1}{L} \quad , \tag{6.2.13}$$

and have used $W(k,t) = \overline{W}_k \exp[2\gamma_k t]$. The diffusion tensor (6.2.12) is of first order in the energy density of waves, but, being the product of $E_1 f_1$, it is not linear, hence *quasi-linear*. It suggests that quasi-linear diffusion is the result of energies, i.e. quadratic quantities. We shall introduce the concept of plasma phonons in Section 6.3 analogous to photons of light. The spectral wave energy density is proportional to the density of phonons of the energy $\hbar\omega$. Equation (6.2.12) can be interpreted that diffusion is proportional to the number of wave quanta (phonons) and the rate of their interaction with particles. The appearance of the wave energy density is a most remarkable aspect showing that particle diffusion and waves are coupled: The presence of waves causes resonant particles to diffuse. Equation (6.2.12) can be generalized to a three dimensional distribution of waves by replacing $W(k,t)$ with an integral in k-space over the waves in resonance with particles of a given velocity $\mathbf{v}$.

6.2.2. Strong turbulence

If the wave energy density is able to grow beyond the levels considered so far, the oscillating wave fields exert an average pressure, or *ponderomotive force*, altering the initial plasma parameters. This is the regime of *strong turbulence*. The theory of strong Langmuir turbulence presented here briefly is still under development, and its importance is unclear under astronomical circumstances. It is here not a matter of deriving accurate formulas, but of identifying some major physical processes.

Assuming pressure equilibrium, a local enhancement of electrostatic waves of total energy density $W_{\rm tot}$, given in Equation (5.2.42), produces a slow adjustment of the average density of both ions and electrons

$$\Delta n = -\frac{W_{\rm tot}}{k_B T} \quad . \tag{6.2.14}$$

The decreasing density reduces the local plasma frequency, and the dispersion relation of Langmuir waves (5.2.24) changes accordingly,

$$\omega_r^2 = (\omega_p^0)^2(1 + 3k^2\lambda_D^2 - \frac{W_{\rm tot}}{n_0 k_B T}) \quad , \tag{6.2.15}$$

where subscript 0 refers to the undisturbed values. A first characteristic of strong turbulence is immediately clear – the wave vector k of a wave at constant ω_r

The above shift of the dispersion relation from smaller to larger k is strongly enhanced by what is called *modulational instability.* The wave distribution can collapse into *cavitons.* The collapse is simply an effect of refraction. Imagine that, in a part of the space occupied by the wave, the density is slightly smaller, and therefore k is larger than in the rest, to match the dispersion relation. The phase velocity there is smaller, thus the wave is focused into the region of low density, where W_{tot} increases and enhances the ponderomotive force (see also Fig. 6.4). It has to push out even more plasma to remain in pressure equilibrium, closing the feedback loop of the instability. Unlike the resonant decay process of the previous section, the modulational instability produces an essentially zero-frequency plasma modulation which does not satisfy a dispersion relation of normal modes of the linearized plasma equations. Accordingly, it is important only if W exceeds some threshold value. Cavitons are often loosely referred to as solitons. For Langmuir waves due to an electron beam with a small range of $\mathbf{\Delta k}$, the best known instability of this type is the so-called *oscillating two-stream instability* having a threshold of

$$\frac{W_{tot}}{n_0 k_B T} \geq 3(\Delta k \lambda_D)^2 \tag{6.2.16}$$

(Sagdeev, 1979). We do not want to give the mathematical details, but point out that this result can be expected from Equation (6.2.14), as it is the requirement for W_{tot} to substantially influence k. The threshold for $W_{tot}/(n_0 k_B T)$ can be estimated from measured quantities in the example of Figure 6.3. It is in the range $2 - 15 \cdot 10^{-7}$, about two orders of magnitude above the observed maximum density of wave energy. Energy densities near threshold have been found in other cases.

The half-width Δk of the resonant Langmuir wave spectrum at half-maximum is difficult to measure. It can be inferred indirectly from its relation to the range of initially unstable phase velocities Δv. Assuming that the growth rate $\gamma(\omega/k)$ of Langmuir waves, as derived from Equation (5.2.25), has a triangular shape with height γ_0 and a base $\Delta v \ll V_b$, one finds

$$\frac{\Delta k}{k_m} \approx \frac{\Delta v}{V_b} \frac{\ln 2}{2\gamma_0 t} \quad , \tag{6.2.17}$$

where $k_m \approx \omega_p / V_b$ is the minimum wave number of the growing Langmuir waves, and V_b is the electron beam velocity. Note the time in the denominator! The wave energy increases, and the spectrum becomes narrower at half-maximum the longer the linear instability has time to grow. This is a property of exponential growth producing an ever sharpening peak, assuming that Δv remains unchanged. The longer W grows and Δk decreases, the more probable is that W exceeds the threshold. Once the modulational instability starts, the growing wavenumber decelerates the waves until they are Landau damped by thermal electrons. This finally quenches the instability. Numerical computations yield, as an order of magnitude estimate for the saturation level of Langmuir waves,

$$\frac{W_{\text{sat}}}{nk_BT} \approx 3\frac{\gamma_0}{\omega_p} \quad , \tag{6.2.18}$$

and typical values are of the order 10^{-5}. The saturation level according to Equation (6.2.18) is very high. For the parameters observed in interplanetary beams (using Eq. 5.2.29 for an estimate of γ_0), it is sometimes found comparable or exceeding the free energy density of the beam, approximately $\frac{1}{2}m_e V_b^2 n_b$. In such cases the beam energy may be exhausted before saturation, and it is advisable to use the beam energy as an upper limit of wave energy.

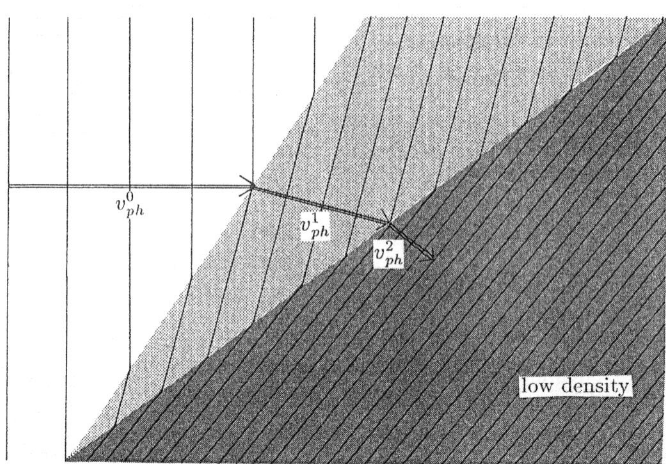

Fig. 6.4. Wave fronts (represented by parallel lines) of a wave with frequency ω are bent toward the region of low phase velocity, $v_{ph} = \omega/k$. Since the phase velocity of Langmuir waves depends strongly on density through k in the dispersion equation, the waves are concentrated into low density regions (dark area).

6.2.3. Deflection of Electrostatic Waves

Langmuir waves have a peculiar dispersion relation (5.2.24), in which the frequency depends only slightly on wave number (assuming $kv_{te} \ll \omega_p^e$). If a wave propagates in a weakly inhomogeneous medium, the wave frequency remains constant, but the wave number must change by a large fraction to satisfy the dispersion relation. This, in turn, affects the phase velocity, and the wave loses resonance with the beam particles. Lateral inhomogeneity deflects the wave (see Fig. 6.4) and is usually even more efficient in removing the wave from resonance. Density inhomogeneities thus tend to isotropize beam produced Langmuir waves. Given a spectrum of density fluctuations, a diffusion coefficient of wave quanta in velocity space can be calculated. It reduces exponentially the phonons in the beam direction and is an effective damping. In fact, its rate can exceed the growth rate and inhibit further growth of the instability.

6.2.4. SUMMARY

Recent developments in both observations and theory of interplanetary electron beams have drastically changed our view of particle streams in the universe. Measurements like Figure 6.3 are detailed enough to test the beam evolution and study the saturation of the bump-on-tail instability.

- Particle beams are stable toward the bump-on-tail instability in a large fraction of space where density variations in the source limit growth or where scattering of electron plasma waves off ion sound waves constitutes an effective damping rate larger than the growth rate (see next paragaph).
- Where the local conditions allow the waves to grow, the wave energy reaches very high levels close to the mean free energy of the beam.
- Localized regions of enhanced wave energy density (cavitons) result.
- Quasi-linear particle diffusion in velocity space is therefore not a continuous process, but occurs in steps when the beam passes a region of enhanced growth rate.

Finally, we may add that the absence of the electrostatic bump-on-tail instability in an electron stream does not imply the stream to be stable. Depending on beam and plasma parameters other instabilities may appear, but usually grow more slowly. They include the resonant electron instability of left hand polarized, low-frequency electromagnetic waves (L-mode) for high β plasmas and intense beams. It is analogous to the electromagnetic ion beam instability of Section 7.2. Furthermore, there may be an instability of lower hybrid waves, the inverse process to electron acceleration by these waves (Chapter 9).

6.3. Plasma Emission

Having studied the initial (*linear*) phase of the growth of electrostatic waves due to a beam in Chapter 4 and the *non-linear* evolution of the instability in Section 6.2, we now turn to the conversion of electron plasma waves into electromagnetic emission. As a rule, the production of radio waves does not substantially influence the evolution of the beam, but is a diagnostic. An exception may be the decay process to be discussed in Section 6.3.3.C.

6.3.1. HARMONICS

Dynamic spectra of type III bursts sometimes show two similar features displaced in frequency by about a factor of two (Fig. 6.5). This important characteristic was already noted by Wild, Murray, and Rowe in 1954, who immediately interpreted it as radiations at the plasma frequency (fundamental, $\nu \approx \nu_p$) and its harmonic ($\nu \approx 2\nu_p$). It will find a natural interpretation in terms of the conversion of electron plasma waves into electromagnetic radiation, and has historically been taken as major evidence supporting the plasma emission hypothesis. Events with harmonic structure have been studied extensively. The frequency ratio is usually found to be smaller than 2 with an average of 1 : 1.8. The ratio being less than 2.0 is interpreted as collisional absorption of the radio waves; the closer the frequency of the electromagnetic wave is to the local plasma frequency, the slower its group velocity and the more time collisions have to randomize the wave energy. Therefore, the low-frequency part of the fundamental is invisible, and the harmonic ratio is below 2. A general discussion of absorption will follow in Section 11.2.

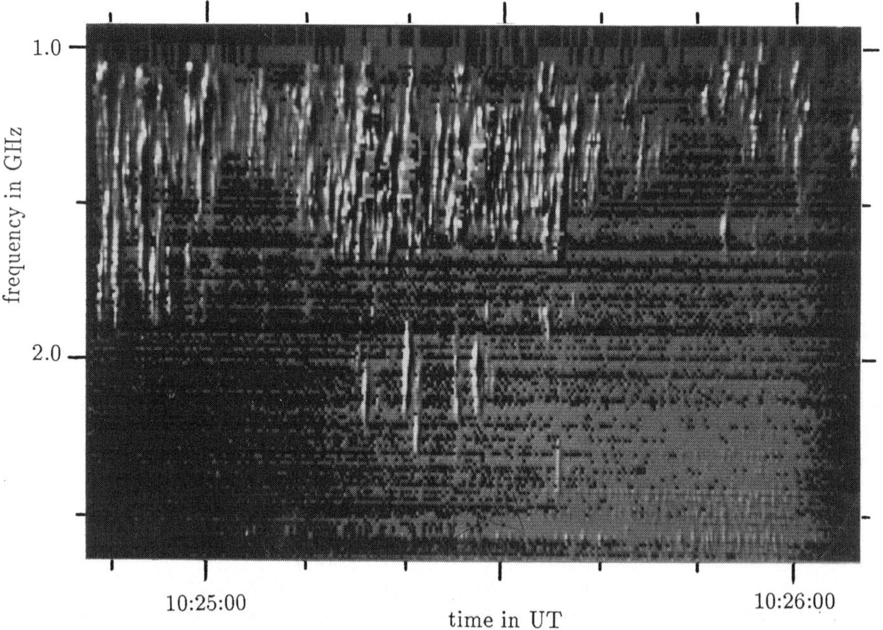

Fig. 6.5. Dynamic spectrogram of solar microwave emission of beams at fundamental and harmonic (narrowband type III bursts). The brightest sources on 1.2 GHz show up at the harmonic frequency around 2 GHz (observed at ETH Zurich).

The emission of the fundamental component at meter wavelengths is more directive than the harmonic. Simultaneous measurements from Earth and from a Sun orbiting spacecraft have allowed direct determination of the cone of emission. The harmonic was found to be practically isotropic, but the fundamental was restricted to about 60° from the vertical. This has been confirmed by spectral observations noting a sharp decline of the number of fundamental-harmonic pairs when the angular distance of the coronal source from the disk center exceeds 50°. The higher directivity of the fundamental will be explained in Chapter 11 by absorption, by beaming toward the vertical direction due to the decreasing refractive index and by ducting.

The recognition of fundamental-harmonic pairs again becomes difficult above about 100 MHz due to the increasing number of bursts per group. In the microwaves ($\gtrsim$ 1 GHz), the diminishing bandwidth greatly facilitates their identification, but the fundamental is often heavily absorbed. Figure 6.5 demonstrates a case of narrowband type III bursts. The fundamental emission is considerably brighter than the harmonics, which are visible only for the brightest bursts.

6.3.2. Phonons and Their Scattering (Wave Conversion)

The conversion of unstably growing waves into other waves is interesting for two reasons: it can limit the growth of the waves, and it can transform them into observable electromagnetic radiation (transverse waves). Since energy and momentum are conserved, the changes in wave energy and momentum between the incident wave and the scattered wave have to be taken up by a third partner. This may be a particle, a group of particles or a third wave (three-wave interaction). The conversion of one wave into another is generally called *scattering* even if the product has a different frequency or wave mode.

The exchange of energy and momentum can be illustrated and simplified by a quantum mechanical approach. Waves in a plasma contained by a conducting box are eigenmodes, like light waves in a vacuum cavity. Quantum mechanics can be applied analogously to a plasma with the result that all the waves also appear as photons (transverse waves) or *phonons* (longitudinal modes) with energy $\hbar\omega$ and momentum $\hbar\mathbf{k}$ ($\hbar$ is the Planck constant divided by 2π). An MHD photon, for example, has an energy of the order of 10^{-30} erg $\approx$ 10^{-18} eV (!). Wave-wave and wave-particle interactions appear as exchanges of quanta governed by conservation laws. The quantum approach is physically correct, but Planck's constant will drop out at the end, demonstrating that it is not necessary. It is sometimes called a semi-quantum mechanical approach. The processes are intrinsically classical.

The elegance and usefulness of the quantum mechanical approach is limited to low wave energy levels, called *weak turbulence*, where the quanta can be considered as individual entities. Strong turbulence conversion will be outlined in the following section.

We start with the scattering of a wave on a particle. In the quantum mechanical picture an incident wave quantum (1) scatters into a quantum (2) of the secondary wave, changing the particle's energy and momentum by

$$\Delta\varepsilon = \hbar(\omega_2 - \omega_1) \quad , \tag{6.3.1}$$

$$\Delta\mathbf{p} = \hbar(\mathbf{k_2} - \mathbf{k_1}) \quad . \tag{6.3.2}$$

The scattered wave may also differ in wave mode. Since only Vlasov's and Maxwell's equations are involved, it is evident that such a process is reversible in time, and wave (2) can therefore also be scattered into wave (1). The process is called *spontaneous*, since one single quantum of type (1) can do it, independent of other quanta of type (1) or (2).

The advantage of the quantum mechanical language is that it allows us to treat scattering as the conversion of a single wave quantum into another. In each such scattering event the number of quanta (phonons or photons) of one wave decreases by unity, and increases for the other. We define a spectral quantum density of bosons, $N(\mathbf{k})$, and an effective *photon temperature*, $T(\mathbf{k})$, using the Rayleigh - Jeans approximation,

$$N(\mathbf{k})\hbar\omega_{\mathbf{k}} := k_B T(\mathbf{k}) := W(\mathbf{k}) \quad . \tag{6.3.3}$$

For propagating transverse waves, $T(\mathbf{k})$ at the observer's site is the observed *brightness temperature* defined in Section 5.1.

A. *Spontaneous Scattering off Ions*

Historically, the first scattering partner envisaged by V.L. Ginzburg and V.V. Zheleznyakov in 1958 to convert electron plasma waves into radiation were thermal ions. Why consider an object as inert as an ion as a scattering partner? Single electrons are indeed well-known to be responsible for Thomson scattering at very high (e.g. optical) frequencies. However, as the wavelength approaches a few times the Debye length, i.e. $\omega \lesssim \omega_p c/v_{te}$, single electrons are effectively shielded and do not scatter. Nevertheless, scattering off a thermal ion is actually accomplished by the Debye shielding cloud of electrons around the ion. The ion is said to be *dressed* by these electrons. Therefore, scattering differs little from Thomson scattering of high-frequency radiation on single electrons. It acts like scattering off a particle with charge $Z_i e$, mass $Z_i m_e$, density $Z_i^2 n_e$, and velocity v_{ti}. Let $w(\mathbf{v}, \mathbf{k_L}, \mathbf{k_t})$ be the differential rate at which a *dressed* ion scatters a Langmuir wave having $\mathbf{k_L}$ into a transverse wave having $\mathbf{k_t}$ with the normalization $\int w(\mathbf{v}, \mathbf{k_L}, \mathbf{k_t}) d^3k_L/(2\pi)^{-3} = w$. An extensive theory (e.g. presented in Melrose, 1980) and some simplifying assumptions yield for solar abundances

$$w \approx w_T/4 \quad , \tag{6.3.4}$$

where the Thomson scattering rate per ion,

$$w_T = \frac{8\pi r_0^2 c\mathcal{N}}{3} \quad . \tag{6.3.5}$$

$\mathcal{N}$ is the refractive index, and r_0 denotes the classical electron radius, $r_0 = e^2/mc^2$. The quantum density of the scattered transverse waves changes according to

$$\frac{\partial N_t(\mathbf{k_t})}{\partial t} \approx \int d^3v \int \frac{d^3k_L}{(2\pi)^3} w(\mathbf{v}, \mathbf{k_L}, \mathbf{k_t}) \; N_L(\mathbf{k_L}) \; f_i(\mathbf{v}) \quad . \tag{6.3.6}$$

Equation (6.3.6) includes only the spontaneous processes (comparable to random collisions between phonons and particles). Often one assumes the spectral quantum density to be constant in $\mathbf{k}$-space from zero to a maximum k. So the differential scattering rate is not needed, and the total scattering rate, w, given by Equation (6.3.4) can be used.

Frequency and wave number of the transverse wave generated by scattering Langmuir waves off ions can be calculated from the dispersion relations and the conservation equations. We consider the simple case of no magnetic field. The dispersion relations have been given by equations (5.2.24) and (4.3.12). Assuming $\omega \gg \Omega_e$,

$$\omega_L \approx \omega_p(1 + \frac{3k_L^2 v_{te}^2}{\omega_p^2})^{1/2} \quad , \tag{6.3.7}$$

$$\omega_t \approx (\omega_p^2 + k_t^2 c^2)^{1/2} \quad . \tag{6.3.8}$$

We claim that the scattered transverse wave has a frequency $\omega_t \approx \omega_L$. Thus Equations (6.3.7) and (6.3.8) give

$$k_t \approx \sqrt{3}\, k_L \frac{v_{te}}{c} \quad . \tag{6.3.9}$$

Under usual coronal conditions Equation (6.3.9) implies $k_t \ll k_L$. The conservation equations (6.3.1) and (6.3.2) require

$$\omega_L - \mathbf{k_L v} = \omega_t - \mathbf{k_t v} \quad , \tag{6.3.10}$$

where $\mathbf{v}$ is the ion velocity assumed to be practically unaltered (since $k_B T \gg \hbar\omega$). The bandwidth of the scattered electromagnetic radiation can be calculated,

$$\Delta\omega \approx |\mathbf{k_t} - \mathbf{k_L}|\, v_{ti} \approx k_L v_{ti} \quad . \tag{6.3.11}$$

Since $k_L v_{ti} \approx \omega_p v_{ti}/V_b \ll \omega_p$, where V_b is the beam velocity, Equation (6.3.11) proves the claimed approximate equality of incident and scattered wave frequency. The scattering process and in particular the bandwidth of the scattered radiation (Eq. 6.3.11) have been demonstrated in a convincing way by radar backscattering from the terrestrial ionosphere. The process will be applied to astrophysical plasmas in Section 6.3.3.

B. *Induced Scattering*

The above spontaneous process was found to be inadequate not only in preventing growth of the bump-on-tail instability, but also in accounting for the observed radio emission. V.N. Tsytovich suggested in 1966 that *induced scattering* by ions could at least explain the latter. Induced scattering occurs in addition to the spontaneous processes analogous to transitions between electron states of an atom (where this process is called stimulated emission). The transition probabilities between atomic states is described by Einstein's coefficients A and B. They allow for spontaneous transition, absorption, and transition induced by the presence of radiation. Induced emission causes an avalanche effect, which is utilized for example in lasers.

Langmuir waves and dressed ions form together a system similar to excited atoms. Induced emission occurs if also a secondary photon (or wave) enters the system and stimulates it to emit more secondary waves. Induced scattering requires the presence of both primary and secondary waves. As the induced process is proportional to both the primary and the secondary wave density, the feedback drives the latter to exponential growth. Needless to say, induced scattering of cosmic plasma waves into observable electromagnetic waves is an attractive possibility.

The scattering process transfers some energy of the primary waves to the ions. The energy absorbed by the scattering particle is always positive for induced processes. If it were negative, one could cool a plasma by irradiation inducing emission from the thermal level of Langmuir waves. This would contradict the second law of thermodynamics. Therefore the scattered wave must be at a lower frequency than the primary wave. For observable radiation this has the important consequence that the primary waves must have a higher frequency than the secondary electromagnetic waves. Electron plasma waves are the only electrostatic mode above the cutoff frequencies of observable radiation. They are the only waves for which induced scattering on ions may be directly observable in coronal plasmas.

Instead of the isolated states of atomic transitions, we have a continuous spectrum of waves. It may be considered as the discrete spectrum of a box whose size we let go to infinity. Einstein's argument for detailed balance of stimulated emission, based on the second law of thermodynamics, implies a rate of induced scattering of $w(\mathbf{v}, \mathbf{k_L}, \mathbf{k_t}) N_t(\mathbf{k}_t)$. Equation (6.3.6) has to be completed to

$$\begin{aligned}\frac{\partial N_t(\mathbf{k_t})}{\partial t} = \int d^3v \int \frac{d^3k_L}{(2\pi)^3} w(\mathbf{v}, \mathbf{k_L}, \mathbf{k_t})[\{N_L(\mathbf{k_L}) - N_t(\mathbf{k_t})\} f_i(\mathbf{v}) - \\ N_L(\mathbf{k_L}) N_t(\mathbf{k_t}) \frac{\hbar(\mathbf{k_L} - \mathbf{k_t})}{m_i} \cdot \frac{\partial f_i}{\partial \mathbf{v}}]\end{aligned} \quad . \tag{6.3.12}$$

The term which is linear in $N_L(\mathbf{k}_L)$ accounts for spontaneous scattering of Langmuir waves into transverse waves weighted by the differential scattering rate $w(\mathbf{v}, \mathbf{k_L}, \mathbf{k_t})$ and the ion velocity distribution f_i (Eq. 6.3.6). The term which is linear in $N_t(\mathbf{k_t})$ is negative, since it represents the inverse process: absorption. The term which is proportional to $N_L N_t$ describes the net effect of induced scattering in both directions differing by the ion distribution at the high and low energy levels. The difference can easily be evaluated using the approximation $f(\mathbf{v} - \hbar(\mathbf{k_L} - \mathbf{k_t})/m_i) \approx f(\mathbf{v}) - \hbar(\mathbf{k_L} - \mathbf{k_t})/m_i \cdot \partial f_i / \partial \mathbf{v}$.

Induced scattering means that radiated photons force the dressed ions to radiate even more. This becomes important when the spectral wave energy density, $W_t(\mathbf{k})$, exceeds the mean thermal energy of ions by a factor of $\omega_t/(\omega_L - \omega_t)$ (Exercise 6.3). For the scattering of Langmuir waves into transverse waves, the threshold is a brightness temperature $T_t \gtrsim 10^9(T_i/10^6)$K. The observed values are frequently much higher and suggest that scattering is generally induced rather than spontaneous, or the emission is of a different nature (see below). Equation (6.3.12) will be applied to plasma emission in Section 6.3.3.

C. *Scattering off Other Waves*

For scattering off a second wave (subscript 2), the conservation equations become

$$\omega_1 + \omega_2 = \omega_3 \quad , \tag{6.3.13}$$

$$\mathbf{k_1} + \mathbf{k_2} = \mathbf{k_3} \quad . \tag{6.3.14}$$

These equations are sometimes referred to as *parametric conditions.* Equation (6.3.13) is the resonance condition for maximum energy transfer of three coupled harmonic oscillators. In fact, the waves are coupled through particles (usually electrons) oscillating in the fields of the waves. If the energy flows from waves (1) and (2) into (3), the scattering is called *coalescence.* It is denoted as (1)+(2) $\rightarrow$ (3), or $a + b \rightarrow c$, where the letters are symbols for waves (L := Langmuir wave, s := ion acoustic wave, t := transverse (electromagnetic) wave, etc.). An important coalescence is the scattering of a Langmuir wave off another Langmuir wave into a transverse wave; $L + L' \rightarrow t$. It is observable as *harmonic emission* at about $2\omega_p$.

The inverse process, $c \rightarrow a + b$, is termed *decay.* Important examples for beams include the decay of a Langmuir wave (L) into an ion acoustic wave and a transverse wave (t), observable as *fundamental radiation*: $L \rightarrow t + s$. Decay into an ion acoustic wave and a secondary Langmuir wave ($L \rightarrow L' + s$) is a competing process.

The conservation conditions (6.3.13) and (6.3.14) greatly reduce the number of possible scattering processes, since the third wave $(\omega_3, \mathbf{k_3})$ must satisfy the dispersion relation of a normal mode. Furthermore, if the outcome should be an observable transverse wave having $\omega_t > \omega_p$, one of the initial waves must be a high-frequency mode, such as a Langmuir wave, z-mode or upper hybrid wave (Section 4.4). The condition on $\mathbf{k_3}$ further reduces the choice of wave modes; Langmuir waves produced by beams have large k_L, and propagating transverse waves have small k_t. They need to be balanced by a wave that is nearly anti-parallel to the incident Langmuir wave and also has a large k. In fact, a major difficulty in wave conversion is to identify suitable waves and plausible physical parameters before calculating their emission. We outline the latter in the following paragraph.

The scattering efficiency off a second wave is evaluated in a way similar to ions. The rate of transitions (1)+(2) $\rightarrow$ (3) per unit $\mathbf{k_1}$ and $\mathbf{k_2}$ is the sum of spontaneous and induced scattering, $u^{123}[1 + N_3]N_1N_2$. It has to be diminished by the rate of the inverse process (3) $\rightarrow$ (1)+(2), namely $u^{123}N_3[1 + N_1][1 + N_2]$. One finds

$$\begin{aligned}\frac{\partial N_3(\mathbf{k}_3)}{\partial t} = \int \frac{d^3k_1}{(2\pi)^3} \int \frac{d^3k_2}{(2\pi)^3} u^{123}(\mathbf{k}_1, \mathbf{k}_2, \mathbf{k}_3)[N_1(\mathbf{k}_1)N_2(\mathbf{k}_2) \\ - N_3(\mathbf{k}_3)\{N_1(\mathbf{k}_1) + N_2(\mathbf{k}_2)\}]\end{aligned} \quad . \tag{6.3.15}$$

The transition probabilities u^{123} result from the non-linear coupling of two normal modes of linear theory. The density perturbation, n_1^1, of one wave interacts with the velocity perturbation, V_1^2, of the other wave to give a non-linear current, $\mathbf{J}_{\rm nl}$, driving the electric field of the third wave. From the linearized Maxwell's equations (4.2.7) and (4.2.10) one derives the general wave equation for the fields of wave (3),

$$\frac{\partial}{\partial t}\left[\frac{|\mathbf{E}_3|^2}{8\pi}+\frac{|\mathbf{B}_3|^2}{8\pi}\right]+\nabla\cdot\left[\frac{c}{4\pi}\mathbf{E}_3\times\mathbf{B}_3\right]=-\mathbf{J}_{nl}\cdot\mathbf{E}_3 \quad . \tag{6.3.16}$$

The first term is the wave field energy density, and the second is the gradient of the Poynting vector. The right side is only different from zero for the non-linear part of the current, $\mathbf{J}_{\rm nl}$. In our case this is the source term for emission. We do not here derive the transition probabilities for particular cases; they are based on the detailed properties of the linear wave modes. Some relevant examples will be given in the following section. A number of them have been tabulated in the literature (e.g. Kaplan and Tsytovich, 1973) for various combinations of waves. The equations for the rate of change of N_1 and N_2 are analogous; for example,

$$\frac{\partial N_1(\mathbf{k}_1)}{\partial t}=-\int\frac{d^3k_3}{(2\pi)^3}\int\frac{d^3k_2}{(2\pi)^3}u^{123}(\mathbf{k}_1,\mathbf{k}_2,\mathbf{k}_3)[N_1(\mathbf{k}_1)N_2(\mathbf{k}_2)- \\ -N_3(\mathbf{k}_3)\{N_1(\mathbf{k}_1)+N_2(\mathbf{k}_2)\}] \quad . \tag{6.3.17}$$

The three-wave process obviously saturates when the bracket in Equations (6.3.15) and (6.3.17) vanish, requiring

$$N_1(\mathbf{k}_1)N_2(\mathbf{k}_2)=N_3(\mathbf{k}_3)\{N_1(\mathbf{k}_1)+N_2(\mathbf{k}_2)\} \quad . \tag{6.3.18}$$

Saturation means that the decay and the inverse coalescence process are in equilibrium. From the point of view of a radiation mechanism, it corresponds to optically thick emission; the intensity cannot be further increased by more scattering in a larger source.

We note that the decay $L \to t+s$ does not satisfy Equation (6.3.18) for saturation unless both N_t and $N_s > N_L$, an unlikely condition. Even without saturation the process has been found to be efficient enough to limit the growth of Langmuir waves and to be compatible with some observed radio emissions (radiation from an optically thin source). It will be applied below to beam radiation.

The coalescence $L+s\to t$ does not work efficiently if ion acoustic waves are strongly damped (Eq. 5.2.39), since, for $N_s \ll N_L$ at saturation (equivalent to maximum emission), $T_t = T_s \ll T_L$. On the other hand, if the ion acoustic waves are excited to $N_s \gg N_L$, they can make weak Langmuir waves observable at a brightness temperature of $T_t \lesssim T_L$. This will be important in Chapter 9, when we consider currents.

For the process $L+L'\to t$ saturation occurs at a brightness temperature

$$T_{\max}(\mathbf{k_t})\approx 2\frac{T_L(\mathbf{k_L})T_{L'}(\mathbf{k'_L})}{T_L(\mathbf{k_L})+T_{L'}(\mathbf{k'_L})} \quad . \tag{6.3.19}$$

The factor 2 arises from $\omega_t = \omega_L + \omega_{L'} \approx 2\omega_p$. For a distribution $T_L(\mathbf{k_L})$ of intense waves collimated along the beam direction and a weaker isotropic population $T_{L'}(\mathbf{k'_L})$ – produced e.g. from the intense waves by induced scattering off ions – the brightness temperature is the temperature of the weaker waves; $T_t = T_L(\mathbf{k'_L})$. In the following section we concentrate further on the processes leading to emission and derive practical formulas.

6.3.3. PLASMA RADIATION EMISSIVITIES

The energy emitted per unit of time, volume, frequency and solid angle Ω is called *emissivity*. It is the rate of change of the photon density per solid angle multiplied by the energy of a photon, thus the emissivity is

$$\eta(\omega) = \hbar\omega \frac{\partial N_t(\mathbf{k_t})}{\partial t} \frac{d^3 k_t}{(2\pi)^3 d\omega \; d\Omega} \quad . \tag{6.3.20}$$

The second quotient converts $\mathbf{k}$ into ω and Ω. In Equation (5.1.1) we have introduced the intensity, I, as the energy of radiation (transverse waves) passing through a unit area per unit time, frequency and solid angle. In the absence of absorption and refraction,

$$I(\omega, \theta) = \eta(\omega, \theta) \; L(\omega, \theta) \quad . \tag{6.3.21}$$

$L(\omega, \theta)$ is the length of the ray path in the source. If the source is inhomogeneous, L is the length over which change becomes essential. For example, if the density changes, the frequency range of the beam excited Langmuir waves moves away form the initial frequency after a small fraction of the scale length. The general transfer of radiation will be discussed in Chapter 11, where the special case of Equation (6.3.21) will be referred to as *optically thin*. In the opposite case, the emission of an *optically thick* source is saturated by its own inverse process. The intensity does not further increase along the ray path, and an outside observer only receives radiation from the surface layer of the source. The intensity at the source then is given by the photon temperature T_t (Eq. 6.3.3),

$$I(\omega, \theta) \approx \frac{2\nu^2 k_B T_t(\omega, \theta)}{c^2} \quad . \tag{6.3.22}$$

The Rayleigh-Jeans approximation for $\hbar\omega / k_B T_t \ll 1$ has been applied in Equation (6.3.22). The factor 2 is usually introduced to account for both electromagnetic modes, assuming equal photon temperatures.

It is accepted that the radio emission associated with coronal and interplanetary electron beams originates from the conversion of beam-excited electron plasma waves into radiation. Yet we have encountered several possibilities in the previous section. A basic classification of the processes distinguishes between: (*i*) weak and strong turbulence (depending on the level of electron plasma waves); and (*ii*) fundamental and harmonic emission. The following subsections summarize the most plausible results for these four cases and outlines their relevance to observations.

A. *Emission at the Harmonic*

We start with the harmonic emission for which there are just two possibilities: weak and strong turbulence.

At a low level of plasma waves (weak turbulence), the coalescence of two waves ($L + L' \rightarrow t$) is a generally accepted emission mechanism. With some simplifying assumptions, the emissivity can be calculated from Equation (6.3.15),

$$\eta \approx \frac{2}{\pi}(k_m\lambda_D)^{-5}\left(\frac{v_{te}}{c}\right)^5 \frac{(W^L_{\rm tot})^2}{n_e k_B T} \quad , \tag{6.3.23}$$

The distribution of Langmuir waves in **k**-space is usually unobservable and has to be estimated from the total wave energy, $W^L_{\rm tot}$. If k_m is the maximum wavenumber of a broadband, isotropic distribution of Langmuir waves with $W(\mathbf{k_L})$ = constant for $k < k_m$, the total wave energy density is given by

$$W_{\rm tot} = W(k)\frac{4\pi k_m^3}{3(2\pi)^3} \quad . \tag{6.3.24}$$

For an optically thin source, the intensity can be evaluated from Equation (6.3.21). It has to be corrected for absorption and refraction in the source and by propagation (Chapter 11). If the source is optically thick, equivalent to saturation of the scattering process (Eq. 6.3.19), the photon temperature in the source becomes

$$T_t = T_L \tag{6.3.25}$$

for isotropic Langmuir waves. The Langmuir waves produced by beams are collimated in the forward direction. A fraction thereof is scattered on ions into the reversed direction and can combine with the former to transverse waves. Since the phonon temperature in the backward direction is lower, the resulting photon temperature is reduced correspondingly.

The photon temperature can be converted into intensity using Equations (6.3.3) and (6.3.22). From Equation (6.3.25) one derives

$$I \approx \frac{12\lambda_D}{(k_m\lambda_D)^3}\left(\frac{v_{te}}{c}\right)^2 W^L_{\rm tot} \quad . \tag{6.3.26}$$

Intensities of $2 \cdot 10^{-13}$ erg s^{-1}Hz^{-1}cm^{-2}sterad^{-1} and brightness temperatures up to 10^{15} K have been reported, bringing $W^L_{\rm tot}$ into the range of strong turbulence.

Strong turbulence packets of Langmuir waves (Section 6.2.2) act as antennas and radiate more efficiently than the coalescence process of weak turbulence. The emissivity of collapsed solitons has been calculated numerically and found between one and two orders of magnitude higher than given in Equation (6.3.23).

The observation of two bands of emissions of type III radio bursts separated in frequency by a ratio of about 1 : 2 confirms the general idea of harmonic emission.

B. *Emission at the Fundamental: Scattering off Ions*

The relevant emission process for the fundamental is less clear. One radiation process is the scattering off the electron shielding cloud of thermal ions presented in Section 6.3.2. If a broad spectrum of Langmuir waves is assumed with $N_L \gg N_t$ and a largest wavenumber k_m (Eq. 6.3.24), the emissivity of Langmuir waves scattering off ions results from Equations (6.3.5) and (6.3.12),

$$\eta \approx \frac{\omega_p^4 W_{\rm tot}^L}{32\pi^2 c^4 n_e k_m} \left(\frac{k_t}{k_m}\right)^2 \left[1 + \frac{v_{ti}}{v_{ph}^L} \frac{W_t(k_t)}{k_B T_i}\right] \quad , \tag{6.3.27}$$

where $v_{ph}^L = \omega_L/k_L \approx \omega_t/k_L$ is the phase velocity of the Langmuir waves. The two terms in the brackets refer to spontaneous and induced scattering, respectively. Induced scattering is negligible at the far end of the source where $W_t(k_t) \approx 0$, but dominates after the ray has passed a critical distance, s_c, where the total energy density of transverse waves has grown to the threshold value $k_B T_i v_{ph}^L/v_{ti}$. At this point induced emission equals spontaneous emission and dominates thereafter. For typical parameters of the solar corona and $W_{\rm tot}^L \approx 6 \cdot 10^{-9} n k_B T$ (measured in the case of Fig. 6.3), the threshold distance becomes $s_c \approx 2 \cdot 10^9$ cm (Exercise 6.4). A necessary condition for induced emission to dominate is source homogeneity on such distances.

C. *Emission at the Fundamental: Decay*

Two more processes have also received attention. One of them proposes that Langmuir waves are directly converted into radio waves on steep density gradients (e.g. Melrose, 1980). Here we consider the second process proposing that the beam produced Langmuir waves decay directly into a transverse wave and an ion acoustic wave ($L \rightarrow t + s$). This process may have been observed in the laboratory and in interplanetary electron beams by its strong radio emission at the fundamental. The parametric equations (6.3.13) and (6.3.14) require that $\omega_L = \omega_t + \omega_s$. Under plausible conditions $k_L \approx k_s \gg k_t$. The ion acoustic waves must have similar wave numbers as the Langmuir waves, but in the opposite direction. Waves with $\omega_s \approx c_s k_L$ have indeed been observed in interplanetary beams, as will be shown in the following paragraph.

The simultaneous observations of waves and particles shown in Figure 6.6 can be interpreted in the following way: The electron waves are driven into instability during the time of positive gradient in the electron distribution. The fastest electrons arrive first. We note, however, that the electrons with highest energies are apparently too sparse to excite waves. The waves are converted into radio waves at 31 – 100 kHz. The radio waves at higher frequency originate closer to the Sun (see for example Figure 5.8). In addition, ion acoustic waves can be found observationally; the solar wind sweeps them by the spacecraft and Doppler shifts their frequency to $\bar{\omega}_s \approx V_{sw} k_s$, where V_{sw} is the solar wind mean flow velocity. Using $k_s \approx k_L \approx \omega_p/V_b$ and the observed solar wind density and velocity, one expects $\bar{\nu}_s \approx 130$ Hz, the center of the band of low-frequency ion acoustic waves in Figure 6.6, correlating with the electron plasma waves.

The decay $L \rightarrow t + s$ is an efficient process, converting electron plasma waves into transverse emission. However, it is limited by a severe condition requiring the growth rates of both transverse waves and ion acoustic waves to exceed the damping rates. The latter are strongly Landau damped unless $T_e \gg T_i$ (Section 5.2.6). In the case presented in Figure 6.6 the observed ratio was $T_e/T_i = 2.8$, permitting growth. Therefore a decay process is likely. It is tempting to speculate that it may also have produced the radio emission at 31 – 100 kHz as the second daughter

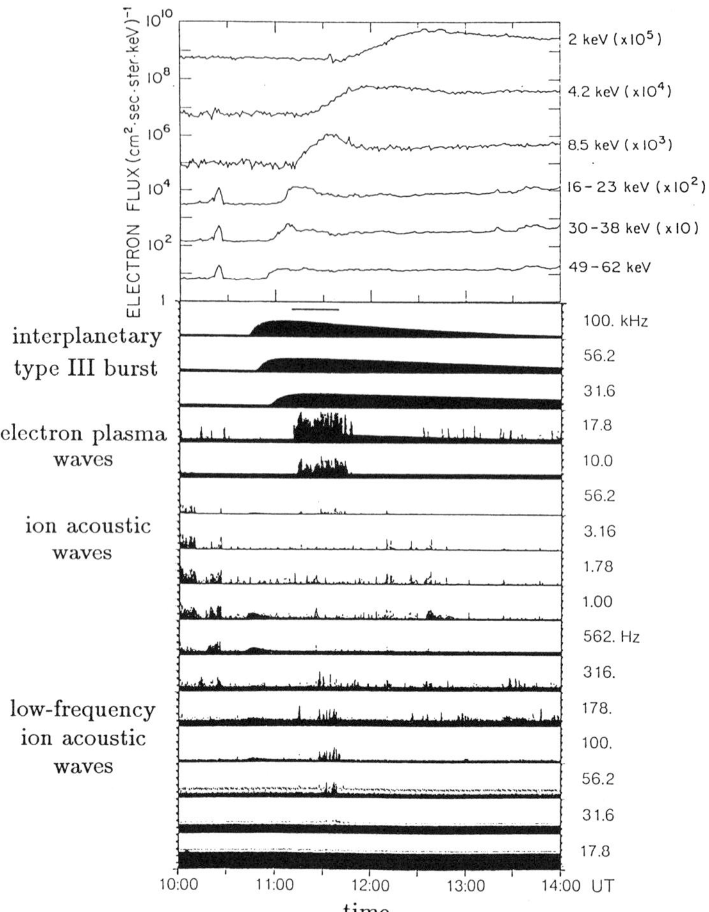

Fig. 6.6. Solar wind particles and waves observed by the ISEE-3 satellite at the Lagrangian point L1 between Sun and Earth. The *top panel* shows non-thermal electrons in different energy channels given at the right. Their velocity distribution has been shown in Figure 6.3. The slower electrons arrive later. The horizontal bar under the top panel indicates the time interval of positive slope in the v_z-distribution of the electrons. The *lower panel* presents the electric field of waves measured in 16 channels. Their frequency is given at the right side. The solid curves give peak intensities, and the black areas are 32 s averages (from Lin *et al.*, 1986).

waves. The observed number of radio photons, however, is an order of magnitude larger than predicted. A quantitative theory of the radio emission of beams is apparently still absent. Nevertheless, decay seems to be an attractive emission mechanism in the interplanetary medium and other non-equilibrium plasmas.

The brightness distribution of fundamental type III radio emission from the solar corona can reach 10^{12} K. If optically thick, weak-turbulence theory requires an equilibrium between photons and Langmuir photons, thus $T_t(\mathbf{k_t}) \approx T_L(\mathbf{k_L})$.

The observed maximum value is compatible with the Langmuir wave energy density at saturation (Eq. 6.2.18). Strong-turbulence effects are likely to occur in the most intense events. The *average* brightness temperature observed in interplanetary space is consistent with optically thick emission by weak-turbulence processes. This greatly simplifies the interpretation, as the details of the conversion process do not need to be known.

There is a general qualitative – and often quantitative – agreement between the observed intensity of plasma emission and the value derived from source parameters. This must be viewed as a remarkable success of more than three decades of painstaking theoretical efforts. The field remains open for new work, and new observations will undoubtedly stimulate progress.

6.3.4. Sense of Polarization

The observed radiation is said to be polarized if one of the two propagating electromagnetic modes predominates. According to Section 4.5 the two modes differ in their physics due to the presence of a magnetic field in the background plasma. Their dispersion relation becomes identical for $\Omega_e/\omega_p \rightarrow 0$ (Eq. 4.5.1). In the presence of a finite magnetic field most emission mechanisms favor one of the modes and yield polarization of characteristic degrees and sign. Propagation, in turn, can also change polarization by preferential absorption or scattering (Chapter 11).

Despite early claims, extensive attempts have not found linear polarization in any type of solar radio emission. This is very likely the result of a propagation effect in the corona: Faraday rotation (Section 11.2.3) changes the orientation of the wave electric field by thousands of radians already during propagation within the source. It obliterates any linear polarization intrinsic to the emission process.

An important diagnostic for the emission mechanism is the question of the preferred wave mode. The observational answer is found by measuring the sense of circular polarization and the prevailing direction (up or down) of the magnetic field in the source. Since the field in the source is not directly observable, one uses the Zeeman effect of optical lines from the underlying photosphere. Several comparisons of magnetic polarity and circular polarization have given a consistent trend. For type III bursts, observed to be left-hand polarized, one finds photospheric fields with the longitudinal components preferentially directed toward the observer, and *vice versa*. According to Section 4.3 such a wave must be *ordinary mode*.

In complex active regions, where the magnetic polarity is not unique, the leading spot (most western part of a spot group and leading in rotation) usually dominates in activity and is often taken as reference. Type III radiation is preferentially polarized in o-mode when referred to the leading spot. The fact that burst polarization is related to the surface magnetic field of the active region is called the *leading spot rule*. It has also been used to infer the radiation mode of other solar radio emissions.

The o-mode emission at the fundamental may be the result of different cutoffs of o(L)-mode (Eqs. 4.3.23 and 4.4.8) and x(R)-mode (Eqs. 4.3.22 and 4.4.10).

Let us assume that the emission mechanism produces a finite spectrum of both modes near the plasma frequency. It will be limited on the low-frequency side by the cutoffs, and on the high side by some common upper limit of Langmuir wave frequencies (to be determined in the following section). Since the o-mode cutoff is lower, its emission is favored. The observed sense of polarization thus follows naturally from the derivations in Chapter 4.

6.3.5. Magnetic Field Strength in the Corona

Polarization of plasma emission is always a consequence of a magnetic field, and it can in principle be used to measure the field strength in the corona. The factors influencing polarization, such as source parameters (viewing angle θ to the magnetic field, angular and spectral distribution of Langmuir waves, etc.), the conversion process and propagation have to be carefully studied before the coronal field strength can be derived from observed circular polarizations.

The polarization of fundamental/harmonic pairs has been studied extensively. In general, the fundamental component is more polarized than the harmonic component, but both are polarized in the same sense. If the degree of polarization depended linearly on the field strength, we would expect the polarizations of the two components to be proportional. In Figure 6.7 this is only the case for small degrees of polarization, where the two polarizations are equal. The harmonic appears saturated at 20% , whereas the fundamental may reach 60% .

Figure 6.7 indicates that the high polarization of the fundamental is determined by an additional factor (causing the scatter) and is non-linear in B. In fact, there are observations supporting both propositions:

- The average ratio of the polarization of fundamental/harmonic pairs has been found to decrease from the center to the limb of the solar disk. Since the path through the corona is longer at the limb, propagation effects are more severe. The observations thus suggest that propagation preferentially reduces the polarization of the fundamental.

- Polarization of the fundamental component has been found to correlate with drift rate and thus with beam velocity. It can be interpreted by the dispersion relation of Langmuir waves (Eq. 5.2.24). For higher beam velocities, the phase velocity, ω/k, of the excited waves is larger. Since the frequency of Langmuir waves varies only slightly, k must be smaller. The dispersion relation then yields a smaller difference, $\omega - \omega_p$, and more waves are produced between the cutoffs of the two modes, where only the o-mode can originate. This effect introduces an additional variable, the beam velocity V_b, that influences polarization. Moreover, this mechanism for polarization at the fundamental is non-linear in magnetic field strength.

The polarization of harmonic radiation seems to be consistent with a simpler relationship to the field strength of the source making it better suited as a diagnostic tool. Furthermore, the conversion processes are well known (Section 6.3.3). If the

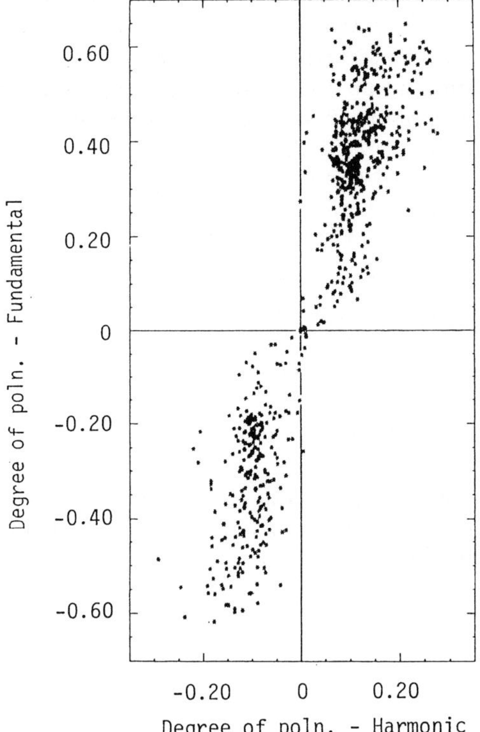

Fig. 6.7. Degree of circular polarization of fundamental component vs. harmonic component of 714 pairs at 80 and 43 MHz (from Dulk and Suzuki, 1980).

combining Langmuir waves are highly collimated in forward and backward direction to the magnetic field, if $\Omega_e \ll \omega_p$, and for weak turbulence, the polarization of harmonic radiation is in the sense of o-mode with a degree

$$P_H \approx \frac{11}{48} \frac{\Omega_e}{\omega_p} |\cos \theta| \quad , \tag{6.3.28}$$

where θ is the viewing angle to the magnetic field. In the other extreme of isotropic Langmuir waves, the polarization would be in the x-mode with

$$P_H \approx -\frac{85}{48} \frac{\Omega_e}{\omega_p} |\cos \theta| \quad . \tag{6.3.29}$$

Harmonic type III bursts of a pair are invariably polarized in the same sense as the fundamental component, thus in o-mode. It is a strong argument for the Langmuir waves in type III sources to be highly collimated. The observations and a detailed theoretical investigation limit them to a cone with an opening angle of about 10° or less in the magnetic field direction. The polarization of the harmonic was observed to vary from the disk center to the limb as $0.14|\cos\phi|$ by Dulk and Suzuki (1980), where ϕ is the aspect angle from the center of the disk. The average field is radial, thus $<\phi>\approx\theta$. Equation (6.3.28) represents both the observed angular behavior and the sense of polarization.

Using the above measurement, Equation (6.3.28) requires an average magnetic field with $\Omega_e/\omega_p \approx 0.61$ at a plasma frequency level of 80 MHz ($\nu_p \approx \nu_{\rm obs}/2$), corresponding to an electron density of $8 \cdot 10^7$ cm^{-3} and typically about half a solar radius above the photosphere. The definition (2.1.4) of the gyrofrequency yields a magnetic field strength of

$$B = \frac{\nu_p}{2.80 \cdot 10^6} \frac{\Omega_e}{\omega_p} \text{ G} . \tag{6.3.30}$$

and, for the above observations, $B = 17$ G. The Alfvén velocity is given by

$$c_A = 6.52 \cdot 10^8 \frac{\Omega_e}{\omega_p} \tag{6.3.31}$$

(Eq. 3.2.13, using solar abundances), becoming $1.1 \cdot 10^8$ cm s^{-1} in our case. The ratio of thermal pressure to magnetic pressure was given in Equation (3.1.51),

$$\beta := \frac{p}{B^2/8\pi} = \left(\frac{\omega_p}{\Omega_e}\frac{2v_{te}}{c}\right)^2 \quad . \tag{6.3.32}$$

For a temperature of $2 \cdot 10^6$ K, one calculates $\beta \approx 3.6 \cdot 10^{-3}$.

We conclude that the theory of polarization of plasma radiation is well developed for the case of weak harmonic radiation. The reader is reminded that the derived values refer to the sources of type III bursts. In fact, the Alfvén velocity is higher than expected for the average corona. The interpretation of these results is a topic of active current discussion.

6.4. Hard X-Ray Emission of Beams

Hard X-rays are important evidence of energetic particles and have played a major role for the appreciation of flares as high energy phenomena. We define hard X-rays as photons having energies in the range from 10 keV to 1 MeV. Since they do not necessarily indicate directed particle motions, their information on beams has to be carefully assessed (cf. also Chapter 8 on trapped particles). Hard X-rays from the solar corona are well observed, but little is known about stellar hard X-ray emission, which will not be considered here.

6.4.1. Emission Process

Collisions of energetic electrons with background particles are a well known mechanism to produce hard X-rays. In these collisions the electrons are deflected, hence they emit electromagnetic radiation. As the energetic particle loses energy in the process, the radiation is called *bremsstrahlung* ('braking radiation'). For particles moving at velocities $v \ll c$ the loss of energy by radiation is very small compared with the energy transfer to the collision partner. It can be the dominant energy loss for highly relativistic electrons. Bremsstrahlung emission depends on the velocity distribution of the collision partners.

The incoherent emission during a collisional interaction follows from procedures similar to those applied in Section 2.6. and from Larmor's equation for the radiation of accelerated charged particles. Quantum mechanical corrections are important. For details, the interested reader is referred to the references at the end of this chapter. The essential properties of the emission process are:

(1) The emitted photons have energies in the range from zero up to the energy of the energetic particle.

(2) In the small angle deflections characteristic for a 'collision' in a plasma (Section 2.6.), the probability of the emission of a photon is roughly constant in the above energy range.

The flux density of photons of energy $h\nu$ is the number of photons per unit time, photon energy, and area. For a stationary source at a distance D it is

$$F_{\rm hxr}(h\nu) = \frac{1}{4\pi\, D^2} \int n_i(\mathbf{x}) d\mathcal{V} \int F_e(\varepsilon, \mathbf{x})\, Q(\varepsilon, h\nu) d\varepsilon \quad [\text{photons s}^{-1}\ \text{cm}^{-2}\ \text{keV}^{-1}]. \tag{6.4.1}$$

The first integral is over the source volume $\mathcal{V}$, the second integral is over particle energy ε, where F_e is the flux of energetic electrons at $\mathbf{x}$ and t per unit time, area, and energy. Q is the differential cross-section for electrons with energy ε to produce photons with energy $h\nu$. The Bethe-Heitler approximation is usually used (e.g. Jackson, 1999). It must be corrected for heavy ions and electron-electron interactions.

The spectrum of the observed hard X-ray flux decreases with energy. It is often possible to approximate it by a power law,

$$F_{\rm hxr}(h\nu) = A \cdot (h\nu)^{-\gamma} \; . \tag{6.4.2}$$

The spectral index, γ, and the constant A are determined by the best fit to the data. The number and energy of the bremsstrahlung producing electrons can then be inferred using appropriate approximations on the source. In the context of beams, there are the following two.

(1) In the *thick-target* approximation a beam encounters a dense layer and stops completely. The kinetic energy of the particles is randomized and lost within the spatial resolution of the observation. The energy distribution in the target, F_e, is different from the injected distribution. Equation (6.4.1) can still be applied to the target-averaged distribution. The impinging flux, F_b, is changed to F_e by the collision process producing $F_e \propto F_b \varepsilon^2$, and is related to the observed X-ray flux by

$$F_b(\varepsilon) \approx 3.28 \cdot 10^{33}\ b(\gamma)\ \frac{D_\odot^2}{\pi r^2}\ A\ \varepsilon^{-\xi}\ \text{[electrons s}^{-1}\text{ cm}^{-2}\text{ keV}^{-1}\text{]}. \quad (6.4.3)$$

(e.g. Hudson *et al.*, 1978) with

$$\xi = \gamma + 1 \quad . \quad (6.4.4)$$

The denominator πr^2 represents the beam area (in cm^2) hitting the target. $D_\odot$ is the source distance in astronomical units, and $b(\gamma) = \gamma^2(\gamma-1)^2\ B(\gamma - \frac{1}{2}, \frac{3}{2})$ is of the order of unity, $B(x, y)$ being the beta function. Note that the electron flux has the form of a power law in energy.

(2) The *thin-target* approximation assumes that the beam propagates without essential changes through a dilute region. An example is a beam accelerated by a flare and escaping into interplanetary space. Since every volume element radiates, the relevant parameter is the energy distribution f_b of the beam electrons per unit *volume* and energy. The exponents of f_b and F_b are related by $\delta = \xi + 1/2$. Equation (6.4.1) can be inverted to

$$f_b(\varepsilon) \ \approx\ 1.05 \cdot 10^{42}\ c(\gamma)\ \frac{D_\odot^2}{\mathcal{V}}\ A\ \varepsilon^{-\delta}\ \text{[electrons cm}^{-3}\text{ keV}^{-1}\text{]}. \quad (6.4.5)$$

with

$$\delta = \gamma - 1/2 \quad , \quad (6.4.6)$$

and $c(\gamma) = (\gamma - 1)/B(\gamma - 1, 1/2)$. $\mathcal{V}$ is the volume of the target. The electron energy distribution again is a power law, but with a different slope, namely $\delta_{\text{thin}} = \delta_{\text{thick}} - 2$, for the same observed γ and in the same units.

6.4.2. Observations

Hard X-ray detectors, brought above the Earth's troposphere by balloons or rockets, count photons in specified energy bands. In the absence of spatial resolution, the count rate is proportional to the integral of the flux density over the channel width, $\int F(h\nu)d(h\nu)$, where $h\nu$ is the photon energy.

Figures 6.8 *left* and *right* present the hard X-ray emissions of two solar flares. They show two typical cases called *impulsive* (duration less than 10 minutes, often less than 1 minute) and *gradual* (or extended). A third type – named *thermal* since its spectrum is comparable with a Maxwellian velocity distribution of 3 to $5 \cdot 10^7$ K – is rare. Impulsive bursts are most frequent and are sometimes followed by a gradual burst. Thermal bursts are observed at low energies (around 20 keV). Of course, the three hard X-ray burst types are extreme situations, and in reality we may observe a mixture of two, or other complications.

The spectral index is observed to decrease in the rise phase of both impulsive and gradual bursts. For impulsive events, the spectral index, γ, decreases again in the decay phase and has a characteristic minimum at peak flux (Fig. 6.8 left). On the other hand, the spectral index of the gradual bursts (Fig. 6.8 right) continues to decrease throughout the event. The evolution of the spectral index is often

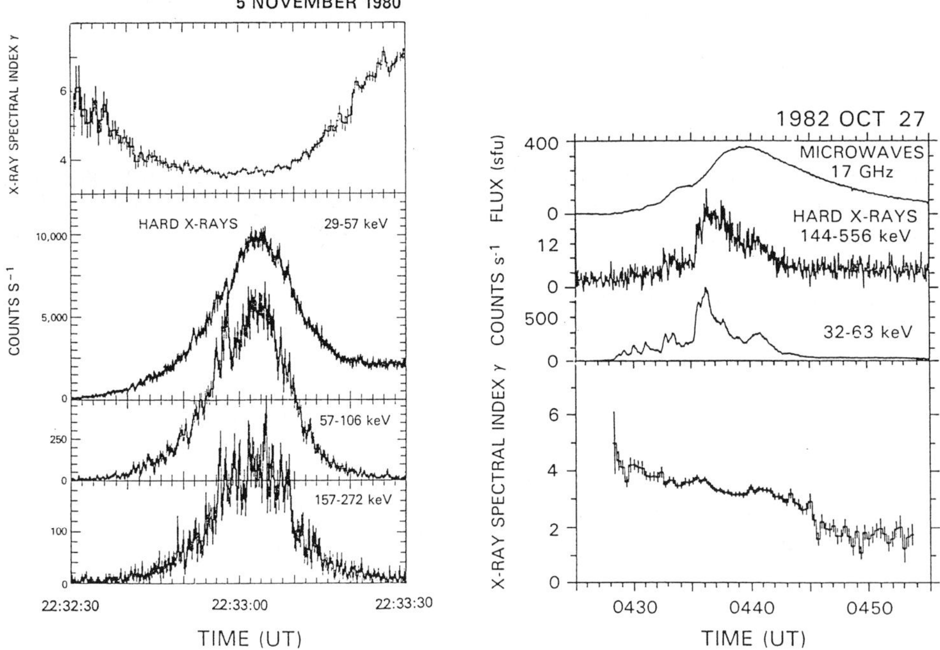

Fig. 6.8. Hard X-ray count rates and spectral index vs. time for, *left*, an impulsive flare (Dennis, 1985), and, *right*, a gradual flare (Kosugi, Dennis, and Kai, 1988). Note the different time scales!

described as 'soft-hard-soft' in impulsive flares and 'soft-hard-harder' in gradual flares. More complicated variations have been reported (Lin and Schwartz, 1986).

6.4.3. X-RAYS FROM BEAMS

Spectral observations alone do not allow us to determine the directivity of the hard X-ray producing electrons. Spatial information from imaging hard X-ray telescopes suggests that some impulsive hard X-ray events originate from beams. The evidence comes mainly from impulsive flares where two or more hard X-ray sources have been observed to brighten up simultaneously within the instrumental resolution of a few seconds. The sources are apparently related by exciters travelling much faster than an Alfvénic disturbance. A likely interpretation are beams of mildly relativistic electrons, as discussed for an example in the following paragraph.

Figure 6.9 gives a hint of the geometry. The soft X-ray structures in Figure 6.9 coincide with two loops suggested by Hα images and magnetograms. Impulsive hard X-rays observed at A, B, and C seem to originate largely from low altitudes and to be produced by precipitating electrons at the footpoints of magnetic loops. Figure 6.10 interprets the observations in terms of precipitating electron beams.

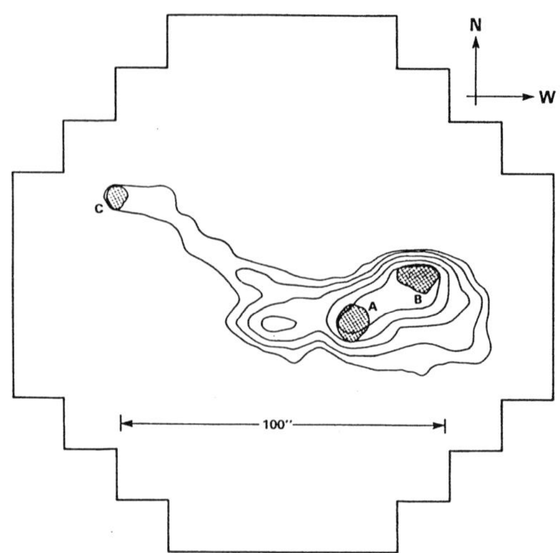

Fig. 6.9. The cross-hatched areas labelled A, B, and C are sources of impulsive hard X-ray emission (16 – 30 keV) observed by HXIS/SMM on the Sun. The contours plot the soft X-ray intensity at 3.8 – 8 keV (from Dennis, 1985).

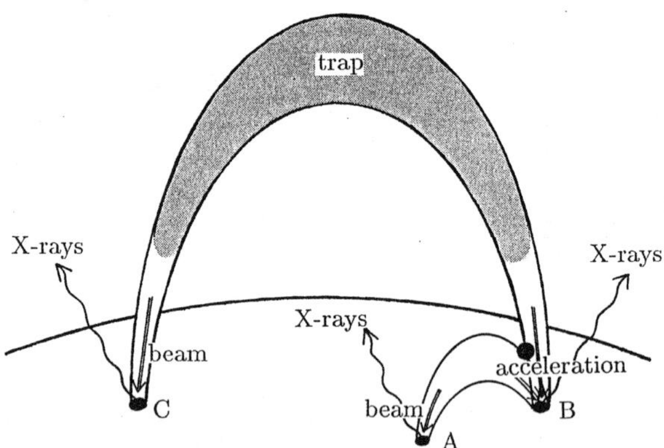

Fig. 6.10. Flares accelerate large numbers of energetic electrons into closed field lines. Some may get trapped, others precipitate into the chromosphere. In an approximately symmetric loop, the directed flux of particles produces two sources of hard X-rays. A second loop was involved in Figure 6.9, producing three sources, labelled A, B, and C.

Note that weak hard X-ray sources have been recently found also on the top of soft X-ray loops (Masuda et al., 1994).

Since the magnetic fields converge, precipitating electrons gain in transverse velocity, conserving the magnetic moment (Section 2.1). If they originate from a nearly isotropic population, the anisotropy of precipitating particles may primarily be transverse to the magnetic field and only minimally in the forward (downward) direction.

6.4.4. Radio – Hard X-ray Association

The most reliable radio emission associated with hard X-ray flares is gyrosynchrotron radiation in microwaves. A close correlation in time was discovered already in 1961 by M.R. Kundu. The two radiations may be produced within the same loop and by the same population of electrons. However, the hard X-rays are preferentially emitted by electrons at the end of their lifetime and near the footpoints, whereas synchrotron radiation is emitted by the spiraling motion anywhere in the loop and also by trapped particles. The microwave – hard X-ray association will be discussed in connection with trapping in Section 8.1.

Hard X-rays during flares suggest electron beams that are many orders of magnitude more intense than the beams necessary to generate the observed coherent radio emission. Because only a small fraction ($\sim 10^{-5}$) of the beam energy appears as bremsstrahlung X-rays in thick-target interactions, the number of electrons in the beam must be very high to produce the observed intense emission. Most of the beam energy is lost in Coulomb collisions with ambient electrons and is converted to heat in the lower corona, transition region, and chromosphere. On the order of 10^{34} electrons s^{-1} above 25 keV ($\approx 10^{27}$ erg s^{-1}) are necessary to make the beam observable to current hard X-ray telescopes. The energy in these particles is a major fraction of the total flare energy. This *thick-target* model is the preferred scenario for impulsive X-ray events.

Type III bursts occur usually in groups. The number of elements per group increases with frequency, and above 300 MHz it can reach several hundreds. Groups of type III bursts are only well associated with flares (seen in X-rays or Hα) at high frequencies ($\gtrsim$ 300 MHz). Since the number of type III bursts is usually much larger than the number of hard X-ray peaks, one cannot expect a one-to-one correlation. Most *metric* type IIIs originate from beams travelling upwards, whereas the X-rays are more probably from beams travelling downwards.

The fraction of reverse drifting type III radio bursts (from low to high frequencies, indicating downward motion) increases with frequency and becomes dominant in microwaves ($\gtrsim$ 1 GHz). Figure 6.11 shows a coincidence of the radio emission of downgoing beams with peaks in hard X-ray emission. The relation between reversed type III radio sources and hard X-ray peaks is rarely one-to-one. The downgoing electrons presumably do not always develop a double hump according to Section 2.3 and thus are not unstable toward growing electron plasma waves.

6.4.5. Diagnostics of the Accelerator

The energy stored in the magnetic field of active regions largely exceeds the requirement of a flare. The build-up of this energy can be observed when new magnetic flux emerges through the photosphere and in the footpoint motion of coronal loops. The evidence is good that the free energy released in coronal flares is magnetic in origin. The conversion of magnetic energy into heat, motion, and energetic particles is most likely initiated by reconnecting magnetic field lines. Through reconnection and subsequent relaxation, the complexity of the magnetic

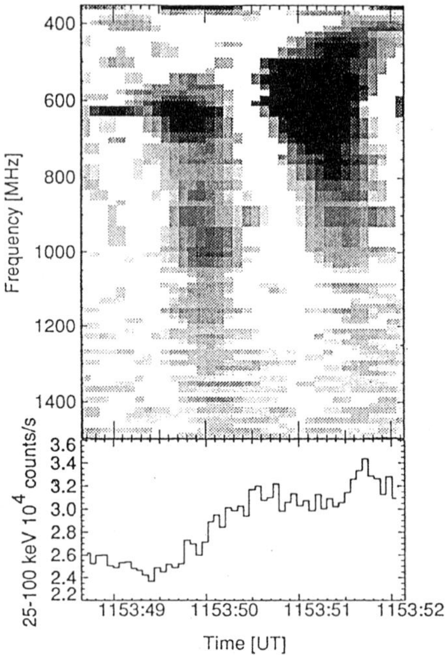

Fig. 6.11. *Top:* Radio flux density of solar type III emission observed by a Zurich spectrometer. Two reversed slope bursts mark two downgoing beams. The second one is accompanied by a simultaneous upgoing beam ($\nu \lesssim 550$ MHz). *Bottom:* Hard X-ray counting rate measured by BATSE on the GRO satellite (from Aschwanden, Benz, and Schwartz, 1993).

field configuration is reduced. Anti-parallel field lines are annihilated, or sheared field lines are untwisted. In physical terms, the free energy stored in the non-potential component of the magnetic field – being equivalent to the presence of a coronal current – is released (Section 9.1). Reconnection (discussed further in Chapter 9) is only fast enough if it takes place on extremely small scales ($\lesssim 1$ km). The accelerator must be much larger to provide enough electrons. It is presently observable only indirectly through fast particles emitting signatures in hard X-rays, radio, and other emissions.

A. *Energy of Flare Electrons*

The energy of a beam hitting a target can be determined in two ways: from the bremsstrahlung emission or from the heating of the target. For an intense beam accelerated in a flare and hitting the chromosphere, the radiation from the beam is visible as bremsstrahlung emission in hard X-rays ($\gtrsim 10$ keV), and from the heated target in soft X-rays ($\lesssim 10$ keV). The energy deposited by the beam heats the plasma. Another source of heat is the accelerator itself, which may not be 100% efficient. If the cooling time of the soft X-ray emitting plasma is significantly longer than the impulsive peaks of electron acceleration and impact, we should have the following relations valid at any given time t during the flare:

$$\int_0^t F_{\rm hxr} dt \; \propto \; \int nk_B T \; d^3x \; \leftrightarrow \; F_{\rm sxr}(t) \quad . \qquad (6.4.7)$$

$F_{\rm hxr}$ and $F_{\rm sxr}$ are the hard and soft X-ray count rates, respectively. The first relation assumes a spectrum constant in time. In practice, it is applied to the flux of an energy range or the full hard X-ray band. The left side (first term) then is proportional to the energy input by the beam. Neglecting heat losses, it is proportional to the increase in thermal energy of the target, or, approximately, to the plasma pressure. The middle (second) term is the thermal energy of the soft X-ray emitting region, heated either by the impacting particles or the waste heat of the accelerator. The second relation ($\leftrightarrow$) assumes thermal soft X-ray emission of an optically thin source. This relation is not linear. Nevertheless, one expects that the derivative of the soft X-ray time profile correlates roughly with the hard X-rays in the thick target scenario. Some correlation should exist even if the plasma of the heated region is allowed to expand (Fig. 6.9) and the two source volumes are not identical.

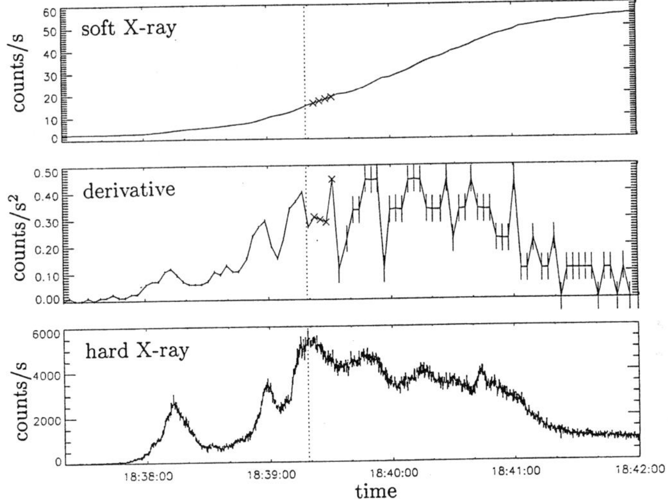

Fig. 6.12. *Top*: Soft X-ray (1–8 Å) flux in units of 10^{-3} erg s^{-1} cm^{-2} observed from an impulsive solar flare by the GOES satellite. *Middle*: Time derivative of top plot. *Bottom*: Hard X-ray counts per second ($\gtrsim$ 30 keV) from HXRBS on the Solar Maximum Mission (from Dennis and Zarro, 1993).

Figure 6.12 clearly shows the expected relationship. It was first pointed out by W.M. Neupert in 1968. The peak times agree to within about 20 s, and the rise and decay times are similar. Nevertheless, the relation between soft and hard X-rays is controversial for gradual flares. Unfortunately, neither the thermal energy nor the non-thermal energy can be determined to better than an order of magnitude, and it is not yet possible to compare the observations in a quantitative way.

B. *Fragmentation of Flares*

Type III radio bursts usually occur in groups related to flares. The electrons of one flare seem to be fragmented into many beams at low altitude, each producing a separate type III burst. The upgoing fraction thereof later combines to one interplanetary type III burst observed also as a particle stream containing up to 10^{33} electrons at 1 AU. This is a small fraction ($\lesssim 10^{-3}$) of the typical numbers of energetic electrons deduced from the hard X-ray bremsstrahlung emission of flares. Apparently, only few of the accelerated electrons escape into interplanetary space. Nevertheless, even upward moving type III bursts and hard X-rays sometimes correlate surprisingly well, although the two emissions do not involve the same electrons.

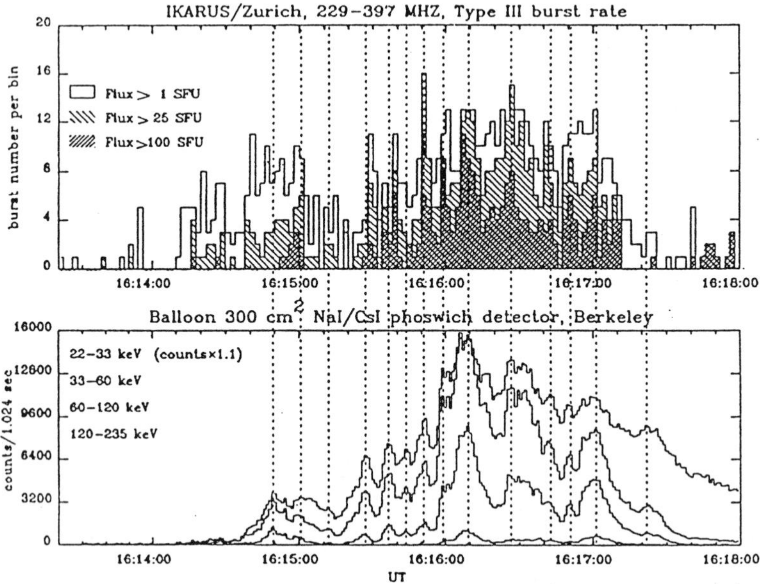

Fig. 6.13. Comparison of the number of beams visible as type III radio bursts with the flux of hard X-ray bremsstrahlung originating from electrons that have precipitated to high densities (from Aschwanden *et al.*, 1991).

As coherent radio emission is not proportional to particle number, its comparison with incoherent hard X-rays, which is proportional to particle number, requires special care. Often, one does not observe a one-to-one correspondence between individual hard X-ray peaks and type III bursts. Figure 6.13 shows the *number* of type III bursts per unit time at different peak flux thresholds in a flare well observed in hard X-rays. The vertical dotted lines mark peaks in hard X-ray flux. They coincide with peaks in type III burst rate. The correlation of the rates of type III bursts and hard X-ray flux suggests that both upward and downward directed beams are generated. Much more importantly, the linear correlation suggests that the total number of electrons is roughly constant per type III producing beam. A flare seems to consist of hundreds of elementary processes of about equal size, each characterized by a type III burst, and each accelerating a similar number of hard X-ray producing electrons. In the case of Figure 6.13, the number is about 10^{32} electrons (>25 keV) per elementary event. It seems to be the result of fragmentation in the primary energy release of flares, consistent with the theoretical notion of small conversion sites.

We may add here that a similar phenomenon is observed in the so-called *narrowband spikes* occurring in groups of more than 10^4 recognizable peaks. They are an extremely short duration (few tens of milliseconds) and narrow bandwidth ($\Delta\nu/\nu \approx 0.02$) type of radio burst in the 0.3 – 8 GHz range (to be discussed in Chapter 8). They may represent an even higher fragmentation in the energy release region. However, since their nature is still not clear, this remains a hypothesis.

Why is the time profile of hard X-ray events much less structured (by an order of magnitude) than the type III radio emission? The time structure of the hard X-ray producing beams may initially be spiky, but may be smoothed out by velocity dispersion as the electrons precipitate down the legs of magnetic loops to the higher density regions near the footpoints where the bulk of the X-rays are produced. The collision time it takes to emit bremsstrahlung is many orders of magnitude longer than the growth time of the bump-on-tail plasma instability. Nevertheless, evidence for occasional time structures down to tens of milliseconds have been reported by Kiplinger *et al.* (1983) for hard X-rays (30 keV) and by Kaufmann *et al.* (1984) for microwaves at 22 and 44 GHz. This represents a powerful constraint on beam models since it represents an upper limit of the collisional stopping time. Using Equation (2.6.29), one derives a lower limit of the hard X-ray source density of $n \gtrsim 3 \cdot 10^{11}$ cm^{-3}.

Independently, laboratory experiments and numerical simulations have shown that the release of magnetic energy by reconnection must be considered as a highly time-dependent and spatially fragmented process. Particle acceleration takes place in localized regions. The process is limited in time, must be impulsive and sometimes explosive. Many such events occurring almost simultaneously may constitute what is observed in hard X-rays as one flare.

Exercises

6.1: Electron beams in a corona are practically neutralized by a return current in the background plasma. Calculate the beam density at which the return current exceeds the instability threshold of ion acoustic waves. Take a background plasma having $T_e = T_i$, and a stable beam distribution $(\partial f(v_z)/\partial v_z \leq 0)$ with a mean velocity of $V_b = 20\ v_{te}$.

6.2: The velocity distribution of electron beams can be measured by spacecraft in the interplanetary medium. Lin *et al.*(1986) have observed an unstable distribution in a range $\Delta v = 10^9$ cm s^{-1} and a background electron temperature of $2 \cdot 10^5$ K. What is the ratio of the saturation level of Langmuir waves to the free energy of the beam?

6.3: Calculate the spectral wave energy density $W_t(\mathbf{k})$ at which induced scattering of Langmuir waves off ions is equally efficient as the spontaneous process, assuming $N_L(\mathbf{k_L}) \gg N_t(\mathbf{k_t})$.

6.4: Calculate the threshold distance s_c after which induced scattering of Langmuir waves dominates. Use

$$s_c \eta_s = W_t(\omega) \mathcal{N} c \quad , \tag{6.4.8}$$

and transform to observable parameters using Equation (6.3.9), $\omega_L \approx \omega_p$, and $k_m \approx \omega_p (3 v_{te})^{-1}$.

Further Reading and References

The beam-plasma system and its instability

Hammer, D.A. and Rostoker, N.: 1970, 'Propagation of High Current Relativistic Electron Beams', *Phys. Fluids* **13**, 1831.

Melrose, D.B.: 1990, 'Particle Beams in the Solar Atmosphere: General Overview', *Solar Phys.* **130**, 3.

Spicer, D.S. and Sudan, R.N.: 1984, 'Beam-Return Current Systems in Solar Flares', *Astrophys. J.* **280**, 448.

van den Oort, G.H.J.: 1991, 'The Electrodynamics of Beam-Return Current Systems in the Solar Corona', *Astron. Astrophys.* **234**, 496.

Radio emission of beams

Akimoto, K., Rowland, H.L., and Papadopoulos, K.: 1988, 'Electromagnetic Radiation from Strong Langmuir Turbulence', *Phys. Fluids* **31**, 2185.

Bekefi, G.: 1966, *Radiation Processes in Plasmas*, J. Wiley and Sons, New York.

Dulk, G.A.: 1990, 'Interplanetary Particle Beams', *Solar Phys.* **130**, 139.

Kaplan, S.A. and Tsytovich, V.N.: 1973, *Plasma Astrophysics*, Pergamon Press, Oxford.

Lin, R.P., Levedahl, W.K., Lotko, W., Gurnett, D.A., Scarf, F.L.: 1986, 'Evidence for Non-Linear Wave-Wave Interactions in Solar Type III Radio Bursts', *Astrophys. J.* **308**, 954.

Melrose, D.B.: 1989, 'The Brightness Temperature of Solar Type III Bursts', *Solar Phys.* **120**, 369.

Wentzel, D.G.: 1982, 'On the Solar Type III Radio Burst Emission Process', *Astrophys. J.* **256**, 271.

Hard X-ray emission

Brown, J.C.: 1975, 'The Interpretation of Spectra, Polarization and Directivity of Solar Hard X-Rays', *IAU Symp.* **68**, 245.

Dennis, B.R.: 1988, 'Solar Flare Hard X-Ray Observations', *Solar Phys.* **118**, 49.

Jackson, J.D.: 1999, *Classical Electrodynamics*, 3rd edition, Chapter 15 on bremsstrahlung, J. Wiley and Sons, Inc. New York.

Kane, S.R.: 1987, 'Propagation and Confinement of Energetic Electrons in Solar Flares', *Solar Phys.* **113**, 145.

Masuda, S., Kosugi, T. Hara, H., and Tsuneta, S.: 1994, 'A Loop-Top Hard X-Ray Source in a Compact Solar Flare as Evidence for Magnetic Reconnection', *Nature* **371**, 495.

Rieger, E.: 1989, 'Solar Flares: High-energy Radiation and Particles', *Solar Phys.* **121**, 323.

References

Aschwanden, M.J., Benz, A.O., and Schwartz, R.A.: 1993, 'The Timing of Electron Beam Signatures in Hard X-Ray and Radio: Solar Flare Observations by BATSE/CGRO and PHOENIX', *Astrophys. J.* **417**, 790.

Aschwanden, M.J., Benz, A.O., Schwartz, R.A., Lin, R.P., Pelling, R.M., and Stehling, W.: 1990, 'Flare Fragmentation and Type III Productivity in the 1980 June 27 Flare', *Solar Phys.* **130**, 39.

Dennis, B.R.: 1985, 'Solar Hard X-Ray Bursts', *Solar Phys.* **100**, 465.

Dennis, B.R. and Zarro, D.M.: 1993, 'The Neupert Effect: What Can it Tell us about the Impulsive and Gradual Phases of Solar Flares', *Solar Phys.* **146**, 177.

Dulk, G.A. and Suzuki, S.: 1980, 'The Position and Polarization of Type III Solar Bursts', *Astron. Astrophys.* **88**, 203.

Hudson, H.S., Canfield, R.C., and Kane, S.R.: 1978, 'Indirect Estimation of Energy Disposition by Non-Thermal Electrons in Solar Flares', *Solar Phys.* **60**, 137.

Kaufmann, P., Correia, E., Costa, J.E.R., Dennis, B.R., Hurford, G.J., and Brown, J.C.: 1984, 'Multiple Energetic Injections in a Strong Spike-like Solar Burst', *Solar Phys.* **91**, 359.

Kiplinger, A.L., Dennis, B.R., Emslie, A.G., Frost, K.J., and Orwig, L.E.: 1983, 'Millisecond Time Variations in Hard X-ray Solar Flares', *Astrophys. J.* **265**, L99.

Kosugi, T.K., Dennis, B. R., and Kai K.: 1988, 'Energetic Electrons in Impulsive and Extended Solar Flares as Deduced from Flux Correlations between Hard X-Rays and Microwaves', *Astrophys. J.* **324**, 1118.

Lin, R. P. and Schwartz, R.A. : 1987, 'High Spectral Resolution Measurements of a Solar Flare Hard X-Ray Burst', *Astrophys. J.* **312**, 462.

Melrose, D.B. : 1980, 'The Emission Mechanisms for Solar Radio Bursts', *Space Sci. Rev.* **26**, 3.

Sagdeev, R.Z.: 1979, 'The 1979 Oppenheimer Lectures. Critical Problems in Plasma Astrophysics – I. Turbulence and Non-Linear Waves', *Rev. Mod. Phys.* **51**, 1.

CHAPTER 7

ION BEAMS AND ELECTROMAGNETIC INSTABILITIES

Energetic electrons and electron beams have so far received primary attention, since their high emissivity – usually favored by the charge to mass ratio – produces most continuum emissions observed in the high energy and radio range. However, this does not mean that ion beams should be neglected. In circumstances where both energetic electrons and ions can be observed, such as in cosmic rays and in the magnetosphere, ions are energetically more important. Energetic ions may propagate over large distances and deposit their energy far from the acceleration site. Furthermore, they can excite low-frequency waves, practically invisible to direct observations but with drastic effects. There is no doubt that a great variety of ion beam phenomena, similar to those in planetary magnetospheres, must be expected also in coronae and other tenuous plasmas.

7.1. Observations of Energetic Ions

7.1.1. Solar ion beams

There exist indications that ion beams play a role in various solar processes. The most conspicuous of them is the emission of γ-ray lines during flares. Chupp (1990) has reviewed the current state of observations of γ-ray lines since their discovery in 1972. The widely accepted scenario is sketched in Figure 7.1. High energy protons ($\gtrsim$ 30 MeV) are accelerated simultaneously with electrons in the tenuous corona. They both travel from the acceleration site to the lower, denser layers of the atmosphere where they interact collisionally with the plasma. Ions can excite the nucleons of heavy ions to higher energy levels similar to electron energy levels in atoms. The most dramatic difference between atomic and nuclear excitations is the scale of the energies involved. The binding energy between electron and proton in the simplest atom is only 13.6 eV whereas 2.2 MeV binds a proton and neutron to form deuterium. When the nucleus relaxes to a lower energy state, it may emit a photon with energy equal to that of the transition. At higher excitation energies ($\gtrsim$ 8 MeV) the nuclei may also emit protons, neutrons, or alpha particles. The majority of nuclear γ-ray line emissions have energies of 1 to 8 MeV. Figure 7.2 displays an observed spectrum exhibiting an excess of γ-rays over the continuum, defined by the expected continuation of electron bremsstrahlung to higher energies. The spectrum is composed of nuclear de-excitation lines (like at 4.4 MeV from ^{12}C and at 6.1 MeV from ^{16}O) which occur simultaneously with hard X-rays within the instrumental time resolution of one second. The most notable delayed feature is the 2.223 MeV deuterium formation line caused by neutron capture of hydrogen. The neutrons result from a collision of an energetic proton with a heavy nucleus. The line is not produced until the neutron has slowed sufficiently for the cross-section of the reaction to become appreciable. The ratios of observed line intensities are used to calculate important physical parameters such as background elemental abundances, proton beam energy distribution, etc.

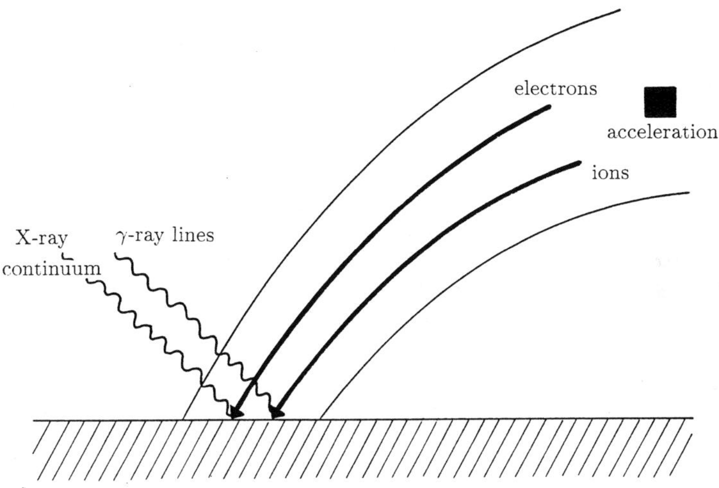

Fig. 7.1. Schematic view of electron and ion acceleration, beam propagation (arrows), and emissions in the solar corona.

Below 2 MeV, the most prominent line is usually at 0.511 MeV, caused by electron-positron annihilation. Positrons result from the decay of positively charged pions, π^+, and the beta decay of unstable nuclei. Both pions and unstable nuclei are interaction products of energetic ions that reach dense layers of the solar atmosphere.

There are other, not yet conclusive, ideas to search for solar ion beams, particularly for those with energies lower than needed to produce nuclear excitation ($<$ few MeV). Such lower energy ions may be of particular significance since a large fraction of the energy released in a flare may go into accelerating these particles.

- High energy protons streaming down may pick up an electron by charge exchange with a background hydrogen atom. The streaming particle becomes a hydrogen atom in an excited state capable of atomic line emissions. Highly Doppler-shifted hydrogen lines (notably Ly α) from such neutralized atoms moving with the beam have been predicted.

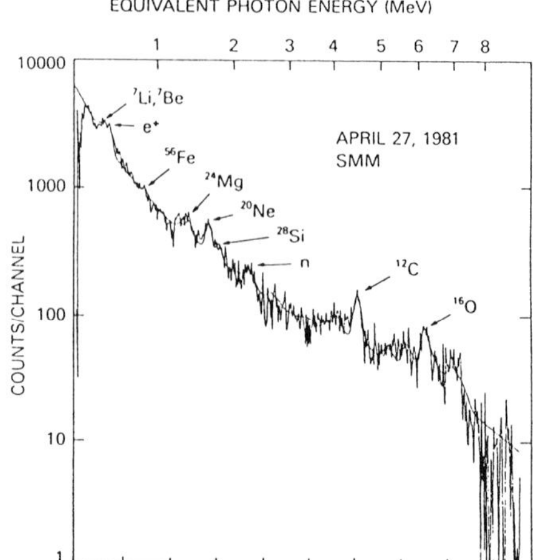

Fig. 7.2. Solar flare γ-ray spectrum taken by the Gamma Ray Spectrometer on board the SMM satellite. The curve is a theoretical fit to the data based on detector response, assumed elemental abundances, ion beam energies, and available nuclear cross sections. Some of the identified radiating nuclei are indicated (from Ramaty and Murphy, 1990).

- Impacting protons excite linearly polarized atomic lines. Chromospheric lines, such as Hα, are a possible diagnostic of proton beams descending from coronal flares. Maximum polarization is predicted in the plane perpendicular to the direction of beam propagation. In fact, linear polarization as high as 2.5% has been reported for bright Hα patches of solar flares.

- Proton beams excite electrostatic waves at low frequencies. It may combine with preexisting Langmuir waves to observable radio emission (Section 7.4).

7.1.2. Cosmic Rays

High energy particles of extraterrestrial origin arrive near the Earth in great numbers. They produce copious secondary particles in the atmosphere, some of which reach the surface. These have been studied since the 1930s by ground-based particle counters. Protons, helium nuclei, and electrons are by far the most numerous particles. The relative abundances of the various nuclei are similar to those in the atmosphere of normal stars. Li, Be, and B are overabundant – mainly the products of collisions between the primary cosmic rays and the interstellar gas, as they travel from the source to the Earth.

There are at least three different sources of cosmic ray particles observed at Earth. (*i*) Supernovae, pulsars, and supernova remnants are probably the major sources of galactic cosmic rays. (*ii*) Stellar activity, in particular the Sun, is known to produce nuclei up to some 10^{10} eV in flares. Since interplanetary energetic protons do not correlate in number with the flare γ-ray line intensities, they appear to be a secondary product of the flare initiated shock wave. (*iii*) Active galactic nuclei are prominent by their powerful X-ray and radio emission of highly relativistic electrons. They may also accelerate ions to high energies. The magnetic field would probably not be strong enough to locally trap the energetic nuclei (see Eq. 7.1.2 below). Once accelerated, they may diffuse into intergalactic space and reach the Earth.

The energy distributions of energetic particles are frequently fitted by a power law of the form

$$f(\varepsilon) = f_0 \varepsilon^{-\delta} \quad . \tag{7.1.1}$$

For solar cosmic ray electrons, one finds – after corrections for propagation – that the energy spectra at the source may be fitted by a power law up to energies around 100 – 200 keV and a second power law with steeper slope beyond and up to the highest energies ($\approx$ 300 MeV). The solar cosmic ray proton and alpha source spectra are also steeper at higher energies. An exponential approximation for the source spectrum is convenient and useful, fitting both power-law ranges by a single spectrum. Often these ion source spectra are even better fit by the difference of two Bessel functions (i.e. a Macdonald function). Such a distribution is predicted by stochastic acceleration (Sections 9.4 and 10.3).

The highest measurable nucleon energies of galactic (i.e. *extra*-solar) cosmic ray ions reach 10^{20} eV. The gyroradius of such particles merits consideration. From Equation (2.1.2) one derives for highly relativistic particles, both electrons and ions,

$$R \;=\; 3.34 \cdot 10^{-3} \varepsilon_{eV} / B \quad [\mathrm{cm}] \quad . \tag{7.1.2}$$

For the highest energies the gyroradius is a few kiloparsecs, exceeding the disk thickness of the Galaxy. Since the galactic magnetic field is curved and inhomogeneous on this scale, the particle orbit is not a spiral, and the extremely high energy particles escape rapidly from the Galaxy. Beams, if formed, would soon randomize, since orbits depend on energy and phase angle.

Galactic cosmic rays have a power-law distribution with an exponent $\delta \approx 2.5$ over most of their range. The spectrum may steepen slightly above 10^{15}eV. Synchrotron radio emission of galactic cosmic ray electrons suggests that the same power law applies throughout the Galaxy and beyond. The remarkable uniformity of the spectrum over such a wide energy range and the universality of the spectral index place stringent limits on theories of galactic cosmic ray acceleration (to be discussed in Section 10.3.2).

The average energy density of galactic cosmic ray particles is of order 10^{-12} erg cm^{-3}. This is comparable to the energy density of typical magnetic fields in interstellar space. Therefore, cosmic ray ions may energetically dominate, particularly near the acceleration sites, and may carry along the magnetic field. Cosmic rays then form an isotropic plasma component capable of MHD-like phenomena.

7.1.3. Ion beams near Earth

Most of our knowledge on ion beams comes from planetary magnetospheres, interplanetary observations near Earth, and comets. A variety of beams with different properties and origins are being studied in the neighborhood of the Earth. A continuous ion stream forms at the bow shock (Chapter 10) and escapes upstream (Sunward) along the magnetic field. The character of the beam depends on the angle between the interplanetary magnetic field and the normal to the shock. This angle varies along the curved bow shock. In the quasi-parallel region (where **B** is nearly parallel to the shock-normal and thus perpendicular to the shock front) the ion speed is low, and the beam has a diffuse velocity distribution. In the quasi-perpendicular region (where **B** is nearly perpendicular to the shock-normal and parallel to the shock front), ions form perpendicular beams. Such beams gyrate in the magnetic field, forming a transverse ring distribution in velocity space. They do not propagate far into the solar wind, except for a small number which eventually move parallel to the field. Nearly monochromatic, high-speed streams are found between the perpendicular and parallel regions. Beams and wave generation have been studied intensively for more than a decade, and are understood in considerable detail.

Other important ion beams have been observed in the Earth's magnetotail, mostly travelling away from the Earth. They are bursty and located on magnetic field lines connecting to magnetic X-points (where oppositely directed field lines interact and presumably annihilate). Beam-particle energies can reach 10^5 eV, a high value among all particles for magnetospheric conditions.

In auroras, the well-known precipitating electron beams are frequently accompanied by various kinds of ion beams. They propagate upward in most cases and seem to be the result of electric fields parallel to the magnetic field. Usually such beams interact with waves and other particles in a complex way. Furthermore, a proton ring current develops after a plasma is injected into the inner magnetosphere, similar to the electrons of an E-layer (Fig. 2.2).

Artificial beams generated by spacecraft and injected into the magnetospheric plasma are particularly well suited to investigate the role of the accelerator – the usually unknown initial condition – in beam evolution.

7.2. Electromagnetic Instabilities of Velocity Space Anisotropy

The interaction of ions with waves is similar to the interaction of electrons with waves, but the higher mass and usually lower velocity make ions more sensitive to instabilities of low-frequency waves. Several instabilities are possible. If the conditions were such that both an electrostatic and an electromagnetic wave can grow, the growth rates (diminished by possible damping) have to be compared. More relevant for the dominating waves is often the threshold for instability, i.e. the conditions where instability equalizes damping. If the unstable situation builds up slowly, the wave with lower threshold drains the free energy, and the threshold of the second wave may never be exceeded.

There are two different types of electromagnetic beam instabilities. We shall first discuss fluid-like (also called non-resonant, 'reactive' or 'hydrodynamic') instabilities. They do not depend on the details of the velocity distribution. The plasma thus can be treated as a fluid; and there is no resonance condition for wave-plasma interaction. All particles are involved in wave growth. Second, the resonant electromagnetic interaction will be derived from the basic kinetic equations of Chapter 5. Resonant (or kinetic) instabilities generally have a lower threshold than fluid instabilities since only a fraction of the particles have to fulfill a certain condition. On the other hand, only a small number lose energy to the waves, and hence their growth rate is generally smaller.

7.2.1. Fire-hose Instability

Consider a low-frequency ($\omega^2 \ll \Omega_i^2$) transverse electromagnetic wave in a homogeneous, electrically neutral plasma. The wave is assumed to be L or R mode, or a combination thereof. Such waves are called Alfvén waves, as Alfvén waves extend to frequencies higher than are valid for MHD and can also be represented by L and R modes (Section 4.3). Let a proton beam travel in such waves. Approximate the particle distribution by

$$\begin{aligned} f_p(\mathbf{v}) = & \frac{n_p}{(2\pi)^{3/2} v_{tp}^3} \exp\left[-\frac{(v_z^2+v_\perp^2)}{2v_{tp}^2}\right] \\ & + \frac{n_b}{(2\pi)^{3/2} v_{tbz} v_{tb\perp}^2} \exp\left[-\frac{(v_z-V_b)^2}{2v_{tbz}^2} - \frac{v_\perp^2}{2v_{tb\perp}^2}\right] \end{aligned} \quad , \tag{7.2.1}$$

where n_b is the beam proton density (satisfying $n_b \ll n_p$), v_{tbz} and $v_{tb\perp}$ are the mean thermal velocities of the beam population parallel and perpendicular to the magnetic field, respectively. The electron distribution is assumed to balance the flow of charges so that the total current vanishes.

In many interesting regions of parameter space, the maximum linear growth rate occurs for wave propagation parallel to the magnetic field. An analysis of linear disturbances similar to the Sections 3.2.1 and 3.2.2 on MHD waves introduces an additional term into the dispersion relation of the Alfvén wave (Eq. 3.2.13) yielding

$$\frac{\omega^2}{k^2} = c_A^2 - \frac{n_b}{n_p}[V_b^2 + (v_{tbz}^2 - v_{tb\perp}^2)] \quad . \tag{7.2.2}$$

The additional term (in brackets) is the sum of two non-MHD pressure contributions of the beam. The first – proportional to V_b^2 – is caused by the bulk motion of the beam. The second is due to temperature anisotropy of the beam. The dispersion relation predicts instability of Alfvén waves for

$$\frac{n_b}{n_p}[V_b^2 + (v_{tbz}^2 - v_{tb\perp}^2)] > c_A^2 \quad . \tag{7.2.3}$$

This instability is called the *fire-hose* (or garden-hose) instability, since the driving force is the beam pressure parallel to the magnetic field. It increases the amplitude of the wave in an analogous way to that of water flowing through a loose hose. We note that the instability also occurs for electron beams; in the threshold condition (7.2.3) V_b^2 is simply replaced by $m_e/m_i\ V_b^2$. In an anisotropic plasma without a beam, condition (7.2.3) with $n_b = n_p$ reduces to $p_z > p_\perp + B^2/4\pi$, where p_z and $p_\perp$ are the parallel and perpendicular particle pressures, respectively. This demonstrates that the fire-hose instability is a general result of pressure anisotropy and not limited to beams. It will be applied to cosmic beams in Section 7.3.

7.2.2. Kinetic Instability

A. *Dispersion Relation of Transverse Waves in Kinetic Plasma*

In Chapter 5, on kinetic plasma, we studied only electrostatic waves. Ion beams give us an excellent opportunity to investigate electromagnetic instabilities caused by some anisotropy in velocity space, of which ion beams are just a special case. We shall apply the same physics to a different situation in Chapter 8, on trapped particles, where particles (mostly electrons) with velocity transverse to the magnetic field predominate.

Here we derive a general dispersion relation for linear waves in kinetic plasma. We linearize the Vlasov-Maxwell equations by introducing a small disturbance and by keeping only first-order terms. In particular, the velocity distribution of the relevant particle population is assumed to have the form

$$f(\mathbf{x}, \mathbf{v}, t) = f_0(\mathbf{v}) + f_1(\mathbf{x}, \mathbf{v}, t) \tag{7.2.4}$$

with

$$| f_1 | \ll | f_0 | \quad \text{for all} \quad \mathbf{x}, \mathbf{v}, t. \tag{7.2.5}$$

Then the Vlasov (collisionless Boltzmann) equation (5.2.1) can be approximated by

$$\left[\frac{\partial}{\partial t} + \mathbf{v}\cdot\nabla_{\mathbf{x}} + \frac{q}{mc}(\mathbf{v}\times\mathbf{B}_0)\cdot\nabla_{\mathbf{v}}\right] f_1 = -\frac{q}{m}(\mathbf{E}_1 + \frac{1}{c}\mathbf{v}\times\mathbf{B}_1)\cdot\nabla_{\mathbf{v}} f_0 \quad . \tag{7.2.6}$$

The operator in brackets on the left side acts as a total derivative on f_1 along the undisturbed orbit of a particle having a velocity $\mathbf{v}$ and being located in $\mathbf{x}$ at the given time t. We integrate Equation (7.2.6) along this path and derive a formal solution for the perturbation f_1,

$$f_1(\mathbf{x}, \mathbf{v}, t) = \frac{q}{m}\int_{-\infty}^{t} dt'(\mathbf{E}_1 + \frac{1}{c}\mathbf{v}'\times\mathbf{B}_1)\cdot\nabla_{\mathbf{v}'} f_0(\mathbf{v}') \quad . \tag{7.2.7}$$

The primed variables describe the undisturbed particle orbit of the particle that is found at the (unprimed) time t at the coordinate $\mathbf{x}$. This approach corresponds to the well-known method of characteristics for linear differential equations of first order. Having assumed a homogeneous background plasma (Eq. 7.2.1) with the magnetic field in the z-direction, the Fourier transform can be carried out by putting in the wave form

$$f_1(\mathbf{x}, \mathbf{v}, t) = \bar{f}_1(\mathbf{v}) \exp\left[i(k_z z + k_x x - \omega t)\right] \quad . \tag{7.2.8}$$

Without loss of generality we have put $k_y = 0$. Later in this section we shall restrict ourselves to waves propagating in the z-direction ($k_x = 0$). Most important, the form of (7.2.8) implies Fourier transformation in time as well as space. Chapter 5 has shown that this would exclude the resonance process between certain particles and the wave. We shall later make good this neglect by the Landau prescription for the integration over v_z and so recover the correct result of the Laplace transformation. That this procedure is legitimate can be seen from the general nature of the derivation in Section 5.2. Now take the analogous wave forms (but in $\mathbf{x}'$ and t') for $\mathbf{E}_1$ and $\mathbf{B}_1$ and put them together with Equation (7.2.8) into (7.2.7). After moving all exponentials to the right side, we get

$$\bar{f}_1(\mathbf{v}) = \frac{q}{m} \int_{-\infty}^{t} dt' \exp[i\{k_z(z'-z) + k_x(x'-x) - \omega(t'-t)\}](\bar{\mathbf{E}}_1 + \frac{1}{c}\mathbf{v}' \times \bar{\mathbf{B}}_1) \cdot \nabla_{\mathbf{v}'} f_0(\mathbf{v}') \quad . \tag{7.2.9}$$

Now let us calculate the differences $z' - z$ etc. from the basic particle motion (Section 2.1). It is useful to decompose it into the motion of the guiding center, given by $\mathbf{X}$ and $\mathbf{V}$, and the gyration of the particle $\mathbf{\Delta x}$ and $\mathbf{\Delta v}$ in the plane perpendicular to the magnetic field (x and y directions), where

$$\mathbf{x} = \mathbf{X} + \mathbf{\Delta x} \quad , \tag{7.2.10}$$

$$\mathbf{v} = \mathbf{V} + \mathbf{\Delta v} \quad . \tag{7.2.11}$$

Using Equations (2.1.1) – (2.1.3), we express the gyration by

$$\mathbf{\Delta x} = R(\sin[\psi + \Omega_z t], \cos[\psi + \Omega_z t]) \quad , \tag{7.2.12}$$

$$\mathbf{\Delta v} = R\Omega(\cos[\psi + \Omega_z t], -\sin[\psi + \Omega_z t]) \quad , \tag{7.2.13}$$

where Ω_z denotes the gyrofrequency times the sign of the charge, $\Omega_z := q/|q|\ \Omega$ (Eq. 4.2.16). We neglect relativistic effects here. The gyration phase, ψ, of the particle is relative to the wave at $t = 0$. Since we assume that the acceleration in the z-direction vanishes,

$$v_z' = v_z = \text{const} \quad . \tag{7.2.14}$$

Neglecting particle drifts perpendicular to the magnetic field, we have

$$v'_x = v_\perp \cos(\psi' + \Omega_z \tau) \quad , \tag{7.2.15}$$

$$v'_y = -v_\perp \sin(\psi' + \Omega_z \tau) \quad . \tag{7.2.16}$$

ψ' is the phase angle at t: $\psi' = \psi + \Omega_z t$. The integration over t' suggests the definition $\tau := t' - t$. Integrating Equations (7.2.14) and (7.2.15) from t to t', one gets immediately

$$z' - z = v_z \tau \quad , \tag{7.2.17}$$

$$x' - x = \frac{v_\perp}{\Omega}[\sin(\psi' + \Omega_z \tau) - \sin\psi'] \quad . \tag{7.2.18}$$

The sine functions of Equation (7.2.18) enter the exponential in the integral of Equation (7.2.9). For finite k_x this will require a development into a series of Bessel functions in the next chapter. Here we drastically simplify the mathematics by assuming parallel wave propagation ($k_x = 0$). It is the direction of fastest wave growth in the present case.

The evaluation of the other terms of the integral in Equation (7.2.9) involves many, but mostly straightforward steps, which we only outline.

• The wave field $\mathbf{B}_1$ must be expressed in terms of $\mathbf{E}_1$ by Faraday's equation (4.2.25),

$$i\mathbf{k} \times \mathbf{E}_1 = \frac{i\omega}{c}\mathbf{B}_1 \quad . \tag{7.2.19}$$

• The operation $\nabla_{\mathbf{v}'} \cdot f_0$ can be written

$$\nabla_{\mathbf{v}'} f_0 = 2\left(v'_x \frac{\partial f_0}{\partial v_\perp^2}, v'_y \frac{\partial f_0}{\partial v_\perp^2}, v'_z \frac{\partial f_0}{\partial v_z^2}\right) \quad . \tag{7.2.20}$$

We have introduced $v_\perp^2$ because it is proportional to the magnetic moment (Eq. 2.1.7). Therefore, $v_\perp^2$ and v_z are constants of motion; the terms $\frac{\partial f_0}{\partial v_\perp^2}$ and $\frac{\partial f_0}{\partial v_z^2}$ can be moved out of the integral.

• The velocities v'_x and v'_y are given by Equations (7.2.15) and (7.2.16), and their sine and cosine functions are replaced by exponentials. The phase angle ψ' cancels out as expected.

• Finally, the time dependence can be compressed into one exponential, $\exp[(k_z v_z - \omega \pm \Omega_z)t]$, and the integral of Equation (7.2.9) yields

$$\bar{f}_1(\mathbf{v}) = \frac{q}{m} \frac{2v_z \bar{E}_1}{i(k_z v_z - \omega \pm \Omega_z)} \left[(1 - \frac{k_z v_z}{\omega})\frac{\partial f_{0\alpha}}{\partial v_\perp^2} + \frac{k_z v_z}{\omega}\frac{\partial f_{0\alpha}}{\partial v_z^2}\right] \quad , \tag{7.2.21}$$

The rest of the derivation is similar to cold plasma. The perturbed current density $\mathbf{J}_1$ is calculated from f_1 yielding Ohm's law,

$$\mathbf{J}_1 = \sum_\alpha n_\alpha q_\alpha \int_{\mathcal{L}} \mathbf{v} f_{1\alpha} d^3v := \hat{\sigma}\mathbf{E}_1 \quad . \tag{7.2.22}$$

The index α has been introduced in Equation (7.2.22) for the particle species. The sum goes over all species that contribute to the current. $\mathcal{L}$ denotes the Landau contour. The conductivity $\hat{\sigma}$ can be evaluated using Equation (7.2.21). The dielectric tensor, $\hat{\epsilon} := \hat{1} - 4\pi\hat{\sigma}/(i\omega)$, contains all the interesting physics. It is used in Ampère's equation to derive the dispersion relation as for Equation (4.2.27). Since we have assumed transverse waves, we find only two solutions (one for each circular polarization),

$$\omega^2 = k^2c^2 + 2\pi\omega \sum_\alpha \frac{\omega_{p\alpha}^2}{n_\alpha} \int_{\mathcal{L}} \left[(1 - \frac{k_z v_z}{\omega}) \frac{\partial f_{0\alpha}}{\partial v_\perp^2} + \frac{k_z v_z}{\omega} \frac{\partial f_{0\alpha}}{\partial v_z^2} \right] \frac{v_\perp^3 \, dv_\perp dv_z}{k_z v_z - \omega \pm \Omega_z^\alpha} \quad . \tag{7.2.23}$$

The $\pm$ signs refer to the L(+) and R(−) modes, respectively. In the cold plasma limit ($f_{0\alpha} \propto n_\alpha \delta(v)$) they correspond to the left and right circularly polarized electromagnetic waves at high frequency derived in Chapter 4. We again refer to them as ion or electron cyclotron waves near the respective gyrofrequency, whistler waves at intermediate frequency (R-mode, $\Omega_i \ll \omega < \Omega_e$), and Alfvén waves at low frequency.

B. *Resonance Condition*

For a kinetic plasma, Landau damping or kinetic instability can be important if waves and particles satisfy the resonance condition expressed by setting the denominator of (7.2.23) to zero,

$$\omega - k_z v_z = \pm\Omega_z^\alpha \quad . \tag{7.2.24}$$

The $k_z v_z$ term may again be interpreted as a Doppler shift of the wave frequency in the frame of reference of the guiding center of the resonant particle. The resonance condition requires that in this frame (*i*) the electric field vector of the wave rotates in the same sense as the particle (e.g. counterclockwise for an ion having $\Omega_z > 0$), and (*ii*) with the frequency of particle gyration. Obviously, energy can easily be transferred if particle gyration and Doppler shifted wave frequency are equal; the particle feels a wave electric field having a constant component relative to the instantaneous particle velocity. A resonance between wave and particle gyration is called *gyromagnetic*. Equation (7.2.24) will be generalized to higher harmonics of the gyrofrequency when waves with finite k_x will be studied in the next chapter (Eq. 8.2.14). We may note already here that Ω_z^α refers to the gyrofrequency of the resonating particle. For relativistic particles, the relativistic gyrofrequency (cf. Eq. 2.1.4) has to be used.

Which wave resonates with which particle? Equation (7.2.24) warrants a careful discussion. Remember that we use here the convention of positive wave frequency. The signs of k_z and v_z indicate whether the wave and the particle, respectively, propagate in the positive or negative $\mathbf{B_0}$-direction. Let us now consider as an example a proton moving with $v_z < 0$. A resonant L-wave has a plus sign on the right side of Equation (7.2.24) and satisfies $\omega = \Omega_p + k_z v_z$. Since low-frequency,

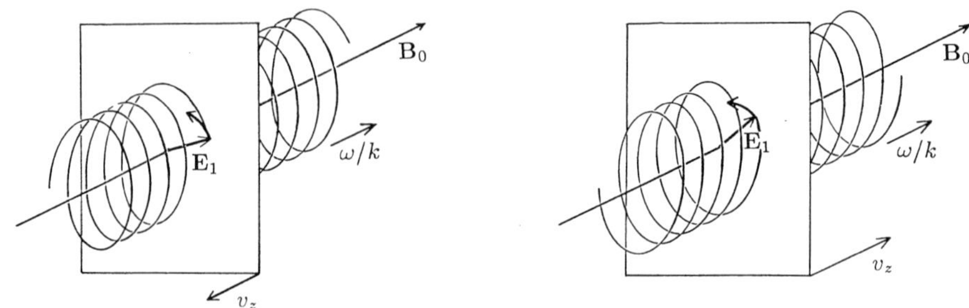

Fig. 7.3. *Left:* The tip of the electric field vector of an L-wave travelling in direction $\mathbf{B_0}$ describes a clockwise (right-handed) spiral if seen along $\mathbf{B_0}$. In the plane moving with (negative) velocity v_z, the $\mathbf{E_1}$ vector rotates counterclockwise. *Right:* R-wave propagating in direction $\mathbf{B_0}$ in anomalous Doppler resonance with a faster moving ion. In the ion frame, the wave propagates in the negative $\mathbf{B_0}$ direction and rotates counterclockwise.

left circularly polarized waves have $\omega < \Omega_p$ (Section 4.1), this requires $k_z > 0$. Therefore, ions moving in the negative direction interact resonantly with L-waves propagating in the positive direction (Fig. 7.3 left) and *vice versa*.

R-waves in resonance with protons must fulfill $\omega = -\Omega_p + k_z v_z > 0$. This implies that particle and wave travel in the same direction. However, in the frame moving with the particle velocity v_z, the wave propagates in the inverse direction, since $|\omega/k_z| < |v_z|$. Figure 7.3 (right) depicts this situation and makes clear that the electric wave vector in the moving frame of reference rotates counterclockwise. Therefore, an R-wave can be Doppler shifted to resonate with ions. This is called *anomalous Doppler resonance.*

The gyroresonant interaction of an electron is similar, except that L and R modes are interchanged. Normal resonance with R-waves requires $\omega = \Omega_e + k_z v_z$; anomalous resonance with L-waves occurs at $\omega = -\Omega_e + k_z v_z$.

C. *Wave-Particle Interaction*

The interaction of particles and waves can be considered as an exchange of photons or phonons. This is a semi-quantum mechanical approach; it does not depend on the value of the Planck constant. For example, let us look at the emission of one quantum. The particle energy, ε, and the particle momentum, $\mathbf{p}$, change by

$$\Delta\varepsilon = -\hbar\omega \quad , \tag{7.2.25}$$

$$\Delta\mathbf{p} = -\hbar\mathbf{k} \quad . \tag{7.2.26}$$

We write $\Delta\varepsilon = \Delta\varepsilon_z + \Delta\varepsilon_\perp$, with

$$\Delta\varepsilon_z := m v_z \Delta v_z = \Delta p_z v_z = -\hbar k_z v_z \quad , \tag{7.2.27}$$

$$\frac{\Delta\varepsilon_z}{\Delta\varepsilon} = \frac{k_z v_z}{\omega} \quad , \tag{7.2.28}$$

$$\frac{\Delta\varepsilon_\perp}{\Delta\varepsilon} = \frac{\Delta\varepsilon - \Delta\varepsilon_z}{\Delta\varepsilon} = 1 - \frac{k_z v_z}{\omega} \quad . \tag{7.2.29}$$

The resonance condition (7.2.24) yields the sign of $v_z k_z$. It is negative for normal resonance and positive for anomalous resonance. A proton interacting with an R-mode wave, for example, requires $\omega - v_z k_z = -\Omega_p$ (anomalous resonance). Thus

$$\frac{k_z v_z}{\omega} = \frac{\omega + \Omega_p}{\omega} > 0 \quad . \tag{7.2.30}$$

Since $\Delta\varepsilon < 0$ (Eq. 7.2.25), Equation (7.2.28) implies

$$\Delta\varepsilon_z < 0 \quad . \tag{7.2.31}$$

From Equations (7.2.29) and (7.2.30) it follows that $\Delta\varepsilon_\perp / \Delta\varepsilon < 0$ and

$$\Delta\varepsilon_\perp > 0 \quad . \tag{7.2.32}$$

In our example (see also Figure 7.4) the proton loses (total) energy, but it gains some perpendicular energy. This is a characteristic of anomalous resonance emission.

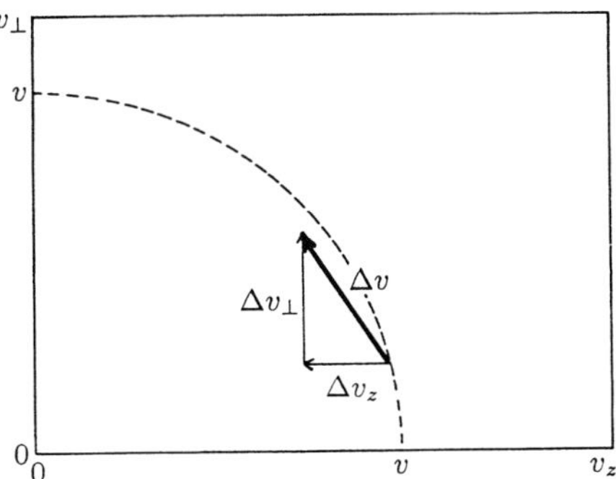

Fig. 7.4. A proton with velocity v emits an R-mode quantum (anomalous resonance) and is displaced in velocity space toward transverse velocity.

The displacement in velocity space is exactly inverse when an R quantum is absorbed by the proton. If absorption and emission occur at random, the particle motion could be described by diffusion. Since the free energy of a beam resides in the parallel direction and can be extracted at the cost of isotropization, it is intuitively clear that emission and absorption are not equally probable. Thus anomalous resonance may drive unstable growth of R-waves. This will be shown in the next subsection.

In a situation of predominant R-mode emission of an ion beam, the general trend is diffusion in the $v_\perp$-direction, decreasing the anisotropy with little energy loss (Exercise 7.1). The result is a randomization of the directions of particle momentum. This property plays an important role in cosmic ray propagation (Section 7.3) and particle acceleration (Section 10.4).

The interaction between proton and L-wave (normal resonance) is similar, except that $\Delta\varepsilon_\perp < 0$ and $\Delta\varepsilon_z > 0$ for emission. The particle gains forward momentum by the emission of a backward photon. Electrons behave similarly with opposite modes: emission of an L quantum (anomalous resonance) increases $v_\perp$, emission of an R quantum decreases it. Note that the emission of quanta brings a particle toward the v_z-axis for normal resonance, but increases the perpendicular velocity for anomalous resonance.

Depending on the gyration phase with respect to the wave, resonant particles experience a force of one sign that either accelerates or decelerates them. Net wave growth or damping thus depends on whether more resonant particles lose or gain energy. Like in the bump-on-tail instability (Section 5.2.4), the appropriate velocity gradient in velocity space determines growth or damping. For electromagnetic waves, particle diffusion is not restricted to the direction of the waves, and the relevant gradients are in pitch angle. A quantitative approach to the phenomenon is quasi-linear diffusion outlined in Section 6.2.1.

D. *Growth Rate*

The previous subsection has given a qualitative picture of electromagnetic wave-particle interaction. The rate at which photons are emitted or absorbed is now calculated from the growth rate, the imaginary part γ, of the frequency given by the dispersion relation (7.2.23). We shall only treat the case $\gamma \ll \omega_r$, where ω_r is the real part of the frequency. This assumption limits us to waves resonant with particles in the high-energy tail, where the number of resonant particles and, therefore, γ is small. It is appropriate for most applications since the free energy resides in relatively few fast particles with long collision time. Then the velocity integral in Equation (7.2.23) can be evaluated using the Plemelj formula (5.2.19),

$$\lim_{\gamma\to 0^+} \frac{1}{v_z - (\omega \mp \Omega_z)/k_z} = \mathcal{P}\left\{\frac{1}{v_z - (\omega \mp \Omega_z)/k_z}\right\} + \pi i \,\mathrm{sign}(k_z)\, \delta(v_z - \frac{\omega \mp \Omega_z}{k_z}) \quad , \tag{7.2.33}$$

where $\delta(v_R)$ is the *delta*-function selecting the residue at the resonance velocity, $v_R = (\omega \mp \Omega_z)/k_z$, given by the resonance condition. The principal part integration, denoted by $\mathcal{P}$, can be solved by expansion into powers of the temperature neglecting the fast particles. We do not need to go into the details, which are similar to the derivation of Equation (5.2.22). Equating the real parts of Equation (7.2.23) one finds that the temperature corrections can be neglected for low-frequency waves ($k^2c^2 \gg \omega^2$) and that ω_r is given by the cold-plasma solutions (4.3.12). The waves (L and R modes) remain identical to Alfvén waves, cyclotron waves, and whistlers in the respective frequency ranges.

The imaginary part of Equation (7.2.23) yields the growth rate

$$\gamma(\omega_r) = \pi \text{ sign}(k_z)\, \omega_r Q \sum_\alpha (\omega_p^\alpha)^2 \eta(v_R^\alpha) \left\{ A_\alpha(v_R^\alpha) - \frac{1}{\pm\Omega_z^\alpha/\omega - 1} \right\} \quad , \tag{7.2.34}$$

having defined for all species α

$$\eta(v_R^\alpha) := \frac{v_R^\alpha}{n_\alpha} \int_0^\infty 2\pi v_\perp dv_\perp f_0^\alpha(v_\perp, v_z = v_R^\alpha) \quad , \tag{7.2.35}$$

$$A_\alpha(v_R^\alpha) := \int_0^\infty v_\perp dv_\perp \frac{v_\perp}{v_z} \left(v_z \frac{\partial f_0^\alpha}{\partial v_\perp} - v_\perp \frac{\partial f_0^\alpha}{\partial v_z} \right) \left[2 \int_0^\infty v_\perp dv_\perp f_0^\alpha \right]^{-1} \Bigg|_{v_z = v_R^\alpha} \quad , \tag{7.2.36}$$

$$\frac{1}{Q} := \sum_\alpha \left(\frac{\omega_p^\alpha}{1 \mp \Omega_z^\alpha/\omega_r} \right)^2 \quad . \tag{7.2.37}$$

The function η has been defined to be proportional to the ratio of particles in resonance to the total particle density. The more particles in resonance, the faster the wave grows. The derivatives in the parentheses of Equation (7.2.36) can be transformed into the derivative in pitch angle, $\partial f/\partial\alpha$. Therefore, A depends only on the gradient of the velocity distribution with respect to pitch angle α and is a measure of pitch angle anisotropy. For an isotropic plasma, Equation (7.2.36) yields $A = 0$, and the growth rate is negative. We have found the analogue of Landau damping for electromagnetic waves. If the velocity distribution is a product of Gaussians with different temperatures for perpendicular and parallel velocities, A is independent of v_R and becomes $(T_\perp - T_z)/T_z$.

Note that in Equation (7.2.34) the gyrofrequency Ω_z has been taken out of the integral as if it were independent of velocity. Of course, this holds only for non-relativistic particles (Eq. 2.1.4). It is relatively uncritical for the low-frequency waves parallel to the magnetic field we are studying in connection with beams. In the non-relativistic approach a particle must have a specific parallel velocity, v_R^α, to be resonant with a wave of frequency ω_r. It can have any perpendicular velocity, and the growth rate only involves the properties of the distribution integrated over all perpendicular velocities. The main change for relativistic beams is the appearance of the relativistic gyrofrequency in place of Ω_z. We shall find in the next chapter that for high-frequency waves, however, the smallest deviation from non-relativistic behavior may be essential.

7.3. Applications to Ion Beams

7.3.1. Instability Threshold

The derivations of wave growth in Section 7.2 are general, and the number of free parameters is large. In many situations only the conditions for growth are of interest, but not its rate. Here we study the threshold for instability of weak ion beam distributions that can adequately be described by Gaussians (Eq. 7.2.1). The anisotropy factor (7.2.36) for the beam component becomes

$$A_b = \frac{T_{b\perp} - T_{bz}}{T_{bz}} - \frac{V_b}{v_R}\frac{T_{b\perp}}{T_{bz}} \quad . \tag{7.3.1}$$

Neglect electrons, since electrons in resonance are rare: they must have energies of the order of $m_e(v_R^e)^2 \approx m_e(\Omega_e)^2/k^2$, exceeding the ion resonance energy by m_i/m_e.

The sign of γ determines growth or damping. The sign of $k\eta$ in Equation (7.2.34) is sign(kv_R): positive for normal resonance and negative for anomalous resonance (Section 7.2.2). A proton beam is unstable for R-waves (anomalous resonance) if

$$A_b < \frac{-1}{\frac{\Omega_p}{\omega} + 1} \quad . \tag{7.3.2}$$

At $\omega \ll \Omega_p$ the right side is approximately $-\omega/\Omega_p$. The combination of Equations (7.3.1) and (7.3.2) demonstrates that a distribution must have a beam-like extension in parallel direction to be unstable. The threshold for instability of a beam with ($T_{b\perp} = T_{bz}$) toward electromagnetic waves can easily be evaluated to be

$$V_b > a\frac{\omega}{k_z} \quad , \tag{7.3.3}$$

where a is unity for a shifted Maxwellian. For a power-law distribution ($f \propto v^{-\delta}$) $a = \frac{1}{3}\delta$. Since at low frequencies ($\omega \ll \Omega_p$) ω/k approaches the Alfvén velocity, c_A (Eq. 3.2.13), a rough estimate and a necessary condition for instability is $V_b \gtrsim c_A$.

Figure 7.5 shows the observed beam velocity versus beam density. The average value decreases from about $2c_A$ at low beam density to $1.1c_A$ at $n_b = n_p$. The figure also displays theoretical thresholds for the kinetic and fire-hose instabilities assuming equal thermal and magnetic pressure in the background plasma ($\beta \approx 1$) and a shifted Maxwellian proton beam having the same temperature as the bulk protons. If $n_b < n_p$, the threshold for the kinetic (resonant) instability is much lower than for the fluid instability. Only for dense beams reaching the order of the background plasma density, does the fire-hose instability threshold become comparable.

It is surprising that, on average, the observed interplanetary beams shown in Figure 7.5 have velocities close to the threshold for kinetic instability. The Alfvén velocity varies from event to event over more than an order of magnitude. Does the accelerator in the Sun know about instability conditions in the solar wind?

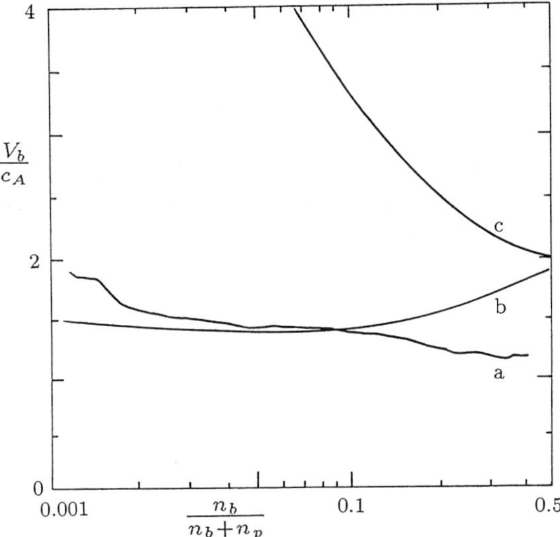

Fig. 7.5. Curve a shows the average beam velocity vs. beam density of several thousand collisionless proton streams observed in the interplanetary medium. The calculated threshold beam velocity (curve b) for electromagnetic instabilities of proton beams having a shifted Maxwellian distribution ($T_{bz} = T_{b\perp}$) is also plotted. R-waves at low frequency in anomalous gyroresonance grow above curve b. The non-resonant fire-hose instability sets in above curve c (after Marsch and Livi, 1987).

The resonant electromagnetic instability apparently slows down the beam to the threshold velocity. Quasi-linear diffusion reduces the anisotropy (Section 7.2.2 c) until the velocity distribution is marginally stable. The beam relaxes to the instability threshold within a few growth times. The beam distribution, on the average, remains near marginal stability. Beam propagation in the presence of low-frequency instabilities is further studied in the following sections.

7.3.2. Wave Growth

Ion beams are frequent in the solar wind, and our analysis suggests that right circularly polarized electromagnetic waves should be abundant. Large amplitude field fluctuations are indeed ubiquitous, in particular upstream of the Earth's bow shock where a super-thermal beam of reflected ions exists most of the time (Chapter 10).

The growth rate is given by Equation (7.2.34), from which one derives for Maxwellian ion components α and for $\omega \ll \Omega_\alpha$,

$$\gamma \approx \sqrt{\frac{\pi}{2}} \sum_\alpha \frac{\rho_\alpha}{\rho} \Omega_\alpha \frac{\Omega_\alpha}{\omega_r} \frac{V_b^\alpha - c_A}{v_{t\alpha}} \exp\left(-\frac{(v_R^\alpha - V_b^\alpha)^2}{2 v_{t\alpha}^2} \right) \quad , \tag{7.3.4}$$

where $\omega_r/k \approx c_A$ has been used, and ρ is the total mass density (Eq. 3.1.26). The wave frequency enters twice in the growth rate (7.3.4): in the denominator and in v_R^α. The resonance condition (7.2.24) yields for R-waves

$$v_R^\alpha = \frac{\omega_r}{k_z}(1 + \frac{\Omega_\alpha}{\omega_r}) \quad . \tag{7.3.5}$$

In the simplest case of one fast, cold beam, the exponential dominates and $v_R \approx V_b$. For $v_R^\alpha \gg V_b^\alpha \gtrsim c_A$, the denominator in Equation (7.3.4) implies maximum growth for $\omega \ll \Omega_\alpha$. Then $v_R^\alpha \gg \omega/k \approx c_A$ (Eq. 7.3.5), indicating that the most rapidly growing waves resonate with particles moving faster than the mean beam velocity. As a rough approximation, $\gamma \approx n_b\Omega_i/n_e$ may be used for the growth rate.

The growth rate of observed quiet-time interplanetary proton beams can reach several percent of the proton cyclotron frequency. Under normal interplanetary conditions this corresponds to growth times of several tens or hundreds of seconds. The instability is fast compared to the solar wind expansion time.

The more energetic proton beams originating in flares (solar cosmic rays) are different. Their speed is high and the density low enough so that the weaker beams reach the orbit of the Earth within less than a growth time. The distribution is still beam-like and highly anisotropic.

The evolution of large proton events on a field line connecting the observer to the flare is most remarkable. They are anisotropic only initially, but become increasingly isotropic near maximum intensity as the beam density increases. The linear growth rate – being proportional to the beam density – is then high enough to drive appreciable quasi-linear diffusion. The observed angular distribution is a measure of beam instability during propagation! Isotropy at 1 AU is the result of well developed instability.

In conclusion, we note that for parameters typical of the interplanetary medium and weak beams the dominant electromagnetic mode is the right-hand resonant ion beam instability. Detailed studies have revealed that extremely hot beams (with beam velocity spread exceeding V_b), as well as anisotropic beams having $T_{b\perp} > T_{bz}$ can also drive L-mode waves unstable at a competitive rate. For extremely dense and fast beams, the fluid fire-hose instability occurs and then grows fastest. A comparative study of all relevant electromagnetic ion-beam instabilities has been published by Gary *et al.*(1984).

7.3.3. Ion Beam Propagation

Most galactic cosmic rays in the energy range 0.1 – 1000 GeV may originate from within our Galaxy and escape from it after a few million years. What keeps them around for so long? In fact, the interaction of ion beams and electromagnetic waves was first proposed for galactic cosmic rays as a means to isotropize streaming particles and to contain them for durations exceeding the free streaming time. The motion of protons may be compared to a situation of enhanced collisions. Scattering by waves deflects the particles by small angles and Fokker-Planck methods (Section 2.6) can therefore be applied. The usual derivation uses the quasi-linear approximation (Section 6.2.1). Here we obtain the same results from a more heuristic approach.

A. *Deflection Time*

The pitch angle of a particle increases or decreases depending on the phase difference between particle gyration and resonant wave. Suppose the phase difference remains nearly constant for a time Δt, called the coherence time. Also assume that the change in pitch angle during this time is small, $\Delta\alpha \ll 1$. The particle then experiences a series of small angular deflections. We expect the pitch angle to execute a random walk described by the diffusion equation (Eq. 6.2.1)

$$\frac{\partial f}{\partial t} = \frac{1}{\sin\alpha}\frac{\partial}{\partial\alpha}\left(D_\alpha \sin\alpha \frac{\partial f}{\partial\alpha}\right) \quad , \tag{7.3.6}$$

neglecting diffusion in energy as would be the case for constant magnetic fields. (Justify this in Exercise 7.1!) The diffusion coefficient in pitch angle is the inverse deflection time, thus $D_\alpha = 1/t_d$. In analogy to Equation (2.6.9) we have defined the deflection time

$$t_d := \frac{\Delta t}{<\Delta\alpha^2>} \quad . \tag{7.3.7}$$

We now estimate t_d from basic physical principles. Particles change their phase Ψ in relation to waves by

$$\frac{d\Psi}{dt} = k_z v_z \pm \Omega_z^\alpha - \omega_r \quad . \tag{7.3.8}$$

Consider waves in a range Δk and a particle resonant with waves at the center k of this range. The time Δt for the particle to get one radian out of phase with the waves at the end of the range, at $k_z + \Delta k/2$, is

$$\Delta t = \frac{2}{v_z \, \Delta k} \quad . \tag{7.3.9}$$

A finite wave range Δk is not the only possible limiting effect on the coherence length. In the interplanetary medium, phase coherence is lost by changes of the wave vector.

During the coherence time, the particle feels the approximately constant Lorentz force, F_1, of the wave magnetic field ($\mathbf{B_1} \perp \mathbf{B_0}$). Therefore, the particle velocity changes by

$$\Delta v_z = -\frac{v_\perp}{v_z}\Delta v_\perp = -\frac{v_\perp}{v_z}\frac{F_1 \, \Delta t}{m_\alpha} = \frac{q_\alpha v_\perp B_1}{c \, m_\alpha}\Delta t \quad . \tag{7.3.10}$$

The first equation assumes energy conservation. From Equation (7.3.10) we derive for the angular deflection

$$\Delta\alpha = \frac{\Delta v_\perp}{v_z} = \Omega_\alpha \frac{B_1}{B_0}\Delta t \quad , \tag{7.3.11}$$

and with Equations (7.3.7) and (7.3.9),

$$t_d \approx \frac{1}{\pi^2}\frac{2\pi}{\Omega_\alpha}\frac{B_0^2}{B_1^2 k}\frac{\Delta k}{} \quad , \tag{7.3.12}$$

where a factor $4/\pi$ has been introduced, making the result agree with the formal derivation. The second factor is the wave period, and the third factor is the ratio of background magnetic field energy to approximate wave magnetic energy.

The deflection time (7.3.12) is exceedingly short on cosmic scales, even for weak waves. For example, cosmic rays at 1 GeV propagating in the interstellar medium ($B_0 \approx 10^{-5}$ G) having a wave energy ratio of say 10^{-6} are scattered within less than a year. Their mean free path, $v_z t_d$, is of the order of 0.06 pc.

A complication may be mentioned: a particle loses coherence with the wave more rapidly if the resonance ceases by its own action on the particle (Exercise 7.2). Resonance is lost for $\Delta v_z \gtrsim v_z \; \Delta k/k$, and the effect could increase the deflection time for exact resonance.

We note that no assumption on the origin of the waves has been made; they may exist independently of the beam or be a result of a beam instability. For waves propagating at an oblique angle to the magnetic field, there is the additional possibility of resonant interaction at the harmonics of the gyrofrequency (Chapter 8). Highly relativistic ions then can scatter on waves driven by lower energy particles of the same beam.

B. *Diffusive Propagation*

Depending on beam density and velocity, we may distinguish two stages of propagation:

(1) The deflection time may be much longer than the propagation time, and the particles arrive *unscattered.*

(2) The deflection time is short for strongly unstable beams, and their velocity distribution becomes nearly, but not exactly isotropic. Such a beam is called *scattered.*

If instability and velocity-space diffusion control the beam, the particles propagate by some residual mean velocity, V_b, from the corona into interplanetary space, or from a supernova to intergalactic space. This propagation is not a free flight, but a diffusion in space given by

$$\frac{\partial f}{\partial t} = \frac{\partial}{\partial \mathbf{x}} \cdot (\widehat{\mathcal{D}} * \frac{\partial}{\partial \mathbf{x}} f) \quad . \tag{7.3.13}$$

The diffusion tensor is diagonal, and its elements are the square of the mean free path ($\ell_{\mathrm{mfp}} = v t_d$) divided by the deflection time (Section 3.1.2.C). Thus,

$$D_{zz} = v_z^2 t_d \quad , \tag{7.3.14}$$

$$D_{xx} = D_{yy} = \frac{R_\alpha^2}{t_d} \quad . \tag{7.3.15}$$

In the perpendicular direction (Eq. 7.3.15) the mean free path, $\ell_{\rm mfp}$, has been replaced by the gyroradius (assuming $\ell_{\rm mfp} \gg R_\alpha$), since one collision also produces a step across the magnetic field of the order of R_α. Therefore, the ratio of the coefficients for diffusion perpendicular and parallel to **B** is

$$\frac{D_{xx}}{D_{zz}} = \left(\frac{\pi}{2}\right)^2 \left(\frac{v_\perp}{v_z}\right)^2 \left(\frac{B_1^2 k}{B_0^2 \, \Delta k}\right)^2 \quad . \tag{7.3.16}$$

Except for the uninteresting singular case $v_z \approx 0$, this ratio is much smaller than unity and tells us that, as expected, diffusion along the field lines is much faster than across as long as the wave field is smaller compared to the background field.

The problem of diffusive propagation is mathematically described by the two coupled equations of wave growth (7.2.34) and diffusion in pitch angle and space, Equations (7.3.6) and (7.3.13). Wave growth depends on the particle distributions f_0^α; particle diffusion is proportional to the wave level B_1. We outline the basic results below.

If the beam velocity exceeds $a\omega/k$, or about c_A, we have an unscattered beam. As it is unstable, Alfvén waves grow at a rate estimated in Equation (7.3.4). For solar proton beams, it may be roughly approximated by $\gamma/\Omega_p \approx n_b/n_e$. A beam density of $10^{-3} n_e$ causes waves to grow at a rate γ of the order of 100 Hz (assuming $B \approx 10$ G). Within about 10 growth times, or less than one second, the deflection time t_d (Eq. 7.3.12) diminishes to a fraction of a second. The anisotropy in velocity space is reduced by quasi-linear diffusion, and the beam velocity decreases. The beam instability develops into a *marginally stable*, quasi-stationary state, in which the proton beam anisotropy is just large enough to excite Alfvén waves to a level that would eliminate any greater anisotropy. The growth rate falls to the order of the inverse deflection time. The mean velocity decreases to about $a\omega/k$ within a few instability growth times. Now the beam appears to be scattered, and propagation is called *diffusive*.

Quiet-time interplanetary proton beams display nicely the characteristic beam velocity of diffusive propagation (Fig. 7.5). The long dwelling time of cosmic rays in the Galaxy can also be explained by diffusive propagation. Similarly, downward moving proton beams accelerated in flares drive growing low-frequency electromagnetic waves during their passage through the corona into the chromosphere, if their density exceeds about 10^{-5} of the background. Quasi-linear diffusion may shorten their penetration depth by orders of magnitude. On a more speculative level, accreting plasma flowing along field lines onto a protostar or compact object may generate a high level of electromagnetic waves sufficient to heat the plasma.

7.4. Electrostatic Ion Beam Instabilities

7.4.1. Low-Frequency Waves

Spacecraft investigations of ion beams near Earth have revealed – in addition to electromagnetic waves – a rich variety of electrostatic waves. An intense broadband spectrum of waves is seen in the magnetotail (where $T_e \gg T_i$) ranging from extremely low frequencies to beyond the local electron plasma frequency. They coincide with fast ion beams (Section 7.1). Most prominent are *ion acoustic waves* at $\omega \lesssim \omega_p^i$, as one would expect from Section 5.2.6, where the growth rate has been evaluated for ions moving against electrons (current). An ion-beam component can be incorporated into Equation (5.2.34) in a similar way. The result is again unstable toward ion acoustic waves. Since ion acoustic waves are Landau damped on the slope of the thermal electron distribution, the threshold is considerably lower for $T_e \gg T_i$. Low-frequency electrostatic waves are of particular interest, if the beam velocity, taken parallel to the magnetic field, is lower than the Alfvén velocity and below threshold for electromagnetic waves.

The *lower hybrid branch*, $\omega \approx \sqrt{\Omega_e \Omega_i}$ (Eq. 4.4.7), has frequently been found at an excited level in the solar wind. The waves are also prominent in the front region of the Earth's bow shock where the arriving solar wind field lines are nearly parallel to the front, and ions are reflected into gyrating perpendicular ion beams (Chapter 10). In contrast to ion acoustic waves, lower hybrid waves exist at all ratios of T_e/T_i, and their role in coronae may be important.

There is a number of possible instabilities of a perpendicular or gyrating ion beam (azimuthal anisotropy). Their importance depends strongly on the ratio of electron to ion temperature, a parameter usually not well known. The *modified two-stream instability* has received particular attention. It is a fluid-like instability between the gyrating beam and the background ions and electrons, much like the two-stream instability in cold plasma (Section 4.6). The instability requires $T_e/T_i \gtrsim 1$ and a beam velocity $V_b > c_{is}$, the ion sound velocity (Eq. 5.2.36), and drives lower hybrid waves.

We have encountered lower hybrid waves in Section 4.4 as space charge waves of ions propagating nearly perpendicular to the magnetic field. Electrons gyrating at a higher frequency do not actively participate if $\omega_r/|k_z| > v_{te}$. In the opposite case, the waves are strongly Landau damped on thermal electrons. If lower hybrid waves are driven by ions and have $\omega_r/|k_\perp| \approx v_{ti}$,

$$\frac{k_z}{k_\perp} \lesssim \frac{v_{ti}}{v_{te}} \ll 1 \quad . \tag{7.4.1}$$

Thus, lower hybrid waves propagate nearly perpendicular to the magnetic field. They enhance the scattering of high-frequency plasma waves (such as Langmuir waves or upper hybrid waves) into observable radio emission. Suitable models have been proposed for coronal currents and shock waves (metric type I and type II radio bursts).

7.4.2. HIGH-FREQUENCY WAVES

Can ion beams radiate plasma emission like the type III bursts of electron beams? In the derivations of Section 5.2 the particle species was not specified. The growth rate (Eq. 5.2.29) is reduced for ion beams by the mass ratio m_e/m_i. For $k \ll k_D$, implying $V_b \gg v_{te}$,

$$\gamma \approx \frac{\pi}{2}\left(\frac{V_b}{v_{tb}}\right)^2 \frac{n_b}{n_e}\frac{m_e}{m_i}\omega_p^e \quad . \tag{7.4.2}$$

How does it compare with the growth of electromagnetic waves? Approximating Equation (7.3.4) with $\gamma \approx n_b\Omega_i/n_e$ and $V_b \approx v_{tb}$, the ratio of the growth rates of Langmuir waves to Alfvén waves is about ω_p/Ω_e and may exceed unity. However, the threshold beam velocity is usually more important. It is about $3v_{te}$ for Langmuir waves and c_A for Alfvén waves. In coronal plasmas $c_A \ll 3v_{te}$. If a beam develops from a super-thermal tail, its mean velocity increases gradually and first meets c_A. Alfvén waves thus diffuse the beam in velocity space and hinder the growth of Langmuir waves. The higher growth rate of Langmuir waves is only effective if the beam moves already initially at high velocity. The interpretation of radio emission by the bump-on-tail instability of ion beams therefore requires a special acceleration process. It has raised little interest for ion beams in coronae.

Exercises

7.1: Derive the ratio $\Delta v_z/\Delta v_\perp$ of the velocity changes of an ion emitting an R-mode quantum. Show that for $\omega \ll \Omega_i$ the ion moves approximately on a circle. This means that it is deflected primarily in pitch angle with only little change in energy.

7.2: The Lorentz force of the wave magnetic field acts on a gyrating particle and changes its phase ψ. This reduces the interaction time between particles and waves. Use $k_z v_z \approx \Omega_i$, and calculate the time the Lorentz force of a wave needs to deflect a particle initially in resonance until the phase difference between particle and wave shifts by one radian.

7.3: Intense beams of ions are accelerated in flares. They are guided along loop-shaped field line into the dense chromosphere (Fig. 7.1). Calculate the growth rate γ/Ω_p of waves due to a beam of protons with the following properties: beam velocity $V_b = 2c_A$, beam thermal velocity $v_{tb} = 100\ c_A$, beam density $n_b/n_p = 10^{-2}$. Assume that $\Omega_p/\omega = 20$ and $B = 10^2$ G.

7.4: Calculate the mean free path of a proton at resonance velocity ($v = 21\ c_A$) in the above example (Exercise 7.3) using the approximation $t_d \approx 1/\gamma$. What is the diffusive propagation time if $L = 10^9$ cm and $c_A = 10^8$ cm s^{-1}?

Further Reading and References

Observations of energetic ions

Chupp, E.L.: 1990, 'Transient Particle Acceleration Associated with Solar Flares', *Science* **250**, 229.

Gary, S.P.: 1985, 'Electromagnetic Ion Beam Instabilities; Hot Beams at Interplanetary Shocks', *Astrophys. J.* **288**, 342.

Duric, N.: 1988, 'The Origin of Cosmic Rays in Spiral Galaxies', *Space Sci. Rev.* **48**, 73.

Simnett, G.M.: 1986, 'A Dominant Role for Protons at the Outset of Solar Flares', *Solar Phys.* **106**, 165.

Electromagnetic and electrostatic ion beam instabilities

Gary, S.P., Smith, C.W., Lee, M.A., Goldstein, M.L., and Forslund, D.W.: 1984, 'Electromagnetic Ion Beam Instabilities', *Phys. Fluids* **27**, 1852.

Kennel, C.F. and Petschek, H.E.: 1966, 'Limit on Stably Trapped Particle Fluxes', *J. Geophys. Res.* **71**, 1.

Tamres, D.H., Melrose, D.B., and Canfield, R.C.: 1989, 'On the Stability of Proton Beams Against Resonant Scattering by Alfvén Waves in Solar Flare Loops', *Astrophys. J.* **342**, 576.

Wentzel, D.G.: 1974, 'Cosmic Ray Propagation in the Galaxy: Collective Effects', *Ann. Rev. Astr. Ap.* **12**, 71.

References

Marsch, E. and Livi, S.: 1987, 'Observational Evidence for Marginal Stability of Solar Wind Ion Beams', *J. Geophys. Res.* **92**, 7263).

Ramaty, R. and Murphy, R.J.: 1990, 'Nuclear Processes and Accelerated Particles in Solar Flares', *Space Sci. Rev.* **45**, 213.

CHAPTER 8

ELECTRONS TRAPPED IN MAGNETIC FIELDS

Until now we have been concerned mainly with beams, the deviations in velocity space from isotropy to distributions enhanced in the direction parallel to the magnetic field. In this chapter we study the opposite: particle distributions with predominantly perpendicular velocities, characteristic of magnetically reflected or trapped particles. The solar and stellar coronae are permeated by loop-shaped magnetic fields forming 'magnetic bottles' for collisionless (fast) particles. After one bounce, particles of pitch angle $\alpha < \alpha_c$ are lost, so they are missing in the reflected population (Section 2.2). If this population is again reflected, it is trapped and has a double loss-cone distribution in velocity space.

The transverse motion of electrons – their gyration in the magnetic field – causes synchrotron emission at high harmonics of the gyrofrequency (to be discussed in Section 8.1). Although not exclusive to trapped electrons, such particles have a long lifetime to emit synchrotron (microwave) radiation before losing their energy by collisions and have a characteristic, high ratio of synchrotron to bremsstrahlung (hard X-ray) flux.

Trapped electrons are important sources for intense *coherent* radio emission. Loss-cone produced emissions include radiation from planets, the most powerful solar decimetric radiations, and probably some stellar radio flares. The driving free energy – the anisotropic velocity distribution of the loss-cone – causes various waves to grow, some of them are observable in radio emission. Section 8.2 shows how trapped particles can be more efficient emitters of coherent radiation than particles in beams, where only the leading front is active. In the best studied case, the Earth's magnetosphere, it is estimated that 10^{-3} of the initial electron energy is converted into radiation. This value greatly exceeds the efficiency of beams, and of incoherent synchrotron and bremsstrahlung emissions from coronae.

Loss-cone instabilities may not only cause intense radio sources visible across the Galaxy, but often control the loss of energetic particles. We assume that the collision rate in a coronal magnetic loop is low. If the particles are deflected by interactions with unstable waves (Section 8.3), the loss-cone instabilities have an important consequence: the particles diffuse into the loss-cone, leave the trap and precipitate into the dense layers at either end of the magnetic loop, where they immediately lose their energy by collisions, emitting hard X-rays (Fig. 6.10). In a low density trap, the particle's lifetime (energy loss) may not be controlled by the local collision time in the trap but by untrapping through waves. Observable effects of trapping will be discussed in Section 8.4.

8.1. Observational Motivation

Trap models are attractive to explain prolonged emissions of non-thermal particles. Since the diffusion in velocity space of particles confined by a magnetic bottle is inevitable, the trapping time is finite, and the particles eventually precipitate into lower layers having a high collision rate. The associated emissions are bound to change in time. The models are usually referred to as *trap-plus-precipitation*.

8.1.1. Incoherent solar emissions

In solar flares, the flux densities of broadband microwave emission ($1 \lesssim \nu \lesssim 30$ GHz) and hard X-rays ($\gtrsim 10$ keV) are usually well correlated. The flare related events in both emissions are classified into impulsive (< 10 minutes) and gradual (> 10 minutes) events (Section 6.4.2), reflecting different injection and trapping conditions. The correlation suggests a common origin (if not identity) of the radiating particles. Yet Figure 8.1 presents the image of a *gradual* flare where the positions of the hard X-ray source and broadband microwave source are well separated. The centroid of the hard X-ray emission is of the order of 10^4 km above the limb, the radio source is about 4 times higher.

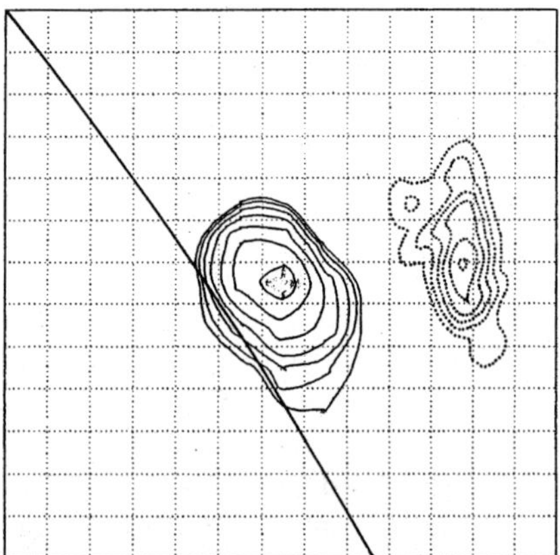

Fig. 8.1. A *gradual* flare near the limb (solid line) was imaged in hard X-rays (solid isophotes, logarithmic steps) and broadband microwaves (dotted isophotes, linear steps). The grid is 10" or 7000 km on the Sun. The hard X-rays were observed by the Hinotori satellite in the 20-30 keV range. The Very Large Array measured the microwave emission at 5 GHz (after Takakura *et al.*, 1985).

As pointed out in Section 6.4.3, the hard X-ray emission of *impulsive* flares seems to originate primarily from the footpoints of loops. The radio source is usually located above the line of zero longitudinal photospheric magnetic field (cf. Fig. 1.4), it is in magnetic loops or an arcade. Figure 8.2 presents an example of 4.9 GHz emission (source B) originating between two Hα flare kernels. The picture is

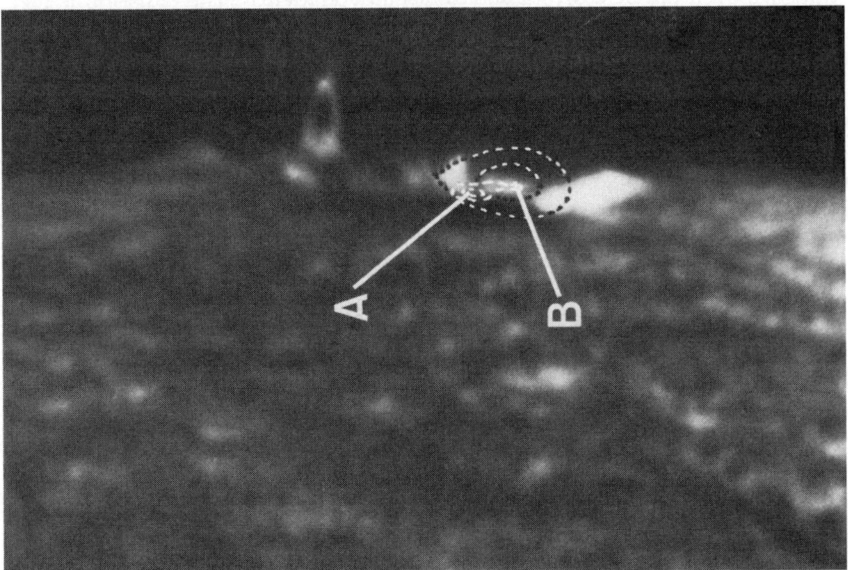

Fig. 8.2. Location of a broadband microwave source observed at 15 GHz (A) and at 4.9 GHz (B) by the Very Large Array near the solar limb during an *impulsive flare*. The contours shown are at the 0.2 and 0.5 levels of the peak brightness. Superposed is an Hα picture from Big Bear Observatory (from Dulk, Bastian, and Kane, 1986).

seen against the solar limb and suggests that the radio source is at a height similar to the Hα features, i.e. probably around 3000 km or below. The curve of zero longitudinal magnetic field passes between the two bright Hα features. Source B appears to be located in a magnetic loop between them. The high-frequency source, A at 15 GHz, is confined to a small core near an Hα bright feature.

In summary, we conclude that hard X-ray and broadband microwave observations are compatible with an interpretation by bremsstrahlung and synchrotron emissions of partially trapped and precipitating energetic electrons. The observations are consistent with emissions from a common loop or interacting loops. Such loops span the neutral line of the photospheric magnetic field and often have their footpoints in Hα-enhanced regions. The microwave radiation comes from a large part of the loop with maxima either near the top or near the feet of the loop; the latter is more likely at short wavelengths (Bastian et al., 1998).

8.1.2. Synchrotron Emission

The magnetic field confines charged particles to a spiraling motion around field lines. The continuous change in the direction of the particle's momentum causes electromagnetic radiation – called *gyromagnetic* emission – according to Larmor's formula. As each particle orbits and emits independently, the emission is incoherent.

Relativistic particle motion strongly enhances the emission, then called *synchrotron* radiation after the man-made accelerators where it was first detected. Its theory has been presented in many textbooks (e.g. Pacholczyk, 1970).

Synchrotron emission depends on the magnetic field and hence provides complementary information not available from X-ray bremsstrahlung. Synchrotron emission occurs at a wide range of harmonics of the (relativistic) gyrofrequency of the emitting particle and has a broad peak at about

$$\omega_c \approx \gamma^2 \Omega_\alpha \quad , \tag{8.1.1}$$

where Ω_α is the non-relativistic gyrofrequency of species α, and γ is the Lorentz factor of the emitting particle. The emission of ions occurs below the electron gyrofrequency by the mass ratio m_e/m_i. More important, the energy loss due to gyration is $(Z_i m_e/m_i)^4$ times smaller for ions and therefore negligible in comparison with that of electrons at the same energy.

The more confined source at high frequency in Figure 8.2 finds an explanation in Equation (8.1.1). The presence of a higher magnetic field strength (and higher Ω_e) near the footpoints shifts the location of the high-frequency source A in relation to the low-frequency source B.

In solar flares, the highly relativistic approximation for synchrotron radiation usually does not apply. More complicated derivations for mildly relativistic electrons yield the so-called *gyrosynchrotron* emission. For a power-law distribution (Eq. 6.4.5) of mildly relativistic electrons with an isotropic pitch angle distribution, Dulk and Marsh (1982) have presented simplified expressions for the emissivity η. For x-mode (where most of the radiation originates), a power-law exponent in the range $2 \lesssim \delta \lesssim 7$, $\theta \gtrsim 20°$ and $\omega/\Omega_e \gtrsim 10$, they derive

$$\eta(\omega, \theta) \approx 3.3 \times 10^{-24} B n_h(> 10\text{keV}) 10^{-0.52\delta} (\sin\theta)^{-0.43+0.65\delta} \left(\frac{\omega}{\Omega_e}\right)^{1.22-0.90\delta}$$
$$[\text{erg s}^{-1}\ \text{cm}^{-3}\text{Hz}^{-1}\ \text{sterad}^{-1}] \quad , \tag{8.1.2}$$

where θ is the emission angle to the magnetic field B [G], and n_h is the density [cm^{-3}] of non-thermal electrons. A power-law cutoff at $\varepsilon_0 = 10$ keV has been assumed in the numerical constant in Equation (8.1.2). If it is at a different but non-relativistic energy, the virtual number density $n_h(> 10\text{keV}) := n_h(> \varepsilon_0)(\varepsilon_0/10)^{\delta-1}$ can be used. Gyrosynchrotron emission becomes optically thick at low frequencies (i.e. it is in equilibrium with its inverse (absorption) process, cf. Section 11.1). For a source depth L [cm] and the above conditions, this occurs at a frequency

$$\nu_{\max} \approx 2.72 \times 10^3\ 10^{0.27\delta} (\sin\theta)^{0.41+0.03\delta} (n_h(> 10\text{keV}) L)^{0.32-0.03\delta} B^{0.68+0.03\delta}\ [\text{Hz}], \tag{8.1.3}$$

Note that both η and $\nu_{\rm max}$ depend on B. In regions of high magnetic field strength, synchrotron emission is stronger and shifted to higher frequency.

In the *optically thin* part of the spectrum ($\nu > \nu_{\rm max}$), Equation (8.1.2) suggests a power-law spectrum for the intensity,

$$I(\omega) \;=\; I(\omega_0)(\omega/\omega_0)^{-\alpha} \quad [\text{erg cm}^{-2}\text{s}^{-1}\text{ Hz}^{-1}\text{sterad}^{-1}] \tag{8.1.4}$$

where $\alpha = 0.90\delta - 1.22$. Using the approximation for highly relativistic electrons, the spectral index $\alpha \approx 0.5\delta - 0.5$.

Equations (6.4.4), (8.1.2), and (8.1.4) can be combined to give a relation between the spectral indices of hard X-rays (thick target model) and synchrotron emission,

$$\gamma = \begin{cases} 1.1\alpha - 0.14 & \text{gyro} - \text{synchrotron approximation} \\ 2\alpha & \text{highly relativistic approximation} \end{cases} , \tag{8.1.5}$$

assuming that both emissions are produced by electrons with the same power-law energy spectrum. This relation can be tested by observations. In view of the frequency dependent positions (Fig. 8.1), it is surprising that the correlation of 2α and γ in time is generally quite good during a flare. In addition to different positions, the observations in microwaves and X-rays often refer to different electron energies. A change in δ with energy may cause a deviation from Equation (8.1.5). Nevertheless, observing approximately the relation (8.1.5) strongly suggests a common origin of the radiating particles.

Impulsive microwave bursts have a higher $\nu_{\rm max}$ (factor of 3 on average) than gradual bursts. This agrees with Figures 8.1 and 8.2, suggesting a lower altitude (a few 10^3 km vs. a few 10^4 km) compatible with higher magnetic field strength for sources of impulsive bursts.

At low frequencies ($\nu < \nu_{\rm max}$), the emission is *optically thick* and – for a power-law distribution of the particle density with exponent δ – also has a power-law form. Its exponent becomes independent of δ at highly relativistic velocities,

$$I(\omega) \approx I(\omega_0)(\omega/\omega_0)^{2.5} \quad . \tag{8.1.6}$$

In reality, the spectral index is rarely 2.5, but in the range from 0 to 10. Low values are generally interpreted by the superposition of multiple sources with different peak frequencies. At the low-frequency end of the spectrum – where emission approaches the plasma frequency in the source – absorption may reduce the observed radiation. In addition, the reduced refractive index near the plasma frequency suppresses the synchrotron emissivity for frequencies $\omega \lesssim \omega_p^2/\Omega_e$, a phenomenon discovered in 1960 by V.A. Razin and V.N. Tsytovich (for a description see e.g. Melrose, 1980). Both collisional absorption and Razin suppression can produce high spectral indices on the low-frequency slope.

Solar flare gyrosynchrotron sources are usually large compared to magnetic scale lengths. Source modelling therefore requires spatial information. Full Sun spectral observations have to be interpreted by complex models. This is of particular importance for unresolved stellar sources (discussed e.g. by Klein and Chiuderi-Drago, 1987).

There remains a fundamental uncertainty in the above observations of incoherent emissions. It is difficult to distinguish particles accelerated or injected into a magnetic loop in one step and then remaining trapped for an extended time, from particles being accelerated continuously, injected, and lost. The two possibilities imply different velocity distributions: (*i*) trapped particles can be identified from their typical loss-cone distribution (cf. Figure 2.6) with a depletion at small pitch angles. (*ii*) Continuously injected particles may have an isotropic or beam-shaped distribution. An equilibrium between particle supply (injection) and depletion (precipitation) may be attractive to simplify theory. Nevertheless, disentangling acceleration and trapping is a formidable observational problem.

8.1.3. Narrowband spikes

Coherent radio emission associated with a loss-cone distribution is direct observational evidence for trapping. Well-known examples are the radio emission of planets, including the auroral kilometric radiation of the Earth and Jupiter's decametric emission. They show that a preferable stay of energetic particles is in magnetic traps. Depletion of the loss-cones by an instability is a necessary process to perceive the trap in coherent emission.

As an example of coherent emission probably caused by a transverse velocity anisotropy, we present narrowband spikes emitted during solar flares. They have first been discovered around 0.3 GHz independently by F. Dröge and P. Riemann, Ø. Elgarøy, and T. de Groot in 1961. However, spikes did not receive much attention until 1978, when C. Slottje also observed them at 2.8 GHz with a much shorter duration and coincident with gyrosynchrotron emission. Narrowband spikes have since been identified by spectrometers in the 0.3 – 8 GHz range and seem to be physically identical (for a review see Benz, 1986). Note that the term 'spike' is also used for short duration hard X-ray and millimeter radio bursts, emissions that are not directly related to the phenomenon under consideration.

Narrowband spikes are a topic of current research, and it is not clear whether their origin is a trap, singly reflected particles, or an acceleration transverse to the field. Figure 8.3 shows a group of hundreds of peaks spread irregularly in time and frequency during the impulsive phase of flares. The number of spikes per second and their frequency-averaged flux correlates well with the flux of impulsive hard X-rays (Fig. 5.1c). This demonstrates that spikes are related intimately to the energetic electrons of flares.

The duration of single spikes is a few tens of milliseconds, and the phenomenon is also known as 'millisecond spikes'. The decay of single spikes can be fitted by an exponential in time. The decay time is surprisingly close to the thermal collision time assuming $T \approx 3 \cdot 10^6$ K and decreases linearly with frequency, as if the emission originated from a source with $\nu_p \approx \nu$.

The narrow bandwidth of 0.2 – 2% of the center frequency is the most remarkable feature. It suggests that the source emits at one of its characteristic frequencies, such as the electron gyrofrequency, the plasma frequency, or the upper hybrid frequency. However, since the source is located in an inhomogeneous

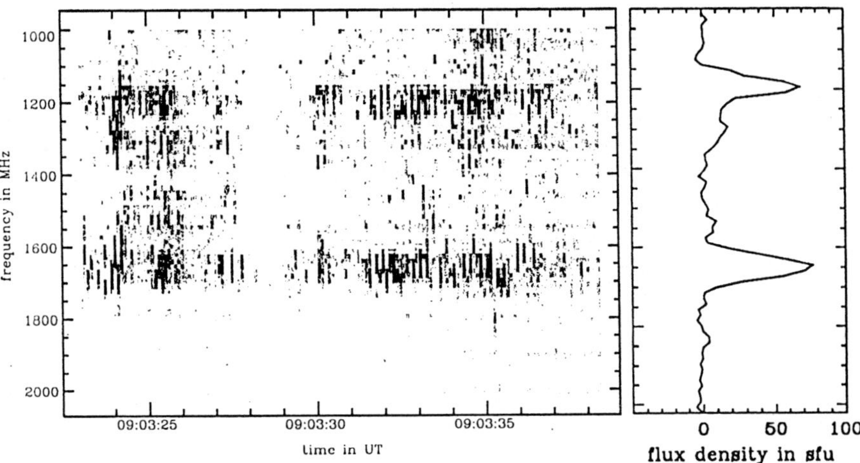

Fig. 8.3. *Left:* The radio spectrogram of narrowband spikes showing two major frequency bands (observed at ETH Zurich). *Right:* Spectrum of the same event at 09:03:32.2 UT (after Krucker and Benz, 1994).

corona, the characteristic frequencies are spread in a finite interval, and the spikes occur all over this band. The bandwidth of a spike, $\Delta\nu$, at an observing frequency ν limits the source dimension, L, along the direction of the gradient,

$$L \leq \frac{\Delta\nu}{\nu} \begin{cases} \times\, H_B & \text{magnetic scale length} \\ \times\, 2H_n & \text{density scale length} \end{cases} \quad . \tag{8.1.7}$$

The magnetic scale length, H_B, applies to emission at a harmonic of the gyrofrequency. For emission at the plasma frequency, $2H_n$ has to be used. With $H \approx 10^9$ cm, a value to be expected for active regions, some source dimensions become smaller than 40 km. The observed flux densities (typically 100 sfu = 10^{-17}erg s^{-1}Hz^{-1}cm^{-2}) and Equation (5.1.1) require a brightness temperature in excess of 10^{13} K. Its large value indicates coherent emission.

Sometimes, when the spikes are confined to a small range of frequencies, several harmonics are observed. Figure 8.3 shows a case with 3 horizontal bands of spikes equally spaced in frequency. They correspond to the harmonic numbers 3, 4, and 5. The modulation of the harmonics correlates in time, indicating a common source.

The observational characteristics of spikes are substantially different from type III bursts caused by beams.

- The instantaneous bandwidth of spikes (spectral width at a given time) is more than an order of magnitude smaller.

- The average duration is a factor of 10 shorter.

- The polarization of spikes follows the leading spot rule and suggests x-mode.

The observational evidence is compatible with a large number of small, transient sources in a stationary volume. This is very different from beams and has led to interpretations by instabilities of transverse velocity space anisotropy in quasi-stationary sources. The observed multitude of sources poses interesting theoretical possibilities and implications on flare fragmentation.

8.2. Loss-Cone Instabilities

Anisotropy in velocity space emerges wherever the magnetic field acts as a particle mirror, an abundant situation in the universe from planets to interstellar space. It involves primarily energetic particles whose mean free path exceeds the length of the magnetic configuration. We study first the linear instabilities driven by a loss-cone distribution of electrons. In the second section the effect of these instabilities on trapping is considered, as well as other non-linear phenomena. Finally, solar and stellar observations are presented exemplifying the theory and demonstrating the practical importance of the subject.

We shall discuss low and high frequency, electrostatic and electromagnetic instabilities. Which of these instabilities dominates depends on the sign of the charge, the velocity distribution of the particles, and the plasma parameters. An introduction such as this cannot exhaust all the possibilities, nor give a complete classification. A first ordering of the possible loss-cone instabilities may separate those which excite waves having $k_\perp v_\alpha/\Omega_\alpha$ much smaller than unity from those where this ratio is much larger than unity (where α refers to the relevant particle species).

- In the former case – studied in Section 8.2.1 – $\lambda_\perp \gg R_\alpha$, where $\lambda_\perp := 2\pi/k_\perp$. The wave fields are homogeneous on the scale of the gyroradius, R_α, of the interacting particle. All particles at a given location feel the wave fields the same way, independent of their energy. The wave then interacts best if its frequency equals the Doppler-shifted fundamental gyrofrequency.

- In the opposite situation ($\lambda_\perp \ll R_\alpha$) the particles interact with the wave fields in a more complex way, depending strongly on the gyroradius (i.e. the perpendicular velocity) of the resonating particle. It leads to interactions with harmonics of the gyrofrequency (Section 8.2.2).

Since the basic equations are the same as for the beam instabilities, we can start at an advanced level.

8.2.1. Low-frequency electromagnetic instability

The interaction of particles with low-frequency electromagnetic waves has been derived in Section 7.2.2 from the Maxwell-Vlasov equations. The dispersion relation (Eq. 7.2.23) relates frequency to wavenumber for small disturbances satisfying the plane wave assumption (7.2.8). Equation (7.2.23) applies to low-frequency, electromagnetic waves propagating parallel to the background magnetic field. The imaginary part of the frequency, the growth rate, has been evaluated in Equation (7.2.34). It can be shown that waves propagating at a finite angle to the magnetic field grow generally at a lower pace. They will be neglected here.

In our case the free energy resides in perpendicular motion. It can be tapped by a process bringing particles toward the v_z-axis. Normal gyroresonance interaction is such a process, transferring particle energy into wave amplification (inverse of Figure 7.4). Therefore, the relevant low-frequency branch for trapped, preferentially transverse moving electrons is the R-mode, in particular the whistler mode (Section 4.3). These right circularly polarized waves have a frequency range of $\Omega_i < \omega < \Omega_e$. Using the dispersion relation of whistlers (Eq. 4.3.19), the growth rate can be derived from Equation (7.2.34) with the result,

$$\frac{\gamma(\omega_r)}{\Omega_e} = \pi \,\mathrm{sign}(k_z)(1 - \frac{\omega}{\Omega_e})^2 \eta(v_R) \left\{ A(v_R) - \frac{1}{\Omega_e/\omega - 1} \right\} \quad . \tag{8.2.1}$$

The resonant velocity, v_R, the fraction of particles in resonance, η, and the measure of pitch angle anisotropy, A, have been defined previously (Eqs. 7.2.24, 7.2.35, 7.2.36, respectively). The resonance velocity – given by the gyroresonance condition (7.2.24) – can be evaluated using the dispersion relation of whistlers (4.3.19) and becomes

$$v_R = c_A \sqrt{\frac{m_p}{m_e} \frac{\Omega_e}{\omega} \left(1 - \frac{\omega}{\Omega_e}\right)^3} \quad . \tag{8.2.2}$$

For $\omega \to \Omega_e$ the resonance velocity approaches zero, and the waves resonate with the thermal particles. In addition, the curly bracket in Equation (8.2.1) becomes infinitely negative for finite A, and the waves are damped. Therefore, only waves with frequencies considerably below Ω_e, being in resonance with sufficiently fast particles, are growing. Typically, they need $v_R \gtrsim c_A\sqrt{(m_p/m_e)} \approx c\Omega_e/\omega_p$. In other words, only the fastest trapped electrons are unstable toward whistler wave growth.

The calculation of the growth rate is only a matter of evaluating the derivatives of the velocity distribution (Exercise 8.1). The growth depends on the sharpness of the loss-cone and the magnetic field. Its maximum is at a real frequency between $10^{-2}\ \Omega_e$ and $10^{-1}\ \Omega_e$ and its dependence on ω_p/Ω_e is shown in Figure 8.4 (curve w). It is small for $\omega_p \lesssim 0.5\ \Omega_e$ and increases to about $10^{-3}\ \Omega_e$ for large ω_p/Ω_e.

Before going on to the next instability, we point out that Figure 8.4 only compares the linear growth rate of infinite plane waves. Which of these modes dominates in reality depends on the global situation: (*i*) If the loss-cone slowly develops out of isotropy, the instability with lowest threshold controls the evolution. (*ii*) The fastest growing instability may develop into a non-linear regime or saturate. (*iii*) The waves escape before growing if the source dimension is smaller than the growth length, $v_{gr}/\gamma(\omega_r)$.

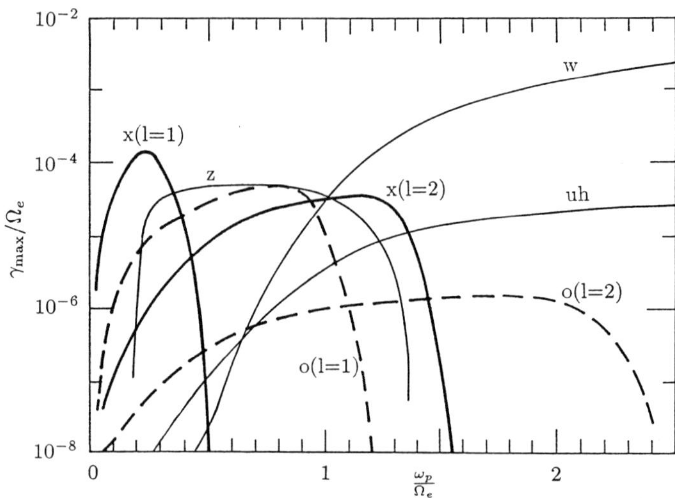

Fig. 8.4. Maximum linear growth rates as a function of ω_p/Ω_e for various loss-cone instabilities discussed in Sections 8.2.1 and 8.2.2. The maximum value in frequency and angle is shown for a background temperature of $2 \cdot 10^6$ K and a Dory-Guest-Harris loss-cone distribution (defined in Exercise 8.1) having $\alpha = 1$, a temperature of $3.5 \cdot 10^7$ K, and a density of 10^{-2} times the background. The letters refer to wave modes, and the numbers refer to harmonics (after Sharma and Vlahos, 1984).

8.2.2. High-Frequency Waves and Cyclotron Masers

A. *Linear Growth Rates*

The velocity distribution of magnetically trapped particles has positive derivatives $\partial f/\partial v_\perp$ near the loss-cone boundaries. One may expect that perpendicular waves can be driven by this anisotropy equivalent to double beams in the v_z-direction. Since $v_\perp$ is involved, the gyrofrequency must enter the physics. It is not surprising that the resonant wave frequency is at the gyrofrequency or higher. The gyroresonance (Eq. 7.2.24 and its generalization to harmonics), in fact, makes it possible for particles to interact with super-luminal waves ($\omega/k > c$). In addition to electrostatic waves dominating the instability of electron beams (Chapter 6), we must include all three high-frequency modes.

We have derived the linear response of a collisionless plasma to a plane wave in Section 7.2.2. The wave amplitude $\bar{f}_1$ of the velocity distribution is the integral in time over the previous history of the acceleration (forces) on the particles and the velocity derivative of the distribution. The integral is along the undisturbed particle orbit (Eq. 7.2.9),

$$\bar{f}_1(\mathbf{v}) = \frac{q}{m}\int_{-\infty}^{t} dt' \exp[i\{k_z(z'-z)+k_x(x'-x)-\omega(t'-t)\}](\bar{\mathbf{E}}_1 + \frac{1}{c}\mathbf{v}'\times\bar{\mathbf{B}}_1) \cdot \nabla_{\mathbf{v}'} f_0(\mathbf{v}') \quad . \tag{8.2.3}$$

The primed variables describe the orbit at time t' of a particle located at the (unprimed) coordinates $\mathbf{x}$ at time t. The motion of the particles of species α can be expressed in guiding center coordinates. We obtained in Equations (7.2.10) – (7.2.18) for $\tau := t' - t$:

$$v'_x = v_\perp \cos(\psi' - \frac{\Omega_z^\alpha}{\gamma}\tau) \tag{8.2.4}$$

$$z' - z = v_z \tau \tag{8.2.5}$$

$$x' - x = \frac{v_\perp \gamma}{\Omega}\left[\sin(\psi' - \frac{\Omega_z^\alpha}{\gamma}\tau) - \sin\psi'\right] \quad . \tag{8.2.6}$$

The gyrofrequency Ω must now be corrected for special relativity by the Lorentz factor γ to the gyrofrequency of the relevant particle, Ω/γ. Ω_z carries the sign of the charge, $\Omega_z := (q/|q|)\Omega$. Equation (8.2.6) brings two sine functions into the exponential of Equation (8.2.3). They are replaced by the identity

$$\exp(i\ A\sin\phi) \equiv \sum_{n=-\infty}^{+\infty} J_n(A)\exp(i\ n\ \phi) \quad , \tag{8.2.7}$$

where J_n is the ordinary Bessel function of the nth order, and A and ϕ are general variables. At this point, the cyclotron harmonics enter through $n\phi$. They are characteristic for waves propagating at finite angles to $\mathbf{B}_0$. The amplitude of the disturbance becomes

$$\bar{f}_1(\mathbf{v}) = \frac{q}{m}\sum_{l,n} J_l(k_\perp R) J_n(k_\perp R)\int_{-\infty}^{0} d\tau \exp\left[i\{(k_z v_z + \frac{l\Omega}{\gamma} - \omega)\tau + (l-n)\psi'\}\right] (\bar{\mathbf{E}}_1 + \frac{1}{c}\mathbf{v}'\times\mathbf{B}_1)\nabla_{v'} f_0 \quad , \tag{8.2.8}$$

where R is the gyroradius (Eq. 2.1.2).

The time integral in Equation (8.2.8) can now be carried out. It allows us to calculate the relation between the wave electric field and the wave current (an Ohm's law, Eq. 7.2.22). The $l \neq n$ terms cancel in the ψ'-integration over velocity space. Ampère's equation gives finally the dispersion relation. We refer to Melrose (1980) for the details of the algebra and discuss only the results. For x-mode and electrons,

$$1 - \frac{c^2k^2}{\omega^2} - \frac{2(\omega_p^e)^2}{\omega^2 n_e}\sum_l \int_{\mathcal{L}} d^3v \left(\frac{l\Omega_e}{\gamma}\frac{\partial f_0}{\partial v_\perp^2} + k_z v_z \frac{\partial f_0}{\partial v_z^2}\right)\frac{v_\perp^2 J'_l(k_\perp R)^2}{k_z v_z + l\Omega_e/\gamma - \omega} = 0 \quad . \tag{8.2.9}$$

The derivative of the Bessel function has been introduced to compress the equation, using $J_l'(x) \equiv -\frac{l}{x}J_l(x) + J_{l-1}(x)$. Each harmonic l produces a singularity of the integrand and thus a solution $\omega_l(\mathbf{k})$ analogous to Equation (5.2.16) for Langmuir waves.

For $\omega_p^2 \ll \Omega_e^2$, the solutions of Equation (8.2.9) have their real parts near the harmonics of the electron gyrofrequency: $\omega_l^r \approx l\Omega_e$ where $l = 1, 2, \ldots$. The corresponding growth rates are of the form

$$\gamma_l = \int d^3v A_l(\mathbf{v},\mathbf{k})\delta(k_z v_z + \frac{l\Omega_e}{\gamma} - \omega_l^r)[\frac{l\Omega_e}{\gamma}\frac{\partial}{\partial v_\perp} + k_z v_\perp \frac{\partial}{\partial v_z}]f_0(\mathbf{v}) \quad . \qquad (8.2.10)$$

The same holds for the o-mode and z-mode. The coefficients A_l can be found in the literature (Wu and Lee, 1979; Winglee and Dulk, 1986). We give as examples some of the more important ones for $k_\perp^2 \gg k_z^2$:

$$A_1^x \approx \frac{\pi(\omega_p^e)^2 v_\perp}{8n_e\omega_r} \qquad (l = 1, \ \mathrm{x-mode}) \qquad (8.2.11)$$

$$A_1^o \approx \frac{\pi(\omega_p^e)^2 v_\perp}{8n_e\omega_r}(\frac{v_z}{c})^2 \qquad (l = 2, \ \mathrm{o-mode}) \qquad (8.2.12)$$

$$A_l^z \approx \frac{\pi(\omega_p^e)^2 v_\perp}{4n_e\omega_r}(\frac{lJ_l}{Rk_\perp})^2 \qquad (l \gg 1, \ \mathrm{z-mode}) \qquad (8.2.13)$$

B. *Particles in Resonance*

Let us now study electrons interacting with the x-mode, an important case in astrophysics. The resonance condition,

$$\omega - k_z v_z = \frac{l\Omega_e}{\gamma} \quad , \qquad (8.2.14)$$

requires the particle gyroperiod, τ_e, to be an integer multiple of the wave period, τ_ω, in the coordinate system moving along the z-axis with velocity v_z. The condition $\tau_e = l\tau_\omega$ is equivalent to $\omega = l\Omega_e/\gamma$. In this frame, the particle moves in a circular orbit (Fig. 8.5) and experiences the same force at the same location for every gyration.

An electron initially located at a when the $\mathbf{E}_1$-field is opposite to $\mathbf{v}$, is accelerated. In the time the particle moves from a to c, the wave electric field changes direction. The particle gains energy until its gyrofrequency decreases due to the increasing Lorentz factor γ. Then its phase in relation to the wave starts to shift. On the other hand, a resonant particle starting at c experiences an $\mathbf{E}_1$-field opposite to Figure 8.5. It loses energy until its phase changes. Figure 8.6 shows the phase ψ between a particle and the wave. The resonant particles oscillate in phase around d.

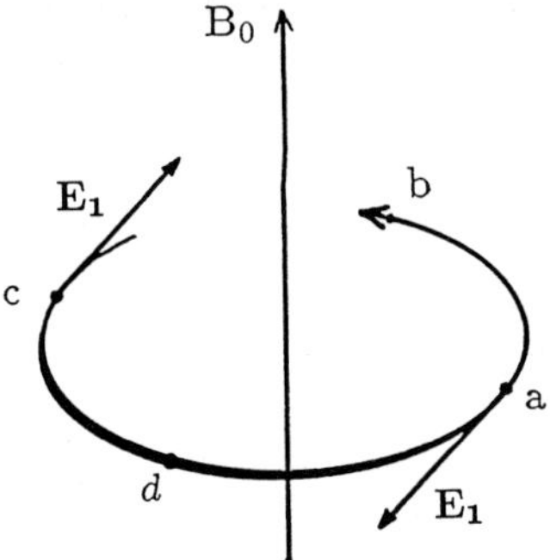

Fig. 8.5. Electron in resonance ($l = 1$) with an x-mode wave ($\mathbf{E}_1 \perp \mathbf{B}_0$).

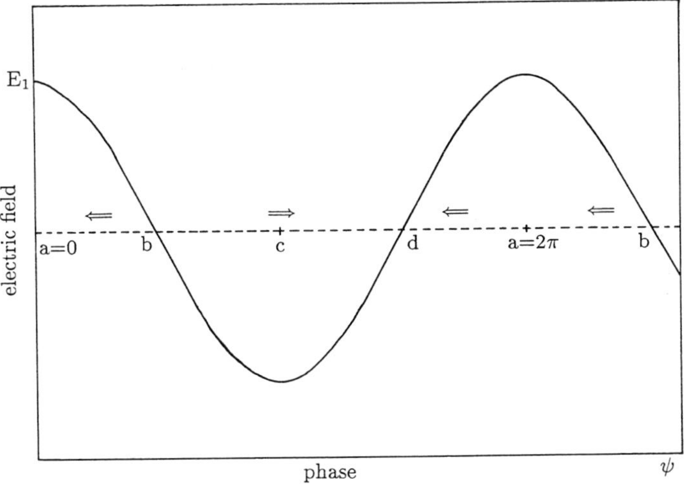

Fig. 8.6. Particles between b, c, and d (defined in Figure 8.5) increase in phase ψ relative to the wave. Particles between d, a, and b decrease in ψ.

The situation is somewhat reminiscent of particles in Čerenkov resonance with Langmuir waves (Section 5.2.5). However, in this case the relativistic effect in the resonance condition plays an important role and changes qualitatively the character of the interaction. Without the variable γ the wave could not trap particles.

If the resulting average particle energy is lower than before the interaction with the wave, the $v_\perp$-component diminishes, and the particle moves in velocity space toward the v_z-axis. If more particles lose energy, the wave amplitude grows, and *vice versa*. The energy gain and loss may be compared to the excitation states of atoms. If there are more electrons in excited states than in equilibrium (inverted population), the states can be depopulated by stimulated emission (like laser or maser action). In analogy, the high-frequency cyclotron instabilities are usually called *masers*, short for microwave amplification by stimulated emission of radiation.

C. Resonance Curve

The resonance condition (8.2.14) defines the surface in velocity space on which the integral for the growth rate is carried out. As may be expected from the discussion of Figure 8.5, the relativistic term plays an important role even for particle velocities $v \ll c$. In the strictly non-relativistic case, the resonance curve in the $(v_\perp, v_z)$-plane is a vertical line (see Fig. 8.7). For small but finite v the resonance depends slightly on $v_\perp$ through γ in the semi-relativistic approximation, $1/\gamma \approx 1 - \frac{1}{2}v^2/c^2$. The resonance condition (8.2.14) then becomes

$$v_z - \frac{v^2}{2c^2}\frac{l\Omega_e}{k}\frac{k}{k_z} + \frac{l\Omega_e}{k_z} - \frac{\omega}{k_z} \simeq 0 \quad . \tag{8.2.15}$$

The relativistic correction (second term in Eq. 8.2.15) may drastically change the resonance for $k_z \ll k$ even for sub-relativistic particle velocities. Equation (8.2.15) can be written as a circle (Exercise 8.2),

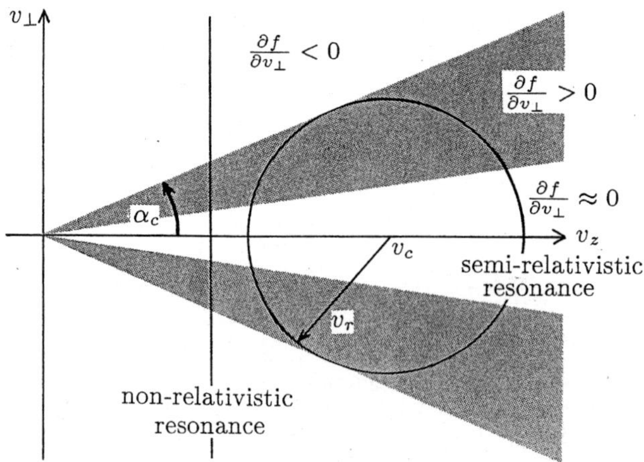

Fig. 8.7. Loss-cone (α_c) and ramp (shaded) velocity distribution of trapped particles with curves of gyroresonance. The resonance circle (semi-relativistic) for a given v_c and maximum wave growth is drawn.

$$(v_z - v_c)^2 + v_\perp^2 = v_r^2 \quad , \tag{8.2.16}$$

with center at v_c and radius v_r, where

$$v_c := \frac{k_z c^2}{l\Omega_e} \tag{8.2.17}$$

$$v_r := \left(v_c^2 - \frac{2(\omega - l\Omega_e)c^2}{l\Omega_e} \right)^{1/2} \quad . \tag{8.2.18}$$

The resonance curve outlines the integration path for the growth rate. The sign of $\partial f/\partial v_\perp$ along the resonance curve determines whether the wave grows or is damped.

In early work on electron cyclotron masers, the term $k_z v_z$ was omitted from the resonance condition. The resonance curve then becomes a circle centered at the origin (Eq. 8.2.17). Others have used the non-relativistic approach (straight vertical line at v_z). Figure 8.7 shows the precarious value of such approximations: Both the $k_z = 0$ and the non-relativistic approach produce lines crossing regions of $\partial f/\partial v_\perp < 0$, and most growth rates of the various modes would be negative for the usual loss-cone distributions.

It was only in 1979, when the semi-relativistic approximation was introduced by C.S. Wu and L.C. Lee, that the full potential of loss-cone instabilities was appreciated. The most unstable waves are the ones with a resonant circle touching the loss-cone boundary (Fig. 8.7). They have the largest number of resonant particles in the region where $\partial f/\partial v_\perp > 0$. Waves with circles penetrating the regime of thermal particles (not shown in Fig. 8.7) are damped. Using the fully relativistic approach, appropriate for truly relativistic velocities, the circles become ellipses.

D. *Loss-Cone Instabilities*

To compare the instabilities and to find the fastest growing mode and harmonic, one has to integrate Equation (8.2.10) along all possible resonant curves, i.e. for all $\omega(k_z, k_\perp)$. Figure 8.8 shows the growth rate in $(k_z, k_\perp)$-space for one mode and one harmonic. Each wave frequency and wave propagation angle defines a resonance ellipse (Eq. 8.2.15). If we now take a smaller angle or a slightly higher frequency, the resonant circle moves to larger v_z (and a smaller number of particles is in resonance). The fastest growing wave is given by the resonance ellipse along which the integral over the value of $\partial f/\partial v_\perp$ weighted by $v_\perp$ (Eqs. 8.2.11 – 8.2.13) has a maximum. In Figure 8.8 it is the wave having $\nu = 999$ MHz and $\theta = 78°$. It will be marked by the subscript *max*.

For $\omega/k \approx c$ and $\omega \approx l\Omega$, the propagation angle $\theta_{\max}$ of the fastest growing wave to the magnetic field can be approximated by

$$\theta_{\max} \approx \arccos \frac{v_{\max}}{c} \tag{8.2.19}$$

(Exercise 8.3), where $v_{\max}$ is the center of the resonance circle of the fastest growing wave. Effective wave growth is confined to a narrow range of propagation angles $\Delta\theta$,

$$\Delta\theta \approx \frac{v_{\max}}{c} \sin \alpha_c \quad , \tag{8.2.20}$$

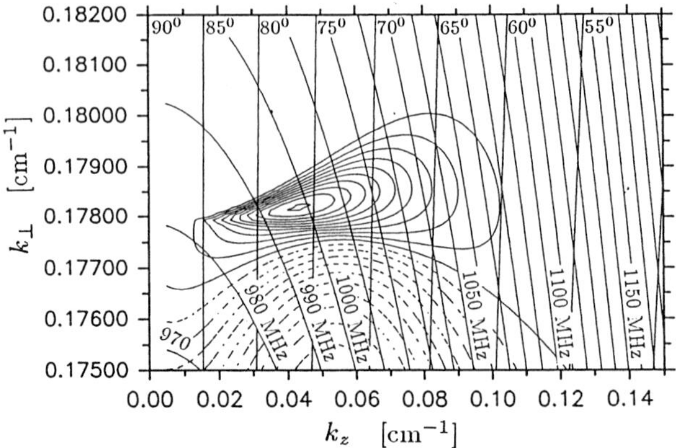

Fig. 8.8. Contours of growth rate in k-space for $l = 1$ (fundamental), x-mode, and $\omega_p/\Omega_e = 0.1$. Positive growth is drawn by solid curves, negative growth (damping) by dashed curves. A $\sin^2\alpha$ distribution ($\alpha =$ pitch angle) of hot particles and a cold population with density $n_h/n_c = 10^{-2}$ was assumed. $\Omega_e = 2\pi \cdot 0.98$ GHz. Curves of constant wave frequency and emission angle form a grid of reference (after Aschwanden and Benz, 1988).

and α_c is the loss-cone angle (Exercise 8.3, using $\Delta v_c \approx v_r$). The frequency range at a particular harmonic l is of the order of

$$\frac{\Delta\omega}{l\Omega_e} \approx \frac{1}{2}\left(\frac{v_{\max}}{c}\right)^2 \Delta\alpha \sin(2\alpha_c) \quad , \tag{8.2.21}$$

where $\Delta\alpha$ (in radians) is the range of pitch angles over which f falls off inside the loss-cone (Exercise 8.3). As a general rule, maximum growth is a few percent above the relevant harmonic of the gyrofrequency and at high pitch angles.

The maximum growth rate of several harmonics and modes is plotted in Figure 8.4 as a function of ω_p/Ω_e. The fundamental x-mode, marked as $x(l = 1)$, grows fastest in plasmas with $\omega_p \lesssim 0.3\ \Omega_e$. The instabilities of higher harmonics, z and o mode become important at higher ratios of ω_p/Ω_e.

In a source of finite length, the number of e-folding growth lengths, v_{gr}/γ, determines how large a wave can grow before it escapes. Waves with only a few growth lengths (say < 10) have little effect on the particle distribution, and their energy is negligible. In particular, we note that the z-mode having a group velocity of only the order of v_{te} may be more important than Figure 8.4 suggests, since the e-folding growth length is relatively small.

The z-mode is important for $\omega_p/\Omega_e \gtrsim 0.3$. Figure 8.9 shows the maximum growth rate at large ω_p/Ω_e in the case where the background component is cold and the energetic electrons have a Dory-Guest-Harris distribution with $j = 1$ (Eqs. 8.4.11 – 8.4.13 and $v_{th} = 0.1c$). With increasing temperature of the background, the harmonic peaks broaden and move to lower ω_p/Ω_e. The growth rate at a harmonic $l \gg 1$ tends to be a maximum when

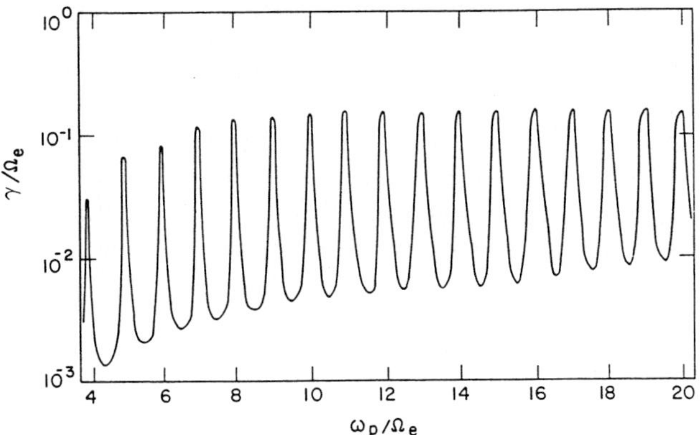

Fig. 8.9. Maximum growth rate of z-mode vs. ω_p/Ω_e. The background population is assumed to be cold (from Winglee and Dulk, 1986).

$$k_\perp v_\perp^m \approx \omega \approx l\Omega_e \quad , \tag{8.2.22}$$

where $v_\perp^m$ is the perpendicular velocity for which $\partial f/\partial v_\perp$ has a maximum. The first approximation is reminiscent of the bump-on-tail instability (Section 5.2.4), particularly in view of the second approximation indicating that the gyration period is much longer than the wave period. Within one wave period the electrons move approximately in linear orbits and thus can be considered unmagnetized.

At a given ω_p/Ω_e growth is possible for harmonics above the resonance of the x-mode at the upper hybrid frequency (Section 4.4),

$$l > l_0 := \omega_{uh}/\Omega_e \quad . \tag{8.2.23}$$

The upper hybrid frequency, $\omega_{uh}^2 = \omega_p^2 + \Omega_l^2$, has been introduced in Equation (4.4.4). Winglee and Dulk (1986) find that for a hot loss-cone distribution – coexisting with a background having a thermal velocity v_{tc} with $(v_{th}/v_{tc})^2 \gg \omega_{uh}/\Omega_e$ – growth at the harmonic closest to l_0 is favored. This seems to be the case for solar continuum bursts (Section 8.4), where the trapped particles have a temperature exceeding the background by 10^2 at $\omega_{uh}/\Omega_e \approx 20$. For $(v_{th}/v_{tc})^2 \lesssim \omega_{uh}/\Omega_e$, growth at more than one harmonic is possible. Electrostatic waves at electron cyclotron harmonics are also called *Bernstein modes* (cf. Fig. 4.3). They are restricted to nearly perpendicular propagation.

The whistler and z-mode waves seem to be of primary importance for interaction with trapped electrons in the solar corona and for the coronae of stars with similar or weaker magnetic fields. The observed non-integer harmonics of solar flare related spikes (Figure 8.3) are consistent with scattering of z-mode (Bernstein) waves into radio emission by $z_1 + z_2 \rightarrow t$ coalescence (Section 6.3.2). The ratio of adjacent modes at frequencies of maximum growth peaks at the observed value for a plasma having $1 \leq \omega_p/\Omega_e \leq 6$ (Willes and Robinson, 1996).

8.3. Precipitation of Trapped Electrons

The lifetime of a fast particle in a magnetic trap can be limited by many competing processes. In the Earth's radiation belts, the trapped electrons are eventually lost by pitch angle diffusion in velocity space. J.M. Cornwall, C.F. Kennel, and H.E. Petschek in the mid-1960s were able to show that a sufficiently populated loss-cone distribution of trapped particles generates whistler waves that drive the particles into the loss-cone. Diffusion in velocity space reduces the anisotropy in velocity space and fills the loss-cone. As the pitch angle is reduced, the altitude of the mirror point decreases, where the particle is reflected. When the pitch angle reaches the loss-cone angle, the particle orbit penetrates into the dense region where collisions thermalize the velocity.

Diffusion into the loss-cone may have several causes:

- Coulomb collisions
- Quasi-linear diffusion due to a loss-cone instability
- Turbulence of other origin

Assume that this diffusion takes a time τ_d. Once in the loss-cone, the particles reach either end of the magnetic trap within about a quarter bounce time and are lost. The maximum escape time for a particle in the loss-cone, τ_e, is at the top of the loop and for the largest pitch angle within the loss-cone, α_c. From Equations (2.2.4) and (2.2.5) one derives

$$\tau_e = \frac{\pi}{4}\frac{L\sqrt{M}}{v_z^{\rm top}} \quad , \tag{8.3.1}$$

where L is the length of the loop, M is the mirror ratio, $B_{\rm sep}/B_{\rm top}$, and $v_z^{\rm top} = v\ \cos\alpha_c$ is the velocity parallel to the magnetic field. In coronal and planetary magnetic loops, particle loss from a magnetic trap is called *precipitation*. The result is a flux of precipitating energetic particles at the footpoints, and this particle beam is what the thick target X-ray model assumes (Section 6.4).

8.3.1. Weak and Strong Diffusion

The lifetime of a particle is a combination of its diffusion time, τ_d, into the loss-cone plus the escape time, τ_e, along the field line. If the diffusion in velocity space is slow and controls the lifetime, diffusion is said to be *weak*. The particle distribution within the loss-cone is strongly reduced (an example can be seen in Figure 2.6). The lifetime, τ_l, can be defined as the total number of trapped particles on a field line divided by the rate they diffuse into the loss-cone. Kennel and Petschek (1966) have approximated it by

$$\tau_l(v) \approx \tau_d \ln[\frac{2}{\rm e}\sqrt{M}] \quad . \tag{8.3.2}$$

It is primarily controlled by the diffusion time, τ_d.

If diffusion is much faster than particles can escape out of the loss-cone (*strong diffusion*), the loss-cone is almost filled up, and the velocity distribution is close to isotropic. The lifetime of a particle in a trap then is given practically by the escape time (8.3.1), multiplied by the portion of the volume of the loss-cone in velocity space. Kennel and Petschek (1966) find

$$\tau_l \approx \tau_e M \quad . \tag{8.3.3}$$

The condition for strong diffusion is $\tau_d < \tau_e M$.

If diffusion is even stronger, such that $\tau_d < \frac{1}{2}L/v_z^{\rm top}$, the particles in the loss-cone do not escape freely but are scattered before they reach the end of the loop. The anisotropy of the trapped particles then is completely lost. In fact, no loss-cone can develop, and precipitation is a diffusion process in space. Such a situation occurs if the particles excite a high level of turbulence and are rapidly deflected by quasi-linear diffusion. The trap then is a 'leaky pail', and the mean lifetime is a statistical value given by *spatial* diffusion. We have encountered a similar case in the diffusive propagation of intense ion beams (Section 7.3). Stochastic acceleration (Sect. 10.4) may produce a situation of trapping by strong diffusion.

8.3.2. Diffusion Time

A. Collisions

Non-thermal particles have a deflection time that is similar to their energy loss time (Eq. 2.6.14 vs. 2.6.29). Energetic particles are deflected at the same rate as they lose energy.

In coronae, the trap is *inhomogeneous* in density and temperature. Collisions are most frequent near the mirror points – the lowest points of the trapped particle's orbit, where the density is highest and the temperature generally lowest. Deflections at the mirror point, however, have a unique property: they always lower the mirror point in whatever direction they occur. (Since the particle is temporarily in an orbit perpendicular to the magnetic field, any disturbance away from this plane decreases the pitch angle.) Therefore, deflections at the mirror points are not a random walk in velocity space as usual, but add linearly. For mirror points near the transition region, this directed mirror point motion can dominate the usual collisional diffusion and reduce the lifetime.

As particles of low velocity have smaller collision times, they are lost first. Collisional diffusion thus predicts that the average particle energy of the trapped population increases with time. This is not generally confirmed by hard X-ray observations. On the contrary, the spectral index of impulsive flares increases after the peak (Fig. 6.8), indicating a declining average particle energy.

B. Quasi-Linear Diffusion

As the loss-cone distribution of trapped particles can drive various instabilities (Section 8.1), wave-particle interactions are an important possibility for deflection. Similarly, wave turbulence produced by other sources – for example by shock waves – may also reduce the effective diffusion time. The coupling between wave growth and particle diffusion has already been studied in Section 6.2.1 on quasi-linear diffusion. The diffusion tensor, $\widehat{\mathcal{D}}$, is proportional to the energy density $W(\mathbf{k})$ of resonant waves (Eq. 6.2.12) and tends to reduce the slope in the particle distribution causing the waves. This derivative, in turn, controls the wave growth rate γ. The mutual coupling can be seen in the equations

$$\frac{\partial f(\mathbf{v})}{\partial t} = \frac{\partial f(\mathbf{k})}{\partial \mathbf{v}} \cdot \left(\widehat{\mathcal{D}}(W(\mathbf{k})) * \frac{\partial f(\mathbf{v})}{\partial \mathbf{v}} \right) \quad , \tag{8.3.4}$$

$$\frac{\partial W(\mathbf{k})}{\partial t} = 2\gamma(\mathbf{k})W(\mathbf{k}) \quad , \tag{8.3.5}$$

valid for an infinite homogeneous source. Typical solutions to Equations (8.3.4) and (8.3.5) are (*i*) the linear phase, when the wave energy is low and growing exponentially, (*ii*) the stationary phase (marginal stability or equilibrium), and (*iii*) periodic deviations from equilibrium called *relaxational oscillations*, when diffusion and wave growth oscillate out of phase around the equilibrium values. They will be discussed in the following section.

8.3.3. Equilibrium of Quasi-Linear Diffusion

Assume that the distribution of energetic particles is initially isotropic. If diffusion is weak, the loss-cone is depleted within the escape time, τ_e. This time is the formation time of the loss-cone. It is at least a few tens of microseconds and may reach several seconds in coronal traps. The growth times calculated from Figure 8.4, however, are less than a microsecond ($B \approx 100$ G). After a few growth times, the wave energy becomes large enough to cause quasi-linear diffusion in velocity space. The diffusion time is usually of the same order as the linear growth time. Therefore, the build-up of the loss-cone only proceeds until an instability sets in, and a balance between build-up and velocity diffusion is established. The distribution remains near the threshold for instability in a so called *marginally stable* state.

If an equilibrium exists, it also must include the waves. The growth of wave energy (Eq. 8.3.5) must be balanced by the energy of escaping waves (assuming a finite source) or by damping. Thus the equilibrium growth rate, γ_e, must satisfy

$$\frac{\partial \omega}{\partial \mathbf{k}} \frac{\partial W(\mathbf{k})}{\partial \mathbf{x}} = 2[\gamma_e(\mathbf{k}) - \gamma_d(\mathbf{k})]W(\mathbf{k}) \quad . \tag{8.3.6}$$

The damping rate, γ_d, includes all wave energy losses within the source as well as non-linear saturation effects. Using the the group velocity, $v_{gr} = \partial\omega/\partial k$, and $\partial W/\partial x \approx 2W/L$,

$$\gamma_e(\mathbf{k}) - \gamma_d(\mathbf{k}) \approx \frac{v_{gr}(\mathbf{k})}{L(\mathbf{k})} \quad . \tag{8.3.7}$$

The integration length, $L(\mathbf{k})$, over which a particular wave with a given **k**-vector can grow, is simply the diameter for a homogeneous source. It may be much smaller for an inhomogeneous source where the most unstable frequency changes with distance, or where the inclination angle to the magnetic field varies. The interaction length then is the distance over which the source parameters change.

The equilibrium growth rate given by Equation (8.3.7) determines the distribution function (e.g. Eq. 8.2.10). The distribution function, in turn, is shaped by diffusion (Eq. 8.3.4). Finally, the diffusion coefficient, $\widehat{\mathcal{D}}$, is connected to the wave energy density (Eq. 6.2.12).

We must note here that the waves cannot grow if the left side of Equation (8.3.6) is smaller than the right side. If it becomes larger, the wave energy accumulates, enhances particle precipitation and thereby reduces the growth rate to the steady value (or oscillates, Section 8.4.3).

If acceleration (or injection) provides continuously new particles, the equilibrium is permanent. It is not sensitive to temporary changes in acceleration, since the particle distribution is a universal function. If acceleration increases, precipitation balances it so that the growth rate, satisfying Equation (8.3.6), remains constant. The precipitation rate and the escaping wave energy will, however, be strongly correlated with acceleration.

8.3.4. Dominant waves

The question of the dominant mode for instability is related to the lowest threshold. For collisionless waves to grow, the growth rate must exceed the thermal collision rate. The wave first satisfying Equation (8.3.6) will establish an equilibrium and stop further evolution. The relevant factor is $(\gamma - \gamma_d)L/v_{gr}$. The bigger it is, the more important the instability.

The dominant wave mode depends on various parameters and can only be determined by detailed calculations. As general rules, we may identify the following regimes:

- The x-mode (maser) may play a role at extremely low ω_p/Ω_e ratios (Fig. 8.4).
- The z-mode tends to dominate at low ω_p/Ω_e values, since its group velocity is much smaller than the x and o modes.
- The whistler mode is important at large ω_p/Ω_e. However, it is overemphasized by the chosen velocity distribution in Figure 8.4 and may involve only the high energy tail of the distribution.

8.4. Observations of Trapped Electrons

Particle trapping must be regarded in view of the relative durations of particle injection (or resupply), trapping, and escape or precipitation. The interpretation of the observations in this scheme, of course, is tentative.

8.4.1. Injection Dominated

If injection is much longer than trapping and precipitation, the signature of trapping may be difficult to observe. This seems to be the case for electrons observed in hard X-rays during impulsive solar flares. The spectrum already changes significantly during the rise phase (Fig. 6.8, left) and continues to change after the maximum with about the same time scale.

Another example of an injection-dominated phenomenon may be narrowband spikes (Section 8.1.3). The build-up time of a loss-cone distribution in conventional coronal loops is comparable or longer than the observed spike duration. Nevertheless, the observation of high harmonics suggests that the emission is a velocity-space anisotropy with a predominantly perpendicular population of electrons leading to a gyroresonance instability.

8.4.2. Trapping and Resupply

Gradual hard X-ray and microwave events appear different from the impulsive bursts; often only one source is visible and located at high altitude (limb observations suggest about 40 000 km). The location and spectrum are suggestive of trapped electrons emitting hard X-rays in a thin target with possibly weak diffusion.

Let us look at the evolution of a set of electrons after their acceleration has stopped. First assume that collisions control the diffusion in velocity space. In the case of *weak diffusion*, the slow particles are deflected and lost first. Thus, the trapped population contains an increasing fraction of high energy particles, and so does the flux of precipitating particles. Therefore, the hard X-ray spectrum would become flatter in time (the technical term *harder* is generally used). This is usually observed (Fig. 6.8, right). A spectrum that *softens* could still be explained under the assumption of weak diffusion, but requires a different diffusion process, preferentially scattering high velocity electrons, such as by whistler waves.

For *strong diffusion*, the precipitation rate is proportional to the escape velocity, and is independent of the diffusion process. We then expect the X-ray spectrum to soften with time, as the fastest particles are lost first. The time scale is dominated by precipitation and is relatively short.

Hard X-ray observations of gradual flares are compatible with Coulomb collisions and weak diffusion. However, acceleration seems to operate throughout the event, and it is not certain that the observed spectral hardening is the result of the longer trapping time with increasing electron energy or of a time dependence of the acceleration process.

Assuming only Coulomb interactions, one finds (Exercise 8.4) that no loss-cone can form at low energies, since the electrons are lost before reaching the end of the loop. At intermediate energies collisional diffusion is strong, and at high energies it is weak. For low solar loops (having for example a length of some 10^4km and a density of 10^{10}cm^{-3}), the three ranges are typically separated at 10 keV and 100 keV, respectively. High coronal loops, 10 times longer and less dense, have the separating energies an order of magnitude smaller. Loss-cone instabilities (in particular whistler waves) can, however, move these ranges to higher energies.

The gradual hard X-ray and microwave emissions of trapped and precipitating electrons are often associated with broadband continua at radio frequencies $\lesssim$ 3 GHz, indicating that other interactions than Coulomb ones are also at work. They form the class of *type IV bursts*, first identified in 1957 by A. Boischot and J.F. Denisse. The broad bandwidth and long duration makes them spectacular in terms of total energy radiated. It is not surprising that these emission processes have also been proposed for stellar radio flares (Section 8.4.4). They have even found applications in radio outbursts of binary stars hundreds of parsecs away.

The solar observations show a large variety of bursts reflecting the many possibilities in altitude, diffusion, instability, and acceleration process. This has led to a confusing number of subclasses. The reader is referred to the end of the chapter for published reviews of the observations. The three major groups are:

A. Moving metric type IV bursts

B. Stationary metric type IV bursts

C. Decimetric type IV bursts

Early interpretations in terms of gyrosynchrotron emission meet the difficulty of the observed bandwidth, which is broad for a coherent emission, but often not enough for the synchrotron process. All type IV emissions are consistent, however, with coherent radiation of loss-cone instabilities. Wherever this can be proved, it is evidence for particle trapping.

A. *Moving Type IV Bursts*

Gradual hard X-ray and microwave bursts are usually accompanied by a meter wave continuum ($\lesssim$ 300 MHz). It lasts between 10 min and >2 h, and is observed to drift slowly to higher altitude and lower frequency. The ratio of maximum to minimum frequency is typically 2 to 4, narrower than for gyrosynchrotron emission. The spectral index at frequencies $\nu > \nu_{\max}$ can reach up to 10, and the observed brightness temperature sometimes exceeds 10^{10}K. There is no doubt that the emission is caused by a coherent process, presumably by electrons trapped in far extending coronal loops. These particles will emit gyrosynchrotron radiation as well, but it seems to be outshone by coherent emission.

The metric source is observed by interferometers to move away from the Sun at roughly constant speed between 200 and 1600 km s^{-1}. The source speed is much smaller than even the thermal electron velocity, implying many bounces and trapping of the non-thermal electrons. The source has been followed out to more than five times the photospheric radius in some cases. Figure 8.10 shows such a moving type IV burst observed with the Culgoora radioheliograph at a constant frequency of 80 MHz. It consists of three sources, typical of many moving type IV bursts. Two – apparently located near the footpoints of the expanding structure – are polarized in the ordinary mode. A third source at the top of the loop is usually weakly polarized. The size of the sources is a large fraction of a solar (photospheric) radius.

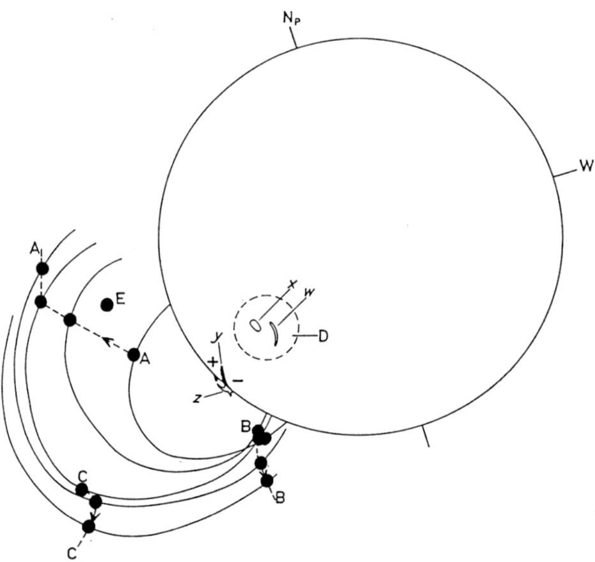

Fig. 8.10. A moving metric type IV burst is represented by the centroid positions of its three components (labelled A, B, and C) at different times separated by a few minutes. D and E mark the source sites of a stationary metric type IV burst appearing at a later time. The letters w and x refer to Hα flares, y and z to a dark filament and an active prominence, respectively, which erupted and reformed during the outburst. The plus and minus signs indicate magnetic polarity. The curves connecting the radio sources are drawn to illustrate the evolution of the expanding magnetic loop, which appears to contain the three moving sources (from Wild, 1969).

Moving type IV bursts are frequently preceded by a faint expanding arch structure seen by white light coronagraphs to move out many solar radii. Below the type IV source, the ejection of filament material may be observed in Hα. Loops with a typical height of $3 \cdot 10^4$ km fill up by hot material evaporating at the footpoints and radiate thermal soft X-rays.

Figure 8.11 attempts to organize the various phenomena in a scenario, which – if not realistic – has now survived for several years. Note that many elements are only tentative. Energy release and acceleration is proposed to occur at an altitude

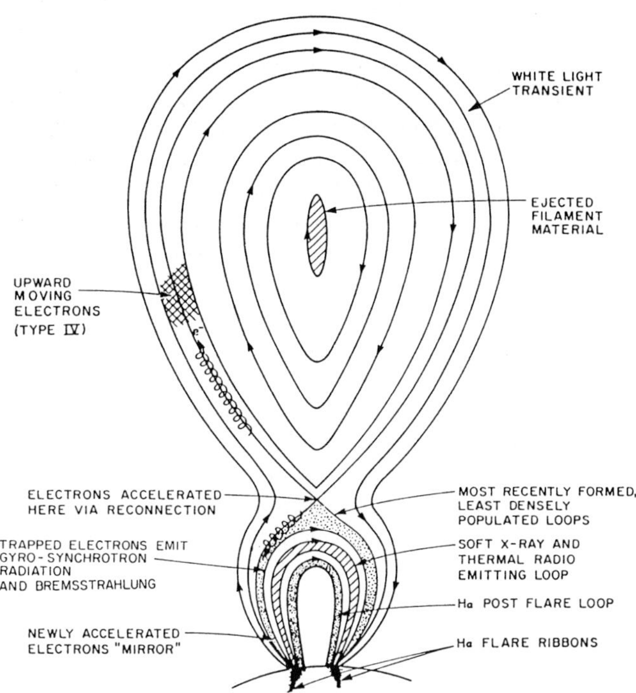

Fig. 8.11. Suggested scenario in which electrons accelerated by reconnection are trapped, and emit gradual hard X-ray and microwave bursts at lower altitude. The geometry also indicates how a concomitant moving type IV event might occur. The drawing is not to scale (after Cliver *et al.*, 1986).

of some 10^4km between the metric type IV and hard X-ray sources. Gradual bursts are associated with eruptive prominences and moving type IV bursts. For this reason, acceleration by reconnection is suggested. In the reconnection process (Section 9.1.2), magnetic field lines are re-ordered and shortened. The changes release magnetic energy and cause acceleration.

The motion of type IV sources strongly suggests radiation by particles trapped in an expanding magnetic arch (Fig. 8.10). Periods of reduced emission have been observed in moving type IV bursts near the onset of secondary peaks of the gradual hard X-ray bursts. One expects an anti-correlation of type IV and X-ray emission from an injection of new particles, temporarily filling up the loss-cone, if the type IV radiation is caused by a loss-cone instability.

At the altitude of moving type IV bursts, the solar corona generally satisfies $\omega_p \gg \Omega_e$. The numerical results presented in Section 8.2 then predict whistlers and z-mode waves as most likely waves for loss-cone instability. The z-mode waves are emitted in a cone with opening angle given by Equation (8.2.20). Two z-waves can scatter to emit radiation ($z + z' \rightarrow t$) at the harmonic frequency,

$$\omega \approx 2 l_1 \Omega_e \quad , \tag{8.4.1}$$

where l_1 is the lowest harmonic number above ω_{uh}/Ω_e (Eq. 8.2.23). The photon temperature at saturation – $T_{\max} \approx T_z$, derived in Equation (6.3.19) – could easily explain the observed high brightness temperature.

The z-mode emission is relatively narrow in bandwidth as the wave frequency is in the range $\omega_p - \frac{1}{2}\Omega_e \lesssim \omega \lesssim \omega_p$. Inhomogeneity in the source, however, broadens the bandwidth of the observed burst in realistic sources. In general, the total bandwidth of emissions by trapped particles is therefore wide. It is not surprising that the location of moving type IV at a given time depends on frequency; the lower the frequency, the higher the sources.

Note that the lifetime of trapped electrons is generally shorter than the duration of a moving type IV burst. At 80 MHz (suggesting a plasma frequency of 40 MHz, Eq. 8.4.1), a 30 keV electron is deflected by collisions within 800 s (Eq. 2.6.20), and its lifetime, according to Equation (8.3.2) for weak diffusion, would be about twice as much. Assuming the loss-cone distribution to be unstable (as required for coherent radio emission), the effective lifetime will be even shorter. The trap needs a continuous supply of energetic electrons to produce emissions over the observed burst durations.

B. *Stationary Metric Type IV Bursts*

Late in large outbursts and usually after the moving type IV burst, a stationary, broadband continuum is sometimes observed for several hours in solar metric radio waves. In Figure 8.10 it was straight above the Hα flare site and labelled D. A weak, oppositely polarized and apparently related source was at E. These sources are stationary and not directly connected to the moving type IV sources. The degree of circular polarization is usually high.

The observational evidence again points to emission by a loss-cone distribution of electrons. The origin of these particles, however, is not clear and seems to require continuous energy release for hours after the flare. Stationary type IV bursts have been observed to turn gradually into noise storms (to be discussed in Section 9.5). Noise storms are interpreted as signatures of substantial magnetic evolution in the corona. If the two phenomena are related, stationary type IV bursts may be considered as the manifestation of magnetic readjustment after the expansion of a coronal arch.

C. *Decimetric Bursts*

For historical reasons related to instrumental developments, solar radio bursts in the decimeter range are the least investigated and least understood of all. This is very unfortunate since the plasma frequencies and electron gyrofrequencies at the site of primary flare energy release are expected to be larger than 300 MHz and possibly in the decimeter range. In addition, the radiated energy of decimetric type IV bursts frequently exceeds their metric counterparts.

Decimetric type IV bursts are continuum events of usually opposite polarization to the gyrosynchrotron emission at higher microwave frequencies. For this reason, they are likely to be o-mode emissions. The total duration sometimes exceeds one hour, but the events are usually modulated with a time constant of typically 20–60 s. Assuming only Coulomb interactions, the deflection time of 30 keV electrons in a plasma with $\nu_p = 400$ MHz (presumably emitting harmonic radiation at about 800 MHz) is only 8 s. For the expected trap parameters, the particle's lifetime is of the same order. Therefore, long duration decimetric type IV bursts require a continuous acceleration process to resupply the trap.

Decimetric type IV bursts occur after the impulsive phase, but are much more irregular than the gradual emissions at meter waves, in hard X-rays and microwave synchrotron radiation. A wide variety of fine structures can be superimposed on the continuum, like broadband absorptions, fast modulations over the whole band, parallel drifting bands of emission, intermediate drift bursts, etc. (catalogued by Bernold, 1980). The tentative interpretation of these structures is generally based on loss-cone emission processes.

8.4.3. Depletion Dominated

Much more frequent than decimetric type IV bursts are *quasi-periodic pulsations*, series of intense, broadband pulses with high drift rate (Fig. 5.1d). They are so abundant that one can also find the name 'decimeter-type' in the old literature. The labels 'type IV pulsations' and 'fast-drift bursts' have been abandoned to avoid confusion. A similar emission for a dMe star is presented in Figure 8.14. Very regular solar pulsations tend to occur after the impulsive flare phase, and the associated hard X-ray flux is often weak. The high ratio of radio to X-ray emissions and the high degree of order of the whole event are suggestive of trapped particles. For this reason they are most likely signatures of trapped electrons that have been impulsively injected. Pulsations seem to occur in the depletion phase, when the distribution of trapped electrons has settled down to a stationary state.

Pulsations are usually polarized in x-mode according to the leading spot rule (Section 6.3.4). They have been interpreted as originating from an unstable loss-cone distribution in a source with high magnetic field ($\omega_p \approx \Omega_e$). Then, x-mode maser emission at the second harmonic or z-mode emission may be dominant (Fig. 8.5). The case for the emission process of pulsations is not yet settled. Nevertheless, they are likely candidates for weak diffusion in self-generated waves. It gives the opportunity to discuss the self-consistent equilibrium between particles and waves, assuming the trapping time to be much longer than the wave growth time.

There are two competing explanations for modulation in pulsations: MHD oscillations in the trap and oscillations in the quasi-equilibrium of the loss-cone instability. The former process is unrelated to the emission mechanism. Consider particles trapped in a high density loop. Fast (magnetoacoustic) waves are also trapped in the loop due to the lower Alfvén velocity than the ambient medium.

The two modes of standing oscillations of fast-mode waves are depicted in Figure 8.12. In the fast kink, the loop oscillates up and down with a period

$$\tau_k = \frac{2L}{j\, c_k} \quad , \tag{8.4.2}$$

where L is the loop length, j the number of nodes, and c_k a mean Alfvén velocity given by

$$c_k := \left(\frac{\rho_i\, c_{Ai}^2 + \rho_e\, c_{Ae}^2}{\rho_i + \rho_e} \right)^{1/2} \quad . \tag{8.4.3}$$

The indices i and e refer to internal and external to the loop, respectively, ρ is the mass density. Since only the modes with low j numbers are easy to excite, the periods given by Equation (8.4.2) are too long to explain the observations.

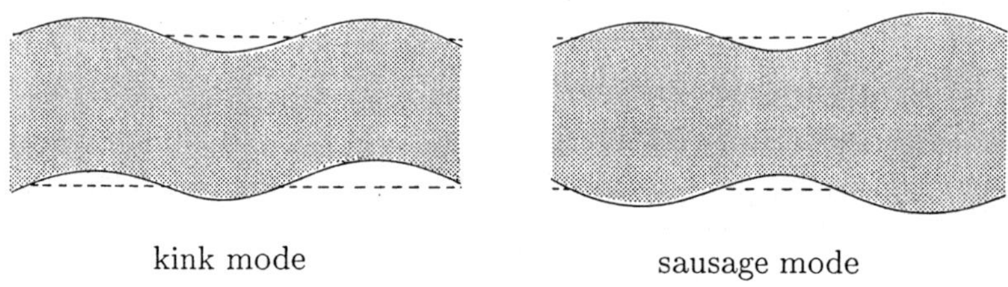

Fig. 8.12. Fast mode oscillations of a dense flux tube.

Another possibility is the fast sausage mode (Fig. 8.12 right), where the loop oscillates in thickness. Its period is fastest as it involves the shortest scale and highest signal speed. It amounts to

$$\tau_s = \frac{2\pi\ell}{c_k} \tag{8.4.4}$$

for a loop diameter ℓ. This period is of order 1 s and compatible with the regular pulsations observed at low decimetric and metric frequencies. The modulation of density and magnetic field may change essential source parameters, modifying in turn the emission process. The details of the theory and more references can be found in Roberts *et al.* (1984).

The alternative interpretation is based on oscillating disturbances of the equilibrium between loss-cone instability (wave growth) and diffusion. Let the perturbation of the velocity distribution and wave energy be given in the form

$$f(\mathbf{v}, t) = f_0(\mathbf{v}) + f_1(\mathbf{v})h(t) \quad , \tag{8.4.5}$$

$$W(\mathbf{k}, t) = W_0(\mathbf{k}) + W_1(\mathbf{k})g(t) \quad . \tag{8.4.6}$$

Equations (8.3.4) and (8.3.5) can then be approximated by equations for the equilibrium (subscript 0) and the temporal parts of the perturbations,

$$\frac{\partial h}{\partial t} = a_{11}g + a_{12}h \quad , \tag{8.4.7}$$

$$\frac{\partial g}{\partial t} = a_{21}g + a_{22}h \quad , \tag{8.4.8}$$

on neglecting quadratic terms. The coefficients have been evaluated by Aschwanden and Benz (1988). The terms a_{11} and a_{12} are proportional to the quasi-linear diffusion rate, $1/\tau_d$, $a_{21} = 0$, and a_{22} is proportional to the growth rate, γ, of the loss-cone driven waves. Equations (8.4.7) and (8.4.8) are linear and form a system of coupled equations of the well-known type first studied by A.Lotka and V.Volterra in the 1920s. If $a_{12} \approx 0$, it has oscillating solutions with a frequency

$$\omega_{ro} \approx \sqrt{a_{11}a_{22}} \approx \sqrt{\frac{2\gamma}{\tau_d}} \quad . \tag{8.4.9}$$

The periodic perturbations of the equilibrium are called *relaxational oscillations*. Elaborate numerical work by Aschwanden (1990) has shown that $\tau_d \approx 10{-}40 \cdot 1/\gamma$. Since γ must exceed the thermal collision frequency for the waves to grow, the relaxational periods must be $\tau_{ro} \lesssim 20\, t_{e,i}$, where $t_{e,i}$ is the electron-ion collision time (Eq. 2.6.32). Relaxational oscillations have short periods and could explain the rapid, irregular pulsations at decimetric frequency. They are a natural element of any self-consistent equilibrium between unstable particle distributions and waves.

8.4.4. Stellar Emissions by Trapped Electrons

Emissions from stellar coronae have also received attention. Yet, how can we identify the stellar processes if we do not fully understand the solar counterparts? Some hope comes from more extreme conditions in the relevant stellar coronae, such as higher magnetic fields or highly relativistic particles. Most of the non-thermal radio emission of stars probably originates from trapped electrons. Gyrosynchrotron radiation is the generally accepted explanation for the quiescent radio emission of early-type 'magnetic' stars (Ap, Herbig Ae, Bp, etc.), binary systems (RS CVn, cataclysmic variables); it is also suggested for active late-type F through M-dwarfs and gradual flares of low polarization in these objects. Coherent emission is generally proposed for highly polarized, impulsive flares.

Here we concentrate on M-dwarfs (dM), an abundant population of small stars $(0.1 - 0.5 M_\odot)$. About 80% of the stars of our galaxy and many nearby stars are of this type. Many of them have optical flares and are called *flare stars*. The majority (85%) shows chromospheric emission lines indicating magnetic activity. They are called dMe stars and are practically identical with the population of flare stars.

A. *Quiescent Radio Emission*

The more active flare stars have a *quiescent* radio emission (Section 1.3.3). By quiescent we mean a component that is always present and varies over time scales longer than one hour. It has a relatively flat spectrum in microwaves and usually has a low degree of polarization. Measurements with very long baseline interferometry give source sizes less than two stellar diameters and brightness temperatures in excess of 10^9 K. The radiation is clearly not a thermal emission of the X-ray emitting coronal plasma. It is generally interpreted as optically thin gyrosynchrotron or synchrotron emission of supra-thermal electrons. The radiation is over a thousand times more intense than the quiet solar radio emission. It has no obvious counterpart on the Sun.

The flares seen on these stars are hardly frequent enough to supply sufficient energetic electrons. It would be puzzling if these particles could be trapped much longer than on the Sun. The scale height (Eq. 3.1.54) is larger than in the solar corona suggesting a homogeneous density in the source region of quiescent emission. Using the observed densities, the lifetime of weak or strong collisional diffusion (Eqs. 8.3.2 and 8.3.3) is much shorter than the interval between flares. The observations suggest either traps at high altitudes with particle densities many orders of magnitude below the X-ray emitting plasma, or quasi-continuous acceleration in flare-like processes.

These small M stars with radii, R_*, between 0.15 to 0.5 $R_\odot$ have strikingly vigorous coronae, as described in Section 1.3.1. Their soft X-ray luminosity exceeds the quiet solar value of about 10^{27} erg s^{-1} by factors up to 1000. As an optically thin bremsstrahlung, it is proportional to the emission measure, $\int n_e^2\, dV$. The observations therefore indicate a much denser corona than on the Sun (average values of several 10^9 cm^{-3} have been suggested). The soft X-ray spectrum is best described by a range of temperatures between 3 and $30 \cdot 10^6$ K. The upper limit, $T_{\rm max}$, suggests that the ratio of kinetic energy to gravitational energy,

$$\frac{\epsilon_{\rm kin}}{\epsilon_{\rm pot}} = \frac{k_B T_{\rm max} R_*}{G M_* m_p} \quad , \tag{8.4.10}$$

frequently exceeds unity. Therefore, a flare star corona is not gravitationally bound to the star. The Zeeman effect in photospheric lines suggests surface fields of several kilogauss covering a large fraction of the star. The occasional observation of a millimetric component ($\nu \gtrsim 10$ GHz) may be interpreted as thermal gyroresonance emission of the X-ray emitting plasma and suggests a ratio of Ω_e/ω_p between 1 and 10, compatible with powerful maser emission in flares at low harmonics or the fundamental of the electron gyrofrequency (next subsection). The magnetic pressure thus dominates the plasma by the factor $1/\beta$ between about 10^5 to 10^2, where β was given by Equation (Eq. 3.1.51). As in solar active regions, the coronal plasma seems to be confined by magnetic forces.

The remarkable coexistence of a corona with a quiescent population of relativistic electrons has already been noted in Section 1.3.3. The radio luminosity of dMe stars can exceed 10^{23}erg s^{-1} outside of flares. The efficiency of gyrosynchrotron emission by solar flare electrons is about 10^{-7} in terms of radiated energy per kinetic particle energy. Applying this value, the input by the non-thermal electrons would exceed the stellar coronal output in thermal soft X-rays. It suggests a better trapping or higher magnetic fields in the stellar case (increasing the gyrosynchrotron efficiency) and a significant role of non-thermal electrons and their acceleration process in coronal heating.

B. *Stellar Flares*

Discovered optically in 1924 by E. Hertzsprung, flares on dMe stars today cover a vast literature. The observers agree that they appear to be qualitatively similar to the Sun, but release 2 – 3 orders of magnitude more energy than the largest solar flares. In addition, they occur so frequently on some stars (such as AD Leonis) that observers can afford to wait for a flare even with large telescopes or satellites with limited access time. As one might expect, the variety of stellar flares is at least as wide as for the Sun. We limit ourselves to the phenomena of trapped particles.

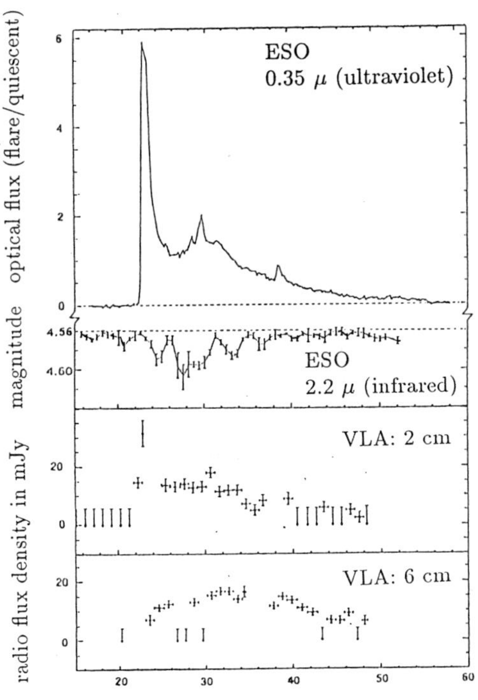

Fig. 8.13. A flare of AD Leo observed at many wavelengths (after M. Rodonò *et al.*, 1989).

Figure 8.13 presents a stellar flare observed at many wavelengths. The top part shows optical continuum emission. It dominates the stellar spectrum, contrary to solar flares, where spectral lines are more important and white light flares are feeble. An impulsive initial peak and a gradual phase are clearly distinguishable. The high-frequency microwave emission (2 cm or 15 GHz) correlates well. A likely interpretation is gyrosynchrotron emission. Its spectral peak occurs at high frequency due to the high magnetic field in the source (Eq. 8.1.3, Exercise 8.5). The 6 cm (5 GHz) emission does not correlate well and may originate from a coherent process. Its peak is delayed by many minutes compatible with a source of trapped particles and similar to solar type IV bursts. The emission and absorption processes of the ultraviolet and infrared flare radiations (0.35 and 2.2 μ, respectively) are not well understood, but seem to be related to incoherent and probably thermal emissions.

The impulsive radio emissions are often, but not always, characterized by complete circular polarization, short time scale and little correlation with optical or soft X-ray emissions. An example is shown in Figure 8.14, presenting a case of quasi-periodic pulsations with a pulse separation of about 0.7 s. The highest intensity of this kind of bursts was measured on the same star with 500 mJy. It is a remarkable factor of 10^4 higher than the strongest solar events at the same frequency and distance.

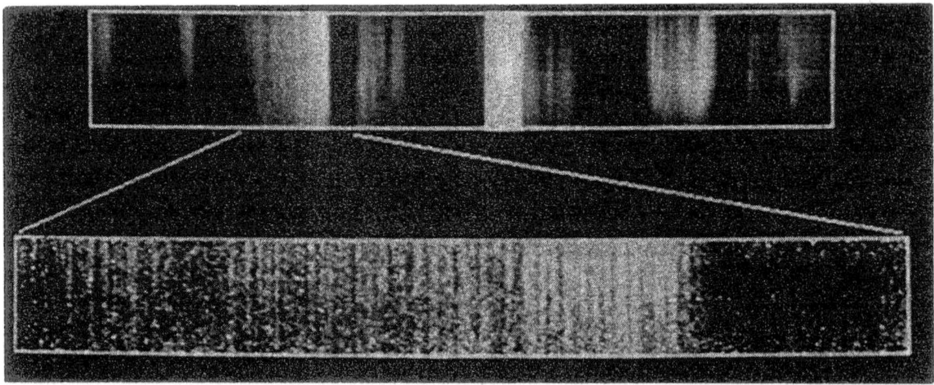

Fig. 8.14. Dynamic spectrum of decimetric radio bursts on AD Leo observed at Arecibo. Both vertical axes represent frequency increasing from 1395 to 1435 MHz. *Top*: 7 minutes of data, *bottom*: enlargement of 50 s. The bright vertical bar in the top panel represents a calibration signal (from Bastian, 1990).

The analogy of the pulsations in Figures 8.14 and 5.1d is striking and calls for an interpretation with the same physics, but other parameters. If the emission process is cyclotron masering of trapped electrons, evoked for the solar counterpart, the radiation frequency is the electron gyrofrequency. The observed frequencies then allow us to derive the coronal magnetic field, about 500 G or 250 G for the case presented in Figure 8.14, assuming fundamental or second harmonic emission, respectively.

More difficult to explain is the 10^5 times higher flux density of the stellar burst. The flux of a maser source is given by the number of e-folding lengths over which the radiation can grow. Since the stellar source is probably not much bigger, the growth rate must exceed the solar value. This may be expected from a high density of trapped particles. Since relaxational oscillations are proportional to $1/\gamma$, but the observed modulations have similar periods, the alternative process, MHD oscillations, seems to be the more likely explanation.

Exercises

8.1: The velocity distribution of magnetically trapped particles can for example be approximated by the Dory-Guest-Harris distribution

$$f(v_z, v_\perp) = n f_z(v_z) f_\perp(v_\perp) \quad , \tag{8.4.11}$$

where

$$f_z(v_z) = \frac{1}{\sqrt{2\pi} v_t} \exp(-\frac{v_z^2}{2v_t^2}) \quad , \tag{8.4.12}$$

$$f_\perp(v_\perp) = \frac{1}{2\pi v_t^2 \, j! \, 2^j} (\frac{v_\perp}{v_t})^{2j} \exp(-\frac{v_\perp^2}{2v_t^2}) \quad . \tag{8.4.13}$$

The number j represents the steepness of the distribution function. Calculate the growth rate of whistler waves for this distribution.

8.2: The resonance condition for gyromagnetic interaction imposes a relation between particle velocity and wave vector. Prove that the resonance curve (Eq. 8.2.15) in the semi-relativistic approximation is a circle and derive its elements (Eqs. 8.2.17 and 8.2.18).

8.3: High-frequency electromagnetic waves grow in populations of trapped particles with loss-cone velocity distributions. The growth rate is given by an integral in velocity space along the curve of resonance, approximately a circle. The resonance conditions relate the center of this circle to the propagation angle θ. Show that the angle $\theta_{\max}$ of the fastest growing wave is given by Equation (8.2.19). Use $\Delta v_c \approx v_r$ to estimate the cone of emission $\Delta\theta$, proving Equation (8.2.20) for $v_{\max} \ll c$. Approximate the frequency range of unstable waves and prove Equation (8.2.21).

8.4: Consider particles injected at the top of a loop and calculate the lifetime of trapped electrons vs. energy. Let the trap be homogeneous with $n_e = 10^{10}$ cm^{-3}, $T_e = 10^7$ K, a loop length $L = 2 \cdot 10^9$ cm, a mirror ratio $M = 10$, and assume that only Coulomb collisions cause diffusion in velocity space.

8.5: At what frequency would the gyrosynchrotron emission peak if a flare occurred on the flare star AD Leonis (distance: 4.9 pc) with a source diameter of 10^9 cm, a fast electron energy index $\delta = 3.5$, a density $n = 10^9$ cm^{-3} and a source field strength of 10^3 G? Put $\sin\theta = 1/2$. What would the approximate flux density of this gyrosynchrotron emission be at this frequency for an observer on Earth?

Further Reading and References

Loss-cone instabilities

Kennel, C.F. and Petschek, H.E.: 1966, 'Limit of Stably Trapped Particle Fluxes', *J. Geophys. Res.* **71**, 1.

Melrose, D.B.: 1980, *Plasma Astrophysics: Non-thermal Processes in Diffuse Magnetized Plasmas*, Vol. 1, Gordon and Breach, New York.

Melrose, D.B.: 1986, *Instabilities in Space and Laboratory Plasmas*, Cambridge University Press, Cambridge, UK.

Vlahos, L.: 1987, 'Electron Cyclotron Maser Emission for Solar Flares', *Solar Phys.* **111**, 155.

Willes, A.J. and Robinson, P.A.: 1996, 'Electron-Cyclotron Maser Theory for Non-integer Ratio Emission Frequencies in solar Microwave Spike Bursts', *Astrophys. J.* **467**, 465.

Winglee, R.M. and Dulk, G.A.: 1986, 'The Electron-Cyclotron Maser Instability as a Source of Plasma Radiation', *Astrophys. J.* **307**, 808.

Wu, C.S. and Lee, L.C.: 1979, 'A theory of the Terrestrial Kilometric Radiation', *Astrophys. J.* **230**, 621.

Observations of trapped particles (Sun)

Bastian, T.S., Benz, A.O., and Gary, D.E.: 1998, 'Radio Emission from Solar Flares', *Annu. Rev. Astron. Astrophys.* **36**, 131.

Benz, A.O.: 1986, 'Millisecond Radio Spikes' *Solar Phys.* **104**.

Dulk, G.A. and Marsh, K.A.: 1982, 'Simplified Expression for the Gyrosynchrotron Radiation from Mildly Relativistic, Non-Thermal and Thermal Electrons', *Astrophys. J.* **259**, 350.

Stewart, R.T.: 1985, 'Moving Type IV Bursts', in *Solar Radiophysics* (ed. D.J. McLean and N.R. Labrum), Cambridge University Press, p. 361.

Emissions of trapped particles (stars)

Bastian, T.S.: 1990, 'Radio Emission from Flare Stars', *Solar Phys.* **130**, 265.

Dulk, G.A..: 1985, 'Radio Emission from the Sun and Stars', *Ann. Rev. Astron. Astrophys.* **23**, 169.

Güdel, M.: 2002, 'Stellar Radio Astronomy: Probing Stellar Atmospheres from Protostars to Giants', *Annu. Rev. Astron. Astrophys.*, in press.

Kuijpers, J.: 1989, 'Radio Emission from Stellar Flares', *Solar Phys.* **121**, 163

Klein, K.-L. and Chiuderi-Drago, F.: 1987, 'Radio Outbursts in HR 1099: A Quantitative Analysis of Flux Spectrum and Intensity Distribution', *Astron. Astrophys.* **175**, 179.

Lang, K.: 1993, 'Radio Observations of Particle Acceleration on Stars of Late Spectral Type', *Astrophys. J. Suppl.* **90**, 753.

Pacholczyk, A.G.: 1970, *Radio Astrophysics*, Chapters 3 and 4 on synchrotron emission, W.H. Freeman and Co., San Francisco.

White, S.M., Kundu, M.R., and Jackson, P.D.: 1989, 'Simple Non-thermal Models for the Quiescent Radio Emission of dMe Flare Stars', *Astron. Astrophys.* **225**, 112.

References

Aschwanden, M.J. and Benz, A.O.: 1988, 'On the Electron-Cyclotron Maser Instability: I. Quasi-linear Diffusion into the Loss-Cone', *Astrophys. J.* **332**, 447.

Bernold, T.: 1980, 'A Catalogue of Fine Structures in Type IV Solar Radio Bursts', *Astron. Astrophys. Suppl.* **42**, 63.

Cliver, E.W., Dennis, B.R., Kiplinger, A.L., Kane, S.R., Neidig, D.F., Sheeley, N.R.Jr., and Koomen, M.J.: 1986, 'Solar Gradual Hard X-Ray Bursts and Associated Phenomena', *Astrophys. J.* **305**, 920.

Dulk, G.A., Bastian, T.S., and Kane, S.R.: 1986, 'Two Frequency Imaging of Microwave Impulsive Flares near the Solar Limb', *Astrophys. J.* **300**, 438.

Krucker, S. and Benz, A.O.: 1994, 'The Frequency Ratio of Bands of Microwave Spikes During Solar Flares', *Astron. Astrophys.* **285**, 1038.

Roberts, B., Edwin, P.M., and Benz, A.O.: 1984, 'On Coronal Oscillations', *Astrophys. J.* **279**, 857.

Rodonò, M., Houdebine, E.R., Catalano, S., Foing, B., Butler, C.J., Scaltriti, F., Cutispoto, G., Gary, D.E., Gibson, D.M., and Haisch, B.M.: 1989, 'Simultaneous Multi-Wavelength Observations of an Intense Flare on AD Leonis', B.M. Haisch and M. Rodonò (eds.), *Proc. IAU Colloq. 104 – Poster Papers*, Reprints Catania Obs., p. 53.

Sharma, R.R. and Vlahos, L.: 1984, 'Comparative Study of the Loss-Cone Driven Instabilities in the Low Solar Corona', *Astrophys. J.* **280**, 405.

Takakura, T., Kundu, M.R., McConnell, D., and Ohki, K.:1985, 'Simultaneous Observations of Hard X-Ray and Microwave Burst Sources in a Limb Flare', *Astrophys. J.* **298**, 431.

Wild, J.P.: 1969, 'Observations of the Magnetic Structure of a Type IV Solar Radio Outburst', *Solar Phys.* **9**, 260.

CHAPTER 9

ELECTRIC CURRENTS

No smoke without a fire, and no magnetic field without a current somewhere in the plasma permeated by the magnetic field. Ampère's equation intimately ties the magnetic field to the current density. In astronomy, the observable parameter is the magnetic field, and currents can be inferred only indirectly. Yet, in everyday life we generally ignore magnetic fields and are more used to the concept of currents describing, for instance, how energy is transported and released in an electric circuit. Currents in coronae do not flow exclusively in well-defined channels, and the circuit analogy is of limited value. In ideal MHD (infinite conductivity, Chapter 3), currents are considered a secondary product of the magnetic field having a non-vanishing curl. Nevertheless, currents in coronae represent large amounts of free energy and are of great interest.

There are many causes of electric currents. Two of them will be given as examples. In the following we study quasi-static currents and their effects (heating and particle acceleration) independent of the origin of the currents and electric fields. The principles of a current in a plasma and its consequences are presented in a straightforward treatment, with little in the way of elegant formalism. Effects that are possibly observable will be discussed in the final section with emphasis on strong, unstable currents.

9.1. Origin of Currents in Coronae

To understand why currents exist in coronae, we have to return to the basic equation of Ampère (Eq. 1.4.2),

$$\nabla \times \mathbf{B} = \frac{4\pi}{c}\mathbf{J} + \frac{1}{c}\frac{\partial \mathbf{E}}{\partial t} \quad . \tag{9.1.1}$$

According to this equation, there are two causes for currents: a non-vanishing curl in B and a variable $\mathbf{E}$ in time. The latter is important for example in waves, but will be neglected here (MHD limit). It is interesting to note that in a plasma a constant electric field is not a primary cause of currents. $\mathbf{E}$ only appears through Ohm's law (Eq. 3.1.45) as a consequence of a current and if the resistivity is finite (cf. Section 3.1.3.B).

If $\nabla \times \mathbf{B} = 0$, the magnetic field is *current free*. Since $\nabla \cdot \mathbf{B} = 0$, the vector identity (A.10) then yields $\nabla^2 \mathbf{B} = 0$. Thus the magnetic field is potential; it means that there exists a scalar magnetic potential $\Psi(\mathbf{x}, t)$ such that $\nabla\Psi = \mathbf{B}$.

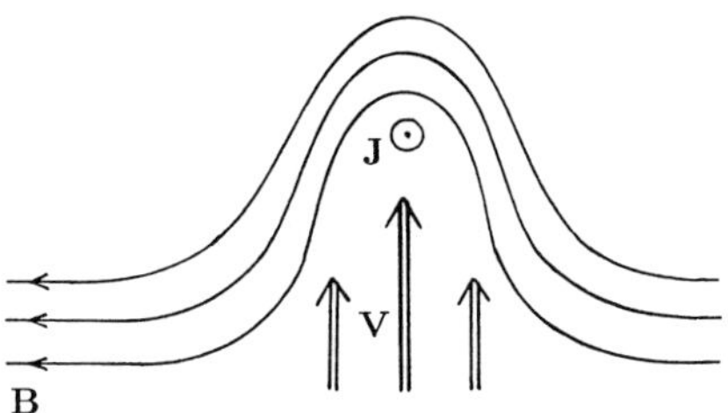

Fig. 9.1. A plasma motion perpendicular to the magnetic field generates a current.

Potential fields have the property that their energy content, $\int B^2 d\mathcal{V}/(8\pi)$, is the minimum for given boundary conditions (minimum energy theorem). In other words, magnetic fields with the same boundary conditions, but carrying a current, always contain more energy. The result is also true for semi-infinite fields. The photosphere is a boundary of a corona, since the photospheric magnetic field is tied to the plasma. The other boundary may be at infinity, where the field is usually assumed to fall off faster than the inverse distance. More realistically, the boundary is interplanetary space where $\mathbf{V} \times \mathbf{B} \neq 0$ and the field forms an Archimedean spiral. A potential field is fully determined by the perpendicular field component at the boundaries. Since the line-of-sight component can be measured by the Zeeman effect in photospheric lines, the perpendicular field component in the center of the disk is well known on the Sun. The potential field is frequently used as a rough estimate of the coronal magnetic field.

If the currents do not vanish, the total magnetic energy is higher than the potential field energy. The difference, called *free magnetic energy*, is due to the presence of currents. It can be released (for example by Ohmic heating) without changing the vertical field components at the boundaries.

9.1.1. MHD GENERATOR

The induction equation (3.1.46),

$$\frac{\partial \mathbf{B}}{\partial t} + \nabla \times (\mathbf{B} \times \mathbf{V}) = \frac{c^2}{4\pi\sigma} \nabla^2 \mathbf{B} \quad , \tag{9.1.2}$$

relates the magnetic field to plasma motion. The functioning of an MHD generator may be imagined as a flow across the magnetic field (Fig. 9.1). The field lines are 'frozen into the matter' (Section 3.1.3) due to the high conductivity of the plasma. They are swept along with the flow like bodily objects.

A current results from the non-vanishing $\nabla \times \mathbf{B}$. The way the current closes depends on the global magnetic field geometry. A rotation (vortex motion) in the photosphere can generate a vertical current into the corona that can be visualized as a twisting of coronal field lines. Independently, an electric field arises from Ohm's law (3.1.45) with $\mathbf{E} \approx -(\mathbf{V} \times \mathbf{B})/c$. Note that the current is not driven by the electric field, as in a laboratory wire!

9.1.2. CURRENT SHEET

Currents are generated wherever magnetic fields of opposite polarity are brought together. In a popular scenario, new magnetic flux emerges from the photosphere into the corona and pushes against the pre-existing magnetic field. The field then must change in strength and direction on a small scale where the two plasmas meet. Figure 9.2 shows a schematic drawing in two dimensions of an idealized case with opposite magnetic fields of the same strength approaching each other. At the contact surface of the two symmetric flows, a current sheet forms due to the finite curl of **B**.

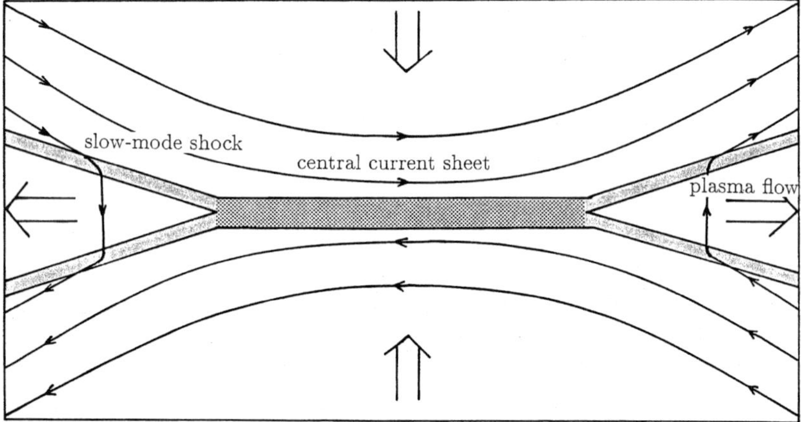

Fig. 9.2. A current sheet forms at the contact surface of two plasmas with opposite magnetic field. For simplicity, a symmetrical, stationary reconnection geometry is shown. Regions of currents parallel and perpendicular to the magnetic field are stippled and hatched, respectively.

Figure 9.2 represents a steady state geometry first proposed in 1964 by H.E. Petschek for solar flares. The counter-streaming inflows meet at a central current sheet where their magnetic fields create a current. The plasma leaves the region at right angles to the incident direction. The outflows carry magnetic field lines that now connect the previously independent fields of the two inflows. The global magnetic field is said to be *reconnected.* Its energy is reduced, and the difference is released into the plasma. Slow-mode shocks (to be discussed in the next chapter) may form at the boundaries between inflowing and outflowing material, linked by the magnetic field. Reconnection does not need to be driven from outside, but may occur spontaneously by a resistive MHD instability such as the tearing mode in a current sheet or in a magnetically sheared structure. The two concepts can be unified (Priest and Forbes, 2000). Furthermore, reconnection is possible without resistivity and collisions as recently evidenced by magnetospheric observations (Øieroset et al., 2001).

The current density in a current sheet is given by Ampère's law,

$$\mathbf{J} \;=\; \frac{c}{4\pi}\,\nabla\times\mathbf{B} \;\approx\; \frac{cB}{4\pi d} \quad . \tag{9.1.3}$$

The half-thickness, d, of the current sheet can be estimated from the electric field in and out of the current sheet

$$\mathbf{E} = \begin{cases} -(\mathbf{V}\times\mathbf{B})/c & \text{outside} \\ 1/\sigma\;\mathbf{J} & \text{inside} \end{cases} \quad . \tag{9.1.4}$$

Faraday's law requires that the electric fields outside and inside the current sheet are the same. Let $V_\perp$ be the perpendicular inflow velocity well outside the current sheet. Equations (9.1.3) and (9.1.4) combine to

$$d \;\approx\; \frac{c^2}{4\pi V_\perp \sigma} \quad . \tag{9.1.5}$$

The conductivity σ, related to the resistivity η by $\sigma = 1/\eta$, will be calculated in the following sections.

Much theoretical work has been devoted to studying the stationary solutions of reconnection. The dynamics of reconnection, variable in time and space, are being explored by laboratory experiments and numerical simulations.

In three dimensions, the inflowing magnetic fields will in general not be in a plane (that is, not anti-parallel), but will approach each other at an oblique angle. In the region of contact the field lines change direction like the steps of a spiral staircase. The current flows parallel to these field lines. At the step where the current is largest, the magnetic field is decoupled from the plasma and can reconnect.

9.2. Classical Conductivity and Particle Acceleration in Stable Currents

What happens if there is a weak electric field *parallel* to the magnetic field? Of course, the charged particles are accelerated. This may continue until the collisional friction between the species equals the electric force. Positive and negative charges move in opposite directions.

Let us study the force equilibrium for the case of an ion. We neglect the effect of the magnetic field (assume parallel motion) and use the frame of reference moving along the z-axis (and $\mathbf{B}$) with the mean electron velocity. For an ion with velocity v, we have

$$q_i E \;=\; m_i \frac{< \Delta v_z >^{i,e}}{\Delta t} \;=:\; \frac{m_i v}{t_s^{i,e}} \quad . \tag{9.2.1}$$

The friction is described by the slowing-down time, $t_s^{i,e}$, due to interactions with electrons. Using Equation (2.6.25) for the ion as the test particle and the background electrons as the field particles,

$$q_i E = m_i(1 + \frac{m_i}{m_e})\frac{A_d}{2(v_{te})^2}G(v/v_{te}) \quad . \tag{9.2.2}$$

A similar equation, (9.2.9), will be derived for electrons. Note that $t_s^{i,e} \neq t_s^{e,i}$.

9.2.1. CONDUCTIVITY

The test particle approach of the previous paragraph is now generalized for particle distributions. The goal is to derive the mean equilibrium velocity (to be called *drift velocity*, V_d) between the ion and electron populations. The conductivity is then defined by the ratio of current density, en_eV_d, to electric field.

It would be natural to use the frame of reference moving with the ions and calculate the friction of the electrons. The G-function in Equation (9.2.2), however, is difficult to integrate since the electron distribution includes velocities with $v \ll v_{ti}$ as well as $v \gg v_{ti}$. Instead we use the electron rest frame as a mathematical trick. The thermal ion velocity distribution is much narrower and usually satisfies an approximation like, for example, $v \ll v_{te}$. The G-function for ions can then easily be approximated and integrated since ions appear as a nearly monoenergetic beam.

For $v \ll v_{te}$, $G(x)$ given by Equation (2.6.11) can be approximated by $x\sqrt{2}/(3\sqrt{\pi})$. Then Equation (9.2.2) predicts a linear relation between electric field and velocity. Such a relation is equivalent to Ohm's law. Putting the ion velocity v in Equation (9.2.1) equal to the drift velocity, V_d, we derive the *conductivity* for weak currents from Equation (9.2.2)

$$\begin{aligned} \sigma := \frac{q_i\ n_i\ V_d}{E} &= \frac{\left(\omega_p^i\right)^2\ t_s^{i,e}}{4\pi} \\ &= \frac{3}{4\sqrt{2\pi}}\ \frac{m_e\ n_i}{e^2 \ln\Lambda\ n_e}\left(\frac{k_BT_e}{m_e}\right)^{3/2} \approx 3.22\cdot 10^6\ T_e^{3/2}\ \frac{20}{\ln\Lambda} \quad [\mathrm{s}^{-1}]\ . \end{aligned} \tag{9.2.3}$$

It is interesting to note that the conductivity rises as $T^{3/2}$; a hotter plasma is a better conductor. On the other hand, it does not change if the plasma density is increased, in spite of the more particles and the larger friction. The reason is the reduced V_d – also a result of the larger number of free charges at constant current density.

It is customary to introduce formally an *electron-ion* collision frequency,

$$\nu_{e,i} := \frac{m_i\ n_i}{m_e\ n_e}\ \frac{1}{t_s^{i,e}} = \sqrt{\frac{2}{\pi}}\ \frac{n_e}{n_i}\ \frac{\ln\Lambda}{\Lambda}\ \omega_p^e \quad , \tag{9.2.4}$$

where we have put $n_iZ_i = n_e$. It is often written in the conventional form,

$$\sigma = \frac{\left(\omega_p^e\right)^2}{4\pi\nu_{e,i}} \quad . \tag{9.2.5}$$

A useful relation is derived from Equation (9.2.3), viz.

$$\frac{V_d}{v_{te}} = 3\sqrt{\frac{\pi}{2}\frac{E}{E_D}} \quad , \tag{9.2.6}$$

where

$$E_D := \frac{q_i \ln \Lambda}{\lambda_D^2} \tag{9.2.7}$$

is called the *Dreicer field.* Its physical meaning will become clear in the following.

The linear relation (9.2.3) between electric field and current density breaks down as the drift velocity between ions and electrons approaches v_{te}. After a maximum of 0.214 at $V_d = \sqrt{2}\, v_{te}$, G decreases, and so does the friction. For $V_d > \sqrt{2}\, v_{te}$ there is no equilibrium between electric acceleration and collisions! At a constant E, the drift increases since Equation (9.2.1) cannot be satisfied, the current grows in time, and the conductivity increases. The critical electric field E_c, for which $V_d = \sqrt{2}\, v_{te}$, is the maximum field for equilibrium. It can be evaluated from Equations (2.6.15) and (9.2.2), and amounts to

$$\begin{aligned} E_c &= 0.214 \frac{q_i}{\lambda_D^2} \ln \Lambda = 0.214\, E_D \\ &\approx 4.32 \cdot 10^{-11} \frac{\ln \Lambda}{20} \frac{Z_i\, n_e}{T} \quad . \end{aligned} \tag{9.2.8}$$

This hypothetical threshold has a simple physical meaning. It is roughly the field for which an average electron increases its velocity by v_{te} within a thermal collision time (2.6.32). Note the $1/T$ dependence! If initially $E < E_c$, Ohmic heating increases the temperature and reduces E_c so that $E > E_c$ may occur. The Dreicer field, $E_D = 4.67\, E_c$, is roughly the electric field above which electrons are practically freely accelerated as in the vacuum.

In reality, ions and electrons do not separate completely in velocity for $E > E_c$, running away into opposite directions. The next section will show that unstable waves inhibit such a 'plasma superconductor'. The waves are driven by instabilities of currents above some threshold density. The waves scatter the directed motion of the current carrying particles and introduce additional friction. Such *anomalous* resistivity can drastically exceed the collisional effects.

9.2.2. Runaway Electrons

The runaway phenomenon nevertheless does exist for certain particles, in particular electrons in one of the wings of the velocity distribution. If their relative velocity to the ions is high ($v \gg V_d$), the friction may be smaller than the accelerating force, qE. These electrons are accelerated electrically until they leave the system. High energy electrons have been found in laboratory plasmas with parallel currents as predicted by this argument. Runaway is an attractive acceleration process for the electrons in coronal currents manifest in flares and other dynamic phenomena.

Consider now a single electron in the frame moving with the ions (Fig. 9.3). Let $E \ll E_D$ and $\mathbf{E} \| \mathbf{B}$. Since collisional friction between the electron and ions

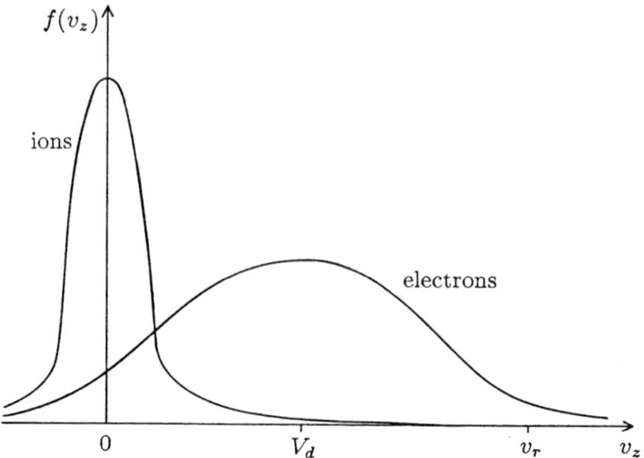

Fig. 9.3. Electron and ion velocity distributions separate by the drift velocity V_d due to a parallel electric field. Electrons beyond v_r run away. The figure is drawn in the rest frame of the ions.

diminishes with increasing electron velocity, there exists a velocity v_r (the runaway velocity) at which friction equalizes the electric force,

$$eE = \frac{m_e v_r}{t_s^{e,i}(v_r)} \quad . \tag{9.2.9}$$

The relevant parameter is now the slowing-down time of a test electron by field ions. We have neglected the interactions with other electrons. Putting in the slowing-down time (Eq. 2.6.25) for $v^2 \gg v_{ti}^2$, one derives (Exercise 9.1)

$$v_r = \left(\frac{m_e A_d}{2eE}\right)^{1/2} = v_{te}\left(\frac{E_D}{E}\right)^{1/2} \quad . \tag{9.2.10}$$

Since ions and electrons have the same v_r but different thermal velocities, ion runaway can usually be neglected.

At a constant electric field, the critical velocity for runaway acceleration does not change through heating. The physical reason is that the collision time in the velocity tail is independent of the background temperature (Table 2.1). More appropriate than constant E may be a constant current density and constant V_d; a temperature increase then reduces E and enhances v_r/v_{te}.

The initial fraction of runaway electrons is given by the condition $v > v_r$. The initial density n_r of runaway electrons is the integral in velocity over the tail of the distribution. For $v_r^2 \gg v_{te}^2$ and using Equations (9.2.6) and (9.2.10),

$$\frac{n_r}{n_e} \approx \exp\left[-\frac{1}{2}\left(\frac{v_r}{v_{te}}\right)^2\right] = \exp\left[-\frac{E_D}{2E}\right] = \exp\left[\frac{-1.88 \cdot v_{te}}{V_d}\right] \quad . \tag{9.2.11}$$

The ratio in the exponent is large by assumption, thus the initial fraction of runaway electrons is small. More accurate calculations do not change this result, unless the drift velocity is close to the thermal electron velocity and the electric field approaches the Dreicer field.

There is also a continuous acceleration of electrons brought beyond v_r by collisions. The particle supply into the runaway regime has about the rate of collisions of electrons with velocity v_r. In a more detailed calculation of the runaway acceleration rate including the effect of the electric field on the distribution at $v < v_r$, Knoepfel and Spong (1979) find for $v_r^2 \gg v_{te}^2$

$$\frac{\dot{n}_e}{n_e} \approx \frac{0.35}{t_t^e}\left(\frac{v_r}{v_{te}}\right)^{3/4} \exp\left[-\sqrt{2}\frac{v_r}{v_{te}} - \frac{1}{4}\left(\frac{v_r}{v_{te}}\right)^2\right] \quad [\mathrm{s}^{-1}] , \tag{9.2.12}$$

where t_t^e is the thermal collision time of electrons (Eq. 2.6.31). Again, acceleration is only appreciable if v_r is close to v_{te}. Note that a temperature increase may drastically enhance the runaway rate! The mechanism has been used to interpret energetic electrons in solar flares (Exercise 9.3).

The flux of runaway electrons constitutes a current in addition to the drifting electrons of the background. If the electric field increases so that v_r/v_{te} is reduced, the current of runaway particles increases. The total current may not necessarily change, since it is tied to the curl of **B** through Ampère's law (Section 6.1.2). Let us assume that the total current in the sheet is not changed by the runaway process. The runaway electrons carry more and more of the current, but the number of electrons traversing a cross-section of the current sheet per second remains constant. The total current therefore sets an upper limit of the runaway rate, $\dot{N}$, that can leave the current sheet,

$$\dot{N} = \dot{n}\,\mathcal{V} \leq \frac{2dwJ}{e} \approx \frac{cwB}{2\pi e} \quad [\text{electrons s}^{-1}] , \tag{9.2.13}$$

where the volume $\mathcal{V}$ of the current sheet is $2dLw$, w is the width of the sheet parallel to **B** and perpendicular to the thickness $2d$, and the L is the length. Ampère's law has been applied in the approximation $J \approx cB/(4\pi d)$.

Even for large values, say $w = 10^9$ cm and $B = 100$ G, Equation (9.2.13) limits the total acceleration rate to about 10^{30} electrons per second. This upper limit is not sufficient for the number of flare electrons derived from solar hard X-ray observations (Section 6.4). Furthermore, the acceleration is limited in time to $N/\dot{N}$, after which a sizable fraction of the electron population in the current sheet has gone into runaway. This time is short, if acceleration is significant. The number problem may, for example, be solved by bringing new electrons into the current sheet.

In addition to electron acceleration, runaway may produce consequences on a global scale. The particles take up energy from the current, decrease resistivity by their smaller friction and possibly change the properties of current instabilities (to be discussed in the next section). The role of the runaway process in the dynamics of reconnection is not yet well understood.

Electron runaway occurs in every current parallel to the magnetic field if $v_r < c$. It is an attractive acceleration process for a number of cosmic phenomena, including solar and stellar flares.

9.3. Instabilities of Electric Currents

The picture of coronal currents would be incomplete without a study of the instabilities of strong currents associated with an appreciable acceleration rate. As we have seen in Sections 5.2.4 and 7.2, moving particle species – such as ions, electrons and runaway electrons – can be in Čerenkov resonance or gyromagnetic resonance with a variety of wave modes.

9.3.1. Parallel currents

Quasi-parallel currents are discussed first. They are important, for instance, in the central current sheet of a reconnection region (Fig. 9.2). The ratio T_e/T_i determines which instability dominates. We treat the likely sequence of phenomena in a cosmic plasma where the electric field and the drift velocity, V_d, rise slowly. Initially let $T_i = T_e$.

A. *Ion Cyclotron Instability*

The ion cyclotron instability has the lowest threshold for $T_i = T_e$. It is driven by moving electrons in anomalous Doppler resonance with the L-mode (Section 7.2.2.B). The resonance condition (Eq. 7.2.24) requires

$$\omega - k_z v_z = -\Omega_e \quad . \tag{9.3.1}$$

Since the low-frequency L-mode has a cutoff at Ω_i, $\omega < \Omega_i \ll \Omega_e$ (Section 4.3). Electrons with resonance velocity $v_z \approx \Omega_e/k_z = v^z_{ph}\Omega_e/\omega$ provide energy for growing waves. For $v_z \approx V_d$, the wave phase velocity along $\mathbf{B}$, v^z_{ph}, must be small. This is the case for resonant L-waves in nearly perpendicular direction with frequencies close to Ω_i, called *electrostatic ion cyclotron waves.* The threshold for Maxwellian distributions is

$$V_d \gtrsim 15 \frac{T_i}{T_e} v_{ti} \quad . \tag{9.3.2}$$

In reality, the electron distribution is skewed by the electric field, and the threshold is somewhat higher. The threshold plotted in Figure 9.4, based on calculations by Morrison and Ionson (1982), takes this into account. It is not well established how the instability saturates, but it is generally agreed and confirmed by experiment that the wave energy level at saturation is low. The waves feed energy primarily to the ions. Therefore, the main effect of the ion-cyclotron instability is ion heating.

B. *Buneman Instability*

If the drift velocity increases and the ion and electron distributions are displaced from each other by more than the thermal velocities ($V_d \gg v_{te}$), they can be considered as two cold streams of particles. This produces one of the oldest known instabilities. In the frame moving with the ions, the dispersion relation of a two-component plasma (Eq. 4.6.5) becomes

$$1 = \frac{(\omega_p^e)^2}{(\omega - kV_d)^2} + \frac{(\omega_p^i)^2}{\omega^2} \quad . \tag{9.3.3}$$

This equation of fourth order in $\omega(k)$ has first been investigated analytically by O. Buneman. The most unstable wave has a frequency $\omega = \omega_r + i\gamma$ with

$$\omega_r = \left(\frac{m_e}{m_i}\right)^{1/3} \omega_p^e \quad , \tag{9.3.4}$$

$$\gamma_k = \frac{\sqrt{3}}{2} \left(\frac{m_e}{2m_i}\right)^{1/3} \omega_p^e \quad . \tag{9.3.5}$$

The waves are of the beam mode (Section 4.6) and purely electrostatic. The frequency given by Equation (9.3.4) is far below the plasma frequency. A kinetic plasma calculation is more realistic and yields a threshold of

$$V_d \gtrsim 1.7\ (v_{te} + v_{ti}) \quad . \tag{9.3.6}$$

Growth is slow near the threshold. In Figure 9.4 the drift velocity for $\gamma = \omega_p^i$ is shown for a fair comparison with the other instabilities.

The non-linear evolution of the Buneman instability saturates by quasi-linear diffusion (Section 6.2.1) of the electrons. This forms a plateau between the original peak of the electron distribution and the ions, bringing the system of the two streams into marginal stability. The electrons are thereby heated to the point where the thermal electron velocity equals the drift velocity ($v_{te} \approx V_d$). Only a small fraction of the energy, $(m_e/m_i)^{1/3}$, is transferred to the ions. After some time, the electron temperature greatly exceeds the ion temperature. This is the starting point of a different instability, the ion acoustic type. It controls the rest of the evolution.

C. *Ion Acoustic Instability*

Ion acoustic (or ion sound) waves are ubiquitous in currents near Earth. They are ion oscillations modified by Debye shielding and have been discussed in Section 5.2.6. The growth rate of ion acoustic waves driven by an electric current has been calculated in Equation (5.2.39). The instability threshold was given by Equation (5.2.40). A more appropriate calculation including the effect of the electric field on the electron distribution is shown in Figure 9.4. Details and references are given in a review by Papadopoulos (1977).

Numerical simulations have followed the non-linear evolution of the instability. They show that both ions and electrons are heated preferentially in parallel direction. For $V_d \gtrsim 2c_{is}$ and $T_e > T_i$, electrons receive a larger fraction of the energy, and T_e/T_i rises. Saturation occurs when unstable waves are scattered on ions into different waves not driven by the current and are damped (a process known as *non-linear Landau damping*). It stabilizes wave growth at a level that has been evaluated using weak turbulence theory. At saturation $T_e \approx 10\ T_i$. The ion acoustic instability can produce an equilibrium between the electric current driving the waves and the wave turbulence causing anomalous resistivity and limiting the drift velocity.

Figure 9.4 compares three instabilities of parallel currents. The following sequence of events may happen in a current sheet in which V_d increases slowly with time. The sheet may start at point A with low current drift velocity and $T_e = T_i$, crosses into the ion cyclotron regime at about $V_d \approx 0.4\ v_{te}$ and moves to the left (region of small T_e/T_i) because of the increasing ion temperature. Then it evolves into the region of Buneman instability where T_e increases and V_d/v_{te} decreases. It settles finally at marginal stability of ion acoustic waves (point B). Note that this scenario is only one of many possibilities, since the number of free parameters is considerable.

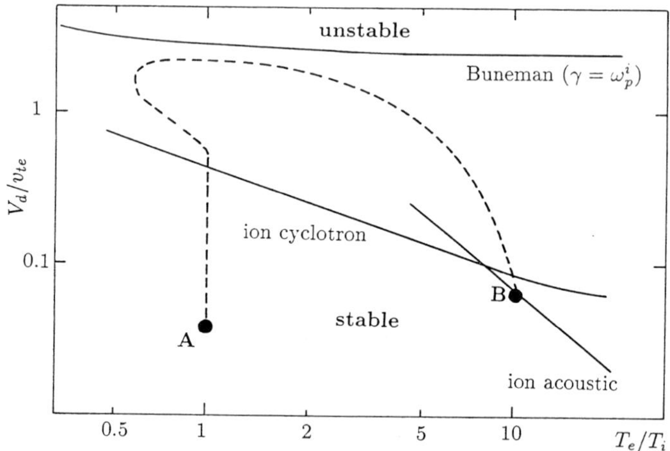

Fig. 9.4. The threshold drift velocity V_d of a parallel current is presented for the three instabilities discussed in the text. It depends strongly on the temperature ratio of electrons and ions. A likely evolutionary scenario for a current system with increasing V_d is indicated by the dashed curve from A to B (see text).

9.3.2. Perpendicular Currents

In a collisionless plasma perpendicular currents are provided by drifting particles. Guiding center motions across the magnetic field lines have been studied in Section 2.1.4. They can produce currents when different species move at different speeds or in different directions. Perpendicular currents are often formed by ions streaming in relation to electrons and the magnetic field.

Perpendicular currents include *gravitational currents* and *field curvature currents* (Eqs. 2.1.28 and 2.1.29). In both cases the ions move faster than the electrons by the mass ratio m_i/m_e. An important example is the curvature of the magnetic field in the slow shocks propagating out of the central current sheet in the Petschek reconnection geometry (Fig. 9.2). This causes a strong perpendicular current. Plasma motions compressing the ambient magnetic field are another and very prolific source of cross-field currents (cf. Fig. 9.1). Such motions also result from the outflows of most reconnection models considered for flares. The perpendicular component of the magnetic field gradient drives the so-called *gradient current* required by Ampère's law, $\mathbf{J} \approx \nabla \times \mathbf{B}c/(4\pi)$.

Note that a perpendicular electric field produces an $E \times B$ drift (Eq. 2.1.26), which does *not* produce currents, but moves the whole plasma transverse to both $\mathbf{E}_\perp$ and $\mathbf{B}$.

Which of the instabilities is most important for cross-field ion streams and currents depends on the ratio T_e/T_i. The *modified two-stream instability* is possible for $T_e \gtrsim T_i$ above a threshold $V_d > c_{is}$. It is like a two-stream instability, but with a modified role of the particles; the electrons are bound to the field and the streaming ions drive the instability. The instability excites waves having a maximum energy density at $k_z \approx k_\perp (m_e/m_i)^{1/2}$. For further information the reader is referred to the review by Huba (1985).

The Lorentz force prohibits runaway particles accelerated in perpendicular currents. Nevertheless, the wave turbulence of these instabilities can heat both ions and electrons to high temperatures. This will be discussed in Section 9.4.3.B.

9.4. Anomalous Conductivity, Heating, and Acceleration

9.4.1. Anomalous conductivity

Fluctuating wave fields can severly increase the frictional effect on drifting particles. We describe it by an effective collision time, $t_{\rm eff}$. Electrostatic waves modulate the ion density. The colliding electron sees bunches of ions with an effective charge-to-mass ratio much larger than q_i^2/m_i. Therefore, the diffusion constant, A_d, defined in Equation (2.6.15) as proportional to this ratio, is enhanced and the slowing-down time (Eq. 2.6.25) is reduced correspondingly.

Current instabilities tend to heat electrons and develop ion acoustic turbulence (Section 9.3.1). The ion acoustic wave energy density at saturation depends only on T_e/T_i and c_{is}/V_d. The effective collision time at saturation has been calculated analytically by R.Z. Sagdeev using weak-turbulence theory;

$$\nu_{e,i}^{\rm eff} = \frac{\omega_p^e}{32\pi} \frac{T_e}{T_i} \frac{V_d}{v_{te}} \quad . \tag{9.4.1}$$

Upon equating anomalous friction with electric acceleration, Equations (9.2.5) and (9.4.1) yield an anomalous conductivity

$$\sigma_{\text{eff}} = 8 \frac{T_i}{T_e} \frac{v_{te}}{V_d} \omega_p^e \tag{9.4.2}$$

for parallel currents strong enough to saturate the ion acoustic instability. It is generally called *Sagdeev conductivity*. Weak-turbulence theory is often used in estimating conductivity, but it is not beyond doubt. Nevertheless, laboratory experiments and numerical simulations confirm that the drift velocity of currents at saturation is in the range $1 - 3c_{is}$, and the minimum anomalous conductivity is typically

$$\sigma_{\text{eff}}^{\text{min}} \approx 20\omega_p^e \quad . \tag{9.4.3}$$

It agrees with Equation (9.4.2) and $T_e \approx 10\ T_i$. The effective electron-ion collision rate becomes $\nu_{e,i}^{\text{eff}} \approx 0.2\ \omega_p^i$. The minimum anomalous conductivity given in Equation (9.4.3) is about a million times smaller than the classical value (9.2.3).

The collision rates of different origins – or the resistivities, $\eta := 1/\sigma$ – have to be added. The ion cyclotron instability, for instance, causes an anomalous conductivity of about $0.2\ \Omega_i$ at saturation.

Even in a stable current it is possible that the ions are correlated. A well known example is the turbulence created by runaway electrons. They may produce a positive slope in velocity space in case of inhomogeneity and drive Langmuir waves. If the wave energy exceeds the threshold given in Equation (6.2.16), the waves collapse non-linearly and produce ion density fluctuations. The generation of such anomalous resistivity by streaming electrons appears to operate in auroral substorms and has been proposed for cosmic ray electrons in clusters of galaxies. It may play the role of a trigger in flare physics.

9.4.2. Ohmic heating

A finite conductivity – whether classical or anomalous – dissipates energy of the current at a rate $\mathbf{J} \cdot \mathbf{E}$. This is called *Ohmic (or Joule) heating*. The energy heats the current-carrying region on a time scale

$$\tau_J := \frac{n_e\, k_B\, T_e}{\mathbf{J} \cdot \mathbf{E}} = \frac{2}{9\pi} \left(\frac{E_D}{E} \right)^2 \frac{1}{\nu_{e,i}} \quad , \tag{9.4.4}$$

assuming a parallel current with $\mathbf{J} = \sigma \mathbf{E}$ and applying Equations (9.2.5) and (9.2.6). The effective collision rate has to be used for $\nu_{e,i}$ in cases of anomalous conductivity. The time scale for Ohmic heating is surprisingly short even for small currents. When the temperature has risen sufficiently, an equilibrium may be reached by radiative or conductive losses.

What happens when the conductivity suddenly drops from its collisional value to an anomalous value? Let us first envisage a typical situation where this may happen. Assume that the anomalous conductivity is produced by a current's drift velocity exceeding the threshold of the relevant instability, $V_d > ac_{is}$, where a is a constant of order 10 (e.g. Eq. 9.3.2). Ampère's law requires that such a strong current can only exist in narrow sheets or threads. A current sheet, for example, would have a thickness of $d = Bc/(4\pi J)$. For coronal values – say $B = 100$ G, $n = 10^9$ cm^{-3}, $T_e = T_i = 2 \cdot 10^6$ K – the thickness is only $d \approx 4 \cdot 10^3$ cm.

If instabilities of currents occur, they generally grow faster than the inductive time scales given by the diffusion time of magnetic fields (Eq. 3.1.48). Ampère's law then requires that the total current of a current sheet also changes at the slow, inductive time scale. The total current is Jdw, where w is the width of the current sheet and is assumed to be constant. The current density, $\mathbf{J}$, therefore remains approximately constant during the initial decrease of the conductivity, E rises by $\sigma/\sigma_{\text{eff}}$, and Ohmic heating increases accordingly. Later, the current sheet expands, as Equation (9.1.5) requires the thickness, d, to increase by a factor $\sigma/\sigma_{\text{eff}}$. Since $Jd \approx$ constant, the current density, J, decreases, while the electric field regains the initial value. The heating rate per volume decreases, and the total power released by Ohmic heating in the expanded current sheet finally reaches the initial value. The sudden decrease in conductivity thus causes a pulse of enhanced Ohmic heating.

9.4.3. Particle Acceleration

A. *Runaway Particles*

Since low-frequency turbulence increases the collision frequency, runaway acceleration is also affected by anomalous conductivity. For the important case of ion acoustic waves it can be shown that the effective electron-ion collision frequency scales with electron velocity in the same way as Coulomb collisions, that is, $\nu_s^{\text{eff}}(v)$ is proportional to $(v/v_{te})^3$. Therefore, the results on classical runaway simply scale with conductivity. E_D increases with $\sigma/\sigma_{\text{eff}}$ (Eqs. 9.2.3 and 9.2.7). The runaway velocity initially does not change, since E rises by the same factor. As E comes back to the initial value, v_r increases with $\sqrt{E_D}$, that is by the factor $\sqrt{(\sigma/\sigma_{\text{eff}})}$ (Eq. 9.2.10). Figure 9.5 schematically displays the changes due to a rise in effective collision frequency at t_0.

Anomalous conductivity reduces the region of runaway in velocity space. However, the rate of diffusion beyond v_r (and therefore the runaway rate) increases with $\nu_{e,i}^{\text{eff}}$ (Eq. 9.2.12). Moreover, the runaway electrons feel a higher electric field after a sudden conductivity decrease and may reach higher energy. A sudden decrease of conductivity may thus cause a surge of high energy runaway electrons.

How significant is the energy released by acceleration? The energy fed into a runaway particle is eE times the length of acceleration, $L/2$ on the average. The flux of runaway particles is $n_r wd < v >$. Therefore, the total energy released in the form of runaway particles is

$$\mathcal{E} \approx \frac{1}{2} eEL \cdot n_r wd < v >= \frac{1}{2} I_{run} EL \quad , \tag{9.4.5}$$

where I_{run} is the total electric current due to runaways leaving the current sheet, and EL is the potential difference. Note that the total Ohmic heating is IEL.

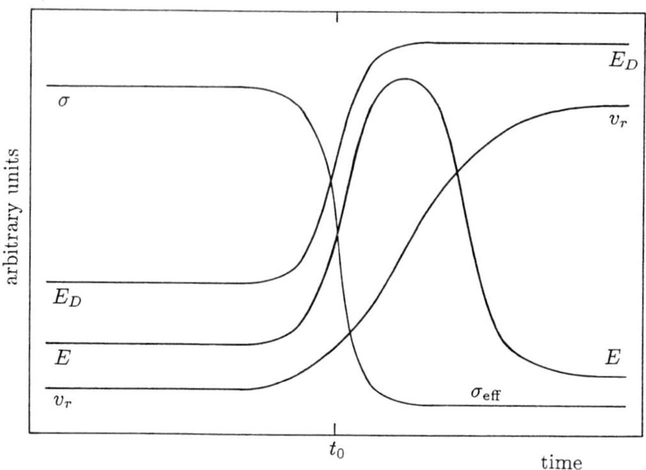

Fig. 9.5. When the conductivity of a current sheet drops to an anomalous value near the time t_0, all the sheet plasma parameters change. The evolution in time is indicated (not to scale).

Therefore, the energy going into runaway acceleration is always smaller than the energy released by Ohmic heating as $I_{run} \leq I$.

B. *Resonance Acceleration*

Instabilities of perpendicular currents have been discussed in Section 9.3.2. Their wave turbulence, primarily lower hybrid waves, can transfer energy to the particles in resonance. Physically, it is equivalent to emissions and absorptions of phonons. The process is of the type of *resonance acceleration* and can be described by a diffusion equation

$$\frac{\partial f_e}{\partial t} + \mathbf{v}\frac{\partial f_e}{\partial \mathbf{x}} = \frac{\partial}{\partial \mathbf{v}} \cdot \left(\widehat{\mathcal{D}} * \frac{\partial f_e}{\partial \mathbf{v}}\right) + \left(\frac{\partial f_e}{\partial t}\right)_{\rm coll} . \tag{9.4.6}$$

Lower hybrid waves propagate at almost perpendicular direction to the magnetic field with $\cos\theta \lesssim (m_e/m_i)^{1/2}$. Their phase velocity is about v_{ti}, and its component in the z-direction,

$$v_{ph}^z = \frac{\omega}{k_z} = \frac{\omega}{k}\frac{k}{k_z} \gtrsim v_{ti}\sqrt{\frac{m_i}{m_e}} \approx v_{te} \quad , \tag{9.4.7}$$

can be sufficiently low to be in Čerenkov resonance ($v_z = v_{ph}^z$) with thermal electrons. Lower hybrid waves – driven for instance by the modified two-stream instability of a perpendicular current – are Landau damped on electrons. Since the electrons are bound to the magnetic field, they are accelerated in the z-direction and diffuse primarily into the v_z tail of the velocity distribution. Equation (9.4.6) can be reduced to one dimension (z-direction) and solved analytically in the limit of quasi-linear diffusion. The result is a Macdonald (modified Bessel) function distribution in v_z, having enhanced tails parallel and anti-parallel to the background magnetic field.

Electron resonance acceleration by lower hybrid waves is in competition with ion heating. Numerical simulations show that for weak currents having $V_d \lesssim 4v_{ti}$ electron acceleration is efficient. The process can produce a large number of $1-40$ keV electrons and has been proposed for flares and in shocks (Section 10.3.3). At relativistic velocities, the resonant wave energy is small, and acceleration is slow. Lower hybrid waves are therefore inefficient in accelerating relativistic electrons.

9.5. Observing Currents

The observation of currents is an important goal of astrophysics. The drift between electrons and ions can be measured directly only in planetary magnetospheres and in interplanetary space. In more distant plasmas, one has to rely on indirect methods. Most important are measurements of $\nabla \times \mathbf{B}$. Since global currents presumably behave close to MHD (Section 9.1.1), they are generally driven by the curl of the magnetic field. A non-vanishing $\nabla \times \mathbf{B}$ indicates the existence of free magnetic energy. We shall also discuss possible detections of current-driven waves by radio observations. Furthermore, electric field measurements using the Stark effect of optical and infrared lines from the Sun suggest occasional large values, reaching up to 700 V cm^{-1} in flares and post-flare loops. They have been attributed to oscillating electric fields of low-frequency (MHD) waves or beam induced electric fields (Section 6.1.2).

9.5.1. Currents in the Photosphere

The longitudinal and transverse magnetic field components can be measured in strong optical lines from the solar photosphere. This completely defines the local magnetic field, and the current density in the photosphere can be determined from Ampère's law (neglecting the displacement current in the MHD limit). Figure 9.6 displays the line-of-sight electric currents, $\mathbf{J}_l = \nabla \times \mathbf{B}_t c/(4\pi)$, flowing at the height where the transverse magnetic field, $\mathbf{B}_t$, was measured. Since the observed region was near disk center, the line-of-sight is close to vertical on the surface. Figure 9.6 (right) represents the sources and sinks of currents flowing above the photosphere. The vertical component of the total current is $5.1 \cdot 10^{21}$ statamp ($1.7 \cdot 10^{12}$ ampères). Note that in a stationary corona ($\mathbf{V} = 0$, $\partial B/\partial t = 0$) obeying *ideal* MHD, these currents are not associated with any electric field. They are driven entirely by the curl of the magnetic field, from which they have been evaluated.

A and B mark the sites in Figure 9.6 (right) where repeated Hα flares started only minutes before the observations. The sites of flare onset were above the maxima of current density within the 2" instrumental resolution. Photospheric currents appear to be closely related to the coronal flare sites.

Fig. 9.6. *Left:* The line-of-sight magnetic field of an active region (4×4 arc minutes field of view) on the solar disk. Solid (dashed) contours represent positive (negative) magnetic polarities. The two thick, closed contours mark the boundaries (umbra) of the leader and follower spot. The thick, open contour separates positive and negative magnetic fields in a small bipolar region. *Right:* Distribution of the line-of-sight electric current density over an enlargement (1.6×1.6 arc minutes). Solid (dashed) contours show currents flowing out of (into) the photosphere. The lowest contour represents a current density of $1.5 \cdot 10^3$ statamp cm^{-2} ($5 \cdot 10^{-3}$ A m^{-2}); successive levels are in increments of the same value. The same magnetic inversion line is shown to aid orientation. A and B refer to sites of repeated flaring (from Hagyard, 1988).

9.5.2. Noise storms

In 1946, shortly after his detection of the radio emission of solar activity, J.S. Hey noted a radiation connected with the appearance of large sunspots on the solar disk. Such long duration radio emission in meter waves ($\nu \lesssim 300$ MHz), composed of narrowband, spiky bursts and a broadband continuum, has become known by the name of *noise storm* (or type I bursts and continuum). The radiation is not related to flares, but occurs for days after the appearance and growth of complex active regions in the photosphere. This association suggests that noise storms are signatures of coronal adjustments to field changes at the photospheric boundary and of coronal evolution. Occasionally, noise storms have been observed to be triggered by a white light transient. In Figure 9.7, the source of a noise storm (observed at a constant frequency of 169 MHz) remains at a roughly constant altitude, but moves with the computed intersection of the white light loop with pre-existing magnetic fields. Noise storms also have been observed to coincide with emerging magnetic flux visible in coronal EUV lines.

The bandwidth of type I bursts is a few percents of the centerfrequency, and the duration is less than one second. A typical observation is shown in Figure 9.8. The bursts occur at rates as high as one per second. Figure 9.7 suggests that their sources are at the contact surface of emerging magnetic flux with the pre-existing magnetic field. In general, type I bursts are interpreted as the signature of many small steps whose cumulative effect is a gradual evolution of the corona.

There are two observational constraints which need consideration. *First*, noise storms are generally accompanied by non-thermal electrons. Their most obvious

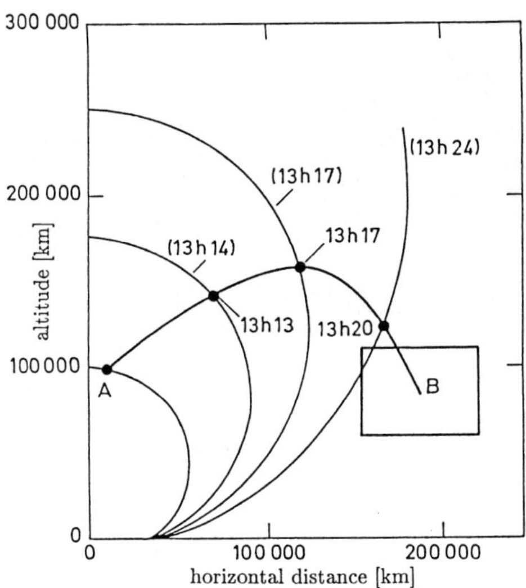

Fig. 9.7. Noise storm source (bullets, observation of Nançay radioheliograph) and white light transient (thin curves, extrapolated from observations of HAO coronagraph on SMM at given times). The origin of the plot is in the center of the associated active region (horizontal) and in the photosphere (vertical). The noise storm follows the thick curve starting at A and becomes stationary in the rectangle marked B (after Lantos, P. and Kerdraon, A., unpublished).

signatures are occasional storms of type III bursts at frequencies below the type I bursts. Also, the noise storm continuum may be caused by electrons trapped in high coronal loops. The number of storm accelerated electrons is insufficient to produce observable hard X-rays. *Second*, and different from type III bursts, type I bursts have high circular polarization, strong center-to-limb effects, and no harmonic structure. An interpretation of the bursts by electron beams, as well as by trapped particles alone, is therefore inappropriate.

Type I bursts have been proposed to be emissions of low-frequency turbulence caused by unstable currents. The waves may coalesce with high-frequency, electrostatic waves (Langmuir, z-mode or upper hybrid waves) into propagating transverse waves at radio frequencies. Such processes have been studied in Section 6.3.2. We abbreviate them by the notation $l + h \rightarrow t$ where l stands for the low-frequency waves, h for the high-frequency wave and t for the observable transverse (radio) wave.

The origin of the high-frequency waves are most likely non-thermal electrons. They could have been accelerated at a type I burst site, later to be trapped (Chapter 8) and to produce high-frequency waves at other burst sites. The storm continuum may originate from the same high-frequency waves. The temporary presence of low-frequency waves in localized regions may enhance the emission and produce the bursts.

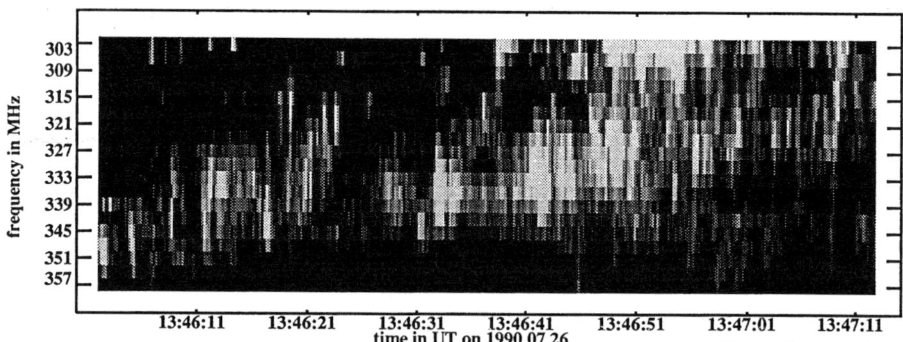

Fig. 9.8. A so-called 'chain' of hundreds of type I bursts traverses a spectral band of a Zurich spectrometer during a noise storm. White is enhanced emission.

The scenario is not yet proved, nor are the details clear. The kind of low-frequency waves depends on the global situation. Proposed are:

- Ion acoustic waves driven by parallel currents. The model appeals to reconnection between the emerging and pre-existing fields.
- Lower hybrid waves in a model suggesting perpendicular currents in the piled-up pre-existing fields, compressed by the weak shocks of emerging flux. The threshold drift for growing lower hybrid waves is considerably lower than for ion acoustic waves at $T_e \approx T_i$.

The essential feature of both instabilities and models are intense low-frequency waves near or at saturation. This has two vital consequences, more general than the theory of type I bursts: (*i*) The emission process becomes optically thick for very thin sources, as is necessary for fast drifting, unstable currents. (*ii*) A high level of low-frequency waves permits the intensity of the high-frequency waves to be relatively weak, consistent with the lack of harmonic emission ($h + h \rightarrow t$). The high-frequency waves then have the role of tracers making the low-frequency turbulence visible.

9.5.3. Radio Emission of Low-Frequency Turbulence

How can current-driven turbulence emit observable radiation? Conversion of one wave mode into another mode has been discussed in Section 6.3.2. The requirements for coupling between low-frequency (l) waves and high-frequency (h) waves and efficient energy transfer $l + h \rightarrow t$ are the parametric conditions,

$$\omega_l \pm \omega_h = \omega_t \qquad (9.5.1)$$

$$\mathbf{k}_l \pm \mathbf{k}_h = \mathbf{k}_t \tag{9.5.2}$$

(Eqs. 6.3.13 and 6.3.14). Since $\omega_l \ll \omega_h$, the emission is at $\omega_t \approx \omega_h$ and is referred to as *fundamental emission*. The dispersion equation of each wave mode relates frequency and wave vector. Considerations similar to those of Equations (6.3.7) – (6.3.9) for the conversion of Langmuir waves into transverse waves suggest that in general $|\mathbf{k}_t| \ll |\mathbf{k}_l|$ and $|\mathbf{k}_h|$. Therefore, $\mathbf{k}_l \approx -\mathbf{k}_h$, thus observable radiation is generated by the coalescence of oppositely directed waves with similar wave numbers.

The rate of photon generation is given by Equation (6.3.15). It also contains absorption losses. The transition probability, $w^{\rm lht}$, is derived from the second-order terms in the wave equations. The emissivity, η, and the absorption coefficient, κ, for transverse waves are then given. They are similar for ion-acoustic waves and for lower hybrid waves interacting with Langmuir waves (the calculations have been done for unmagnetized high-frequency waves for simplicity). The absorption coefficient of this process has been evaluated by Benz and Wentzel (1981) to be

$$\kappa \approx 5 \; \frac{\omega_p^e}{c} \, \frac{W_{\rm tot}^l}{nk_BT} \quad [{\rm cm}^{-1}] \; , \tag{9.5.3}$$

where $W_{\rm tot}^l$ is the total energy density of the low-frequency wave integrated in k-space. With the low-frequency wave energy density at saturation (Section 9.4), optical thickness unity is achieved over a distance of a few meters!

For an optically thick source, the photon temperature, T_t, can be derived similar to Equation (6.3.19) with the result

$$T_t = \frac{\hbar\omega_t}{k_B} \, \frac{N_l N_h}{N_l + N_h} \quad , \tag{9.5.4}$$

N being the spectral quantum density defined in Equation (6.3.3),

$$N(k)\hbar\omega_{\mathbf{k}} \; := \; k_B T(\mathbf{k}) \; := \; W(\mathbf{k}) \quad . \tag{9.5.5}$$

Equation (9.5.4) has a surprising consequence. The photon temperature of the emitted radiation (and therefore the source intensity) is determined by the wave mode with the lower quantum density, although the absorption occurs via the waves with the higher density! In particular for $N_h \ll N_l$, Equations (9.5.4) and (9.5.5) give

$$T_t \;\approx\; \frac{\hbar\omega_p}{k_B} N_h \;=\; T_h \;\approx\; \frac{(2\pi)^3}{k_B \Omega} \, \frac{W_{\rm tot}^h}{k_h^2 \Delta k_h} \quad . \tag{9.5.6}$$

Wave anisotropy (solid angle Ω) and a narrow spectral range (Δk_h) have been assumed on the right side of Equation (9.5.6). Of course, the total wave energy density, $W_{\rm tot}^h$, only includes high-frequency waves that satisfy the parametric conditions (9.5.1) and (9.5.2) and have low-frequency interaction partners.

Equation (9.5.6) provides a simple way to estimate the brightness temperature of a source. Inversely, it says that the observed brightness temperature of type I bursts – typically 10^{11} K – is caused by a high-frequency wave density about 10^4 times the thermal value (given in Eq. 5.2.43 for Langmuir waves). These waves indeed cause no detectable harmonic emission (Exercise 9.5).

Exercises

9.1: The frictional drag of the ions on a fast electron decreases with increasing electron velocity. Calculate the stagnation velocity, v_r, where friction equals the electric force of a field **E** parallel to **B**. Prove Equation (9.2.10) and show that ions and electrons have the same v_r.

9.2: Currents have been proposed to heat the corona. Calculate the electric field driving a parallel current that balances thermal conduction loss by Ohmic heating. Approximate thermal conduction by

$$\frac{\partial \mathcal{E}}{\partial z} v_{te} \approx \frac{n_e m_e v_{te}^4}{L^2 \nu_{e,i}} \ [\mathrm{erg\ s^{-1} cm^{-3}}] \ , \tag{9.5.7}$$

where L is the length of the loop (take 10^9cm); $\nu_{e,i}$ is the classical, thermal electron-ion collision frequency, and $\mathcal{E}$ is the thermal energy density. Let $T = 10^6$K. Note that this scenario is too simplistic for a realistic model of coronal heating!

9.3: Runaway electrons have been suggested for the non-relativistic energetic electrons observed in solar and stellar flares. Let us assume a reconnection model (Fig. 9.2) with an electric field parallel to **B** in the central current sheet. Calculate self-consistently the initial runaway ratio and steady-state acceleration rate of two plasma flows with opposite magnetic fields of 100 G approaching each other with a speed of 10^4 cm s^{-1}. Assume an electron density of 10^{10} cm^{-3} outside the current sheet; 10^{13} cm^{-3} inside, a length and width of 10^9 cm each, a temperature of 10^6 K and use Coulomb conductivity (Eq. 9.2.3). Compare the derived acceleration rate to the observed rate (Chapter 6)!

9.4: An electric field parallel to **B** accelerates runaway electrons. If the electric field is too strong, the current is unstable and drives waves changing the friction between runaways and background. At $T_e = T_i$, the condition for stability is $V_d < 0.4\ V_{te}$. Consider first 100 keV electrons to be accelerated over a length of 10^5 cm in a current channel with density 10^8 cm^{-3}. What is the upper limit on the temperature compatible with stability? Secondly, calculate the energy achieved by runaway acceleration in a current sheet with $V_d = 10^{-4} v_{te}$, $T_e = 2 \cdot 10^6$ K and over the same length.

9.5: Type I radio bursts of solar active regions have been interpreted as emissions of intense low-frequency waves coupling to a low level of high-frequency waves. Use the observed absence of harmonic emission to derive an upper limit on the phonon temperature of the high-frequency waves. Observations show a ratio of harmonic to fundamental flux densities, $F_h/F_f < 10^{-4}$. Harmonic emission is expected to be optically thin, fundamental emission is optically thick. Use the Langmuir wave approximations of Section 6.1 for the high-frequency waves and assume an isotropic distribution with $k_m = k_D$ and $\Delta k = k_m$. Neglect absorption and refraction effects and take for example $T_e = 2 \cdot 10^6$ K, $n_e = 10^8$ cm^{-3} and a source thickness of 10^4 cm.

Further Reading and References

Currents in plasma

Alfvén, H.: 1981, *Cosmic Plasma*, D. Reidel. Dordrecht, Holland.

Foukal, P. and Hinata, S.: 1991, 'Electric Fields in the Solar Atmosphere: A Review', *Solar Phys.* **132**, 30.

Huba, J.: 1985, in *IAU Symp. 107*, 'Unstable Current Systems and Plasma Instabilities in Astrophysics' (eds. M.R. Kundu and G.D. Holman), D. Reidel, Dordrecht, Holland, p. 315.

Knoepfel, H. and Spong, D.A.: 1979, 'Runaway Electrons in Toroidal Discharges', *Nuclear Fusion* **19**, 785.

Morrison, P.J. and Ionson, J.A.: 1982, 'Temperature Gradient and Electric Field Driven Electrostatic Instabilities', *Phys. Fluids* **25**, 1183.

Papadopoulos, K.: 1977, 'A Review of Anomalous Resistivity for the Ionosphere', *Rev. Geophys. Sp. Sci.* **15**, 113.

Acceleration and heating by currents

Benz, A.O.: 1987, 'Acceleration and Energization by Currents and Electric Fields', *Solar Phys.* **111**, 1.

Holman, G.D.: 1985, 'Acceleration of Runaway Electrons and Joule Heating in Solar Flares', *Astrophys. J.* **293**, 584.

Radiation of low-frequency turbulence

Benz, A.O. and Wentzel, D.G.: 1981, 'Coronal Evolution and Solar Type I Radio Bursts: An Ion-acoustic Wave Model', *Astron. Astrophys.* **94**, 100.

Kai, K., Melrose, D.B., and Suzuki, S.: 1985, in *Solar Radiophysics*, (eds. D.J. McLean and N.R. Labrum), Cambridge University Press, Chapter 16 on noise storms.

Reconnection and current sheets

Heyvaerts, J., Priest, E.R., and Rust, D.M.: 1977, 'An Emerging Flux Model for the Solar Flare Phenomenon', *Astrophys. J.* **216**, 123.

Øieroset M., Phan T.D., Fujimoto M., Lin R.P. and Lepping R.P.: 2001, 'IN Situ Detection of Collisionless Reconnection in the Earth's Magnetotail', *Nature* **412**, 414.

Priest, E.R.: 1982, 'Solar Magnetohydrodynamics', D. Reidel, Dordrecht, Holland.

Priest, E.R. and Forbes, T.: 2000, 'Magnetic Reconnection', Cambridge University Press.

Reference

Hagyard, M.J.: 1988, 'Observed Nonpotential Magnetic Fields and the Inferred Flow of Electric Currents at a Location of Repeated Flaring', *Solar Phys.* **115**, 107.

CHAPTER 10

COLLISIONLESS SHOCK WAVES

A shock may be defined as a layer of rapid change propagating through the plasma. In its rest frame, the shock is approximately invariable in time and is marked by a rapid transition of parameters from the medium ahead (also called *upstream*) to the medium behind the shock (*downstream*). A simple example of a shock occurs when a plane piston moves at high velocity into a homogeneous plasma at rest. A shock develops ahead of the piston between the piled-up plasma and the undisturbed plasma. The shock propagates at a velocity similar to or higher velocity than that of the piston. Another example is a wave (or a pulse) with an amplitude so high that the wave velocity of the crest is faster than the velocity of the dip. Small disturbances thus travel faster on the wave crest, overtake it and add in front of it. The wave profile steepens in the front of the wave crest and develops into a shock (like an ocean wave approaching a sloping beach steepens and finally breaks).

A characteristic property of shocks is the dissipation of energy and increasing entropy as it moves through the plasma. The region of energy dissipation, called the *shock front*, is generally extremely thin compared to the lateral extension. It is often approximated by a surface.

The shock structures emphasized in this chapter are those that occur in low-density, fully ionized plasmas – such as coronae – where the collisional mean free path is much longer than the width of the structure. Momentum and energy of a disturbance are transferred across the shock by electric and magnetic fields, and waves. Such collisionless shock waves are produced by numerous violent processes in coronae, and in interplanetary and interstellar space. The eruptions of a coronal prominence or, on a smaller scale, of a surge or a spicule, are likely to generate a disturbance that propagates ahead and into the corona. Shocks are predicted in the outflows expelled by the reconnection process. Solar flares can produce shocks reverberating throughout the heliosphere. Strong collisionless shocks are conspicuous by associated particles, some of them – as in supernova remnants – accelerated to highly relativistic energies. Most of our knowledge on the physics of collisionless shocks, however, comes from much closer to home: spacecraft have gathered a rich collection of data at the Earth's bow shock and in the interplanetary medium over the past three decades.

In the logical sequence from simple to complex plasma phenomena, non-linear waves and shocks claim the final position. The field of collisionless shocks is too vast to be covered adequately in this limited space. We give here some fundamentals and concentrate on the astrophysically most interesting aspects of particle acceleration and heating.

10.1. Elementary Concepts

The basic equations for shocks are the same as for any plasma phenomenon: the Boltzmann equation (1.4.11) – or the Vlasov equation (5.2.1) for the collisionless case – and Maxwell's equations (1.4.2) – (1.4.5). As in Chapter 3 on linear waves, conservation of mass, momentum, and energy densities can greatly facilitate the treatment of shocks if the velocity distribution of the particles is such that the conserved moments can be analytically computed. Here, the theme is not the conservation of an equilibrium state being slightly disturbed by a wave, but the focus is on conservation laws in the large-amplitude transition of plasma parameters at the shock.

10.1.1. Types of shocks

There is an impressive variety of shocks in nature. In the classical *hydrodynamic* shock, the disturbance is propagated by collisions. A conducting gas with a magnetic field and (collision-dominated) Maxwellian particle distributions – described by the *magnetohydrodynamic* (MHD) equations – has three types of shock solutions in analogy with the fast, intermediate (Alfvénic), and slow MHD modes of linear waves. Correspondingly, the MHD shock description classifies fast-mode, intermediate-mode, and slow-mode shocks. They are usually abbreviated to *fast*, *intermediate*, and *slow shock*. Figure 10.1 depicts the three possibilities of the magnetic field orientation behind a shock moving at an oblique angle to the magnetic field of the plasma ahead.

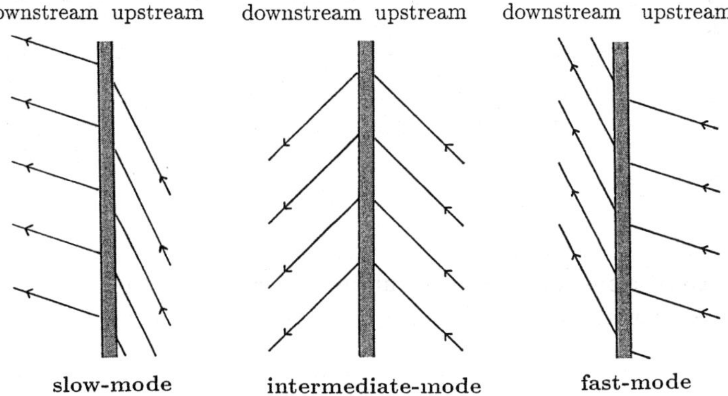

Fig. 10.1. The changes in magnetic field direction (arrow) from ahead (upstream) to behind (downstream) an MHD shock are characteristic for its type.

In all shocks, the component of the magnetic field normal to the shock front is continuous across the shock. This assures continuity of field lines. The component of the magnetic field tangential to the shock front increases across fast shocks, decreases across slow shocks, and remains equal for intermediate shocks.

MHD shocks travel faster than the linear MHD modes. In the limit where the compression ratio of downstream to upstream density approaches unity, the fast, intermediate, and slow shocks reduce to the corresponding linear waves and propagate at their phase velocities (Fig. 3.1). Similar to the linear wave modes, only one type of MHD shock can propagate perpendicular to the magnetic field, namely the fast shock.

The thickness of an MHD shock is limited by the distance to dissipate the energy. We estimate it in two thought experiments. (*i*) In a non-magnetic situation (such as an acoustic (slow) shock moving parallel to the magnetic field ahead), the scale of energy dissipation is roughly the mean free path of thermal particles. Physically, this is the approximate distance the plasma behind the shock can influence the plasma ahead and *vice versa*. For an interplanetary density of $n_e \approx 10$ cm^{-3} and a temperature of $T \approx 10^5$ K, the non-magnetic shock thickness would be more than an astronomical unit. It is reduced by a factor of $\cos\theta$, if there is an angle θ between shock propagation and upstream magnetic field. Even that is many orders of magnitude longer than is actually observed.

(*ii*) Another form of energy dissipation is Ohmic heating due to finite conductivity. It may be relevant for shock propagation nearly perpendicular to the magnetic field (to be called *quasi-perpendicular* shocks). If it dominates, the shock thickness Δx can be estimated in the following way. The heating rate is

$$\frac{\partial \mathcal{E}}{\partial t} = \frac{J^2}{\sigma} \quad . \tag{10.1.1}$$

The current density may be approximated from Ampère's law and the magnetic fields ahead and behind the shock. Typical coronal values suggest an incredible shock thickness of some 10^{-4} cm (Exercise 10.1). This is smaller than important plasma length-scales, such as the ion gyroradius or the Debye length, and cannot be realistic either.

The thought experiments demonstrate that a description using Coulomb collisions is generally inadequate for the physics within the coronal shock layer. Kinetic plasma processes, and in particular collisionless waves, cause faster interactions between the upstream and downstream plasmas. This reduces effectively the mean free path and increases the heating rate (10.1.1). If the dominant dissipation mechanism is known, an order-of-magnitude estimate of the shock thickness may be obtained by equating energy input and dissipation rate in the rest frame of the shock.

The major variety of collisionless shock fronts may be called *turbulent*. They correspond to non-linear structures for which no single wave mode solution (collisional or collisionless) exists. Collisionless waves are driven by small-scale processes, such as unstable electric currents and beams. The interaction of the wave turbulence with particles simulates collisions. In effect, this causes an *MHD-like* shock phenomenon, where the width of the turbulent layer takes the role of the mean free path. It is therefore not surprising that MHD considerations often lead to quantitatively correct results by using some anomalous conductivity. In many coronal shocks, the electron and ion temperatures do not equilibrate within the shock, or the stresses between the media ahead and behind the shock may be transmitted entirely by non-thermal, energetic particles. Such cases can obviously not be approximated by MHD.

The different mobility of electrons and ions being compressed creates charge separations and electric fields. The electric potential of collisionless shocks can reflect ions approaching the front from ahead. Quasi-perpendicular shocks reflect ions at an appreciable number if the Alfvénic Mach number, $M_A := V_1/c_A$, exceeds a threshold (between about 1.1 and 2.2, depending on the upstream plasma parameters). V_1 is the shock velocity relative to the upstream medium. Such strong shocks are called *supercritical.* Their observational characteristic are ion streams in the region ahead of the shock causing intense low-frequency electromagnetic turbulence in a region called the *foot* of the shock (Fig. 10.3). The global shock structure changes at the threshold to supercritical behavior. In subcritical shocks, the energy is mainly dissipated by resistivity and most of the heating goes into electrons. Heating of both ions and electrons takes place in supercritical shocks as the wave profile breaks (i.e. ion reflection sets in) and other dissipation mechanisms, such as viscosity appear.

A further distinction of shocks is frequently made according to their driving agent. The extreme models are called *blast wave* and *piston-driven* shocks. The *blast wave* limit considers an impulsive deposition of mass, energy and momentum that is short compared to the propagation time. The input is an explosion, whose effects then propagate through the corona or interplanetary medium as a shock front. After the explosion, the flow at the site of explosion is soon exhausted and does not propagate to the point of observation. The observer only sees a disturbance passing by. In the *driven shock*, the hot material continues to be ejected from the source and follows the shock front. This case resembles the flows in a classic laboratory device called shock-tube, where a piston is driven into a gas. Astrophysical examples of pistons include destabilized magnetic loops or eruptive filaments, flare material, and supernova ejecta. Observations of the kinetic energy flux following immediately the passage of a shock front can distinguish the two types. At a given point of observation, the energy flux decreases with time behind blast waves; for driven shocks it rises. In interplanetary space, where shocks and energy flux can be observed, intermediate cases have also been reported.

In summary, shocks can be classified according to:

(1) particle velocity distribution: MHD or collisionless;

(2) angle between shock-normal and upstream magnetic field: quasi-perpendicular or quasi-transverse;

(3) ion acceleration: subcritical or supercritical;

(4) driving agent: blast wave or piston driven.

10.1.2. Conservation Equations (MHD Shocks)

Consider a shock in its rest frame (Fig. 10.2). The plasma moving into the shock from upstream (right) leaves the shock layer downstream (left) at a different, smaller velocity. In equilibrium and in absence of sources and sinks, the inflowing mass, momentum, and energy must equal the outflowing values. It is instructive to further simplify the mathematics by assuming the MHD approximations (Section 3.1.3), implying that the velocity distributions ahead and behind the shock are Maxwellian and $T_e = T_i$. We shall show in this section how the three conservation equations then determine the downstream state entirely in terms of the shock velocity and the conditions ahead.

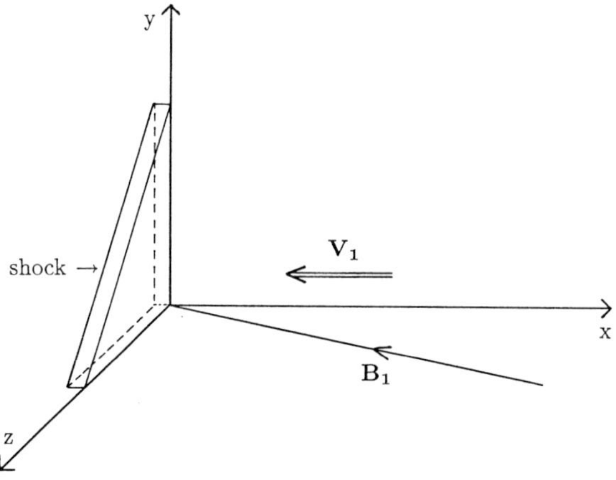

Fig. 10.2. A *normal incidence* frame of reference can be defined, in which the shock is at rest, the upstream velocity is normal ($V_{x1} = V_1$), and the magnetic field is in the (x, z)-plane (implying $B_{y1} = 0$ ahead of the shock).

Let the shock be stationary and homogeneous in the y and z directions of the coordinate system depicted by Figure 10.2. The normal incidence frame is one of the rest frames of the shock, moving along the front such that the inflow is normal to the front. It will furthermore be assumed that the mean velocities of the species are equal and that there is only one ion species, having a density $n_i = n_e$.

The MHD equation of continuity (3.1.40) is integrated in the x-direction from an arbitrary point 1 sufficiently far in the upstream medium through the shock to a point 2 downstream. The result, by partial integration, is

$$\rho_1 V_1 = \rho_2 V_{x2} \quad , \tag{10.1.2}$$

where $V_1 = V_{x1}$ is the shock speed. The equation expresses that the inflowing mass per unit area equals the outflow. An equation like (10.1.2) is called a *jump condition*. Since the regions ahead and behind the shock are assumed to be homogeneous, the subscripts 1 and 2 refer to the general conditions before and after the shock. In a compression shock $\rho_2 > \rho_1$, thus $V_1 > V_{x2}$, and the incoming plasma is decelerated at the shock. The opposite would not be invariable: an expansion shock flattens as it propagates and finally disappears. It is not of interest here.

The conservation of momentum is described by the vector equation (3.1.50). Integrating the x-component over the x-direction, one derives again by partial integration

$$\rho_1 V_1^2 + p_1 + \frac{B_{z1}^2}{8\pi} = \rho_2 V_{x2}^2 + p_2 + \frac{B_{z2}^2}{8\pi} \quad . \tag{10.1.3}$$

The equation expresses pressure balance between the upstream and downstream plasmas. Note how the incoming energy is converted into heat and magnetic energy.

It has been used in Equation (10.1.3) that Ampère's equation requires $B_{y2} = 0$, and that the conservation of the y-component of momentum density results in $V_{y2} = 0$. The conservation of the z-component of momentum yields

$$\frac{B_{x1}B_{z1}}{4\pi} = \frac{B_{x2}B_{z2}}{4\pi} - \rho_2 V_{x2} V_{z2} \quad . \tag{10.1.4}$$

Energy conservation requires that the energy flow through the shock is constant. The flow is composed of kinetic energy, $\sum_\alpha \frac{1}{2} \int m_\alpha v^2 \mathbf{v} f_\alpha(v) d^3v$, and electromagnetic energy, $\mathbf{E} \times \mathbf{B}c/(4\pi)$. The conservation of the x-component requires

$$[5p_1 + \rho_1 V_1^2 + \frac{B_{z1}^2}{2\pi}]V_1 = [5p_2 + \rho_2(V_{x2}^2 + V_{z2}^2) + \frac{B_{z2}^2}{2\pi}]V_{x2} - \frac{B_{x2}B_{z2}}{2\pi}V_{z2} \quad . \tag{10.1.5}$$

From the absence of magnetic monopoles ($\nabla \cdot \mathbf{B} = 0$) follows

$$B_{x1} = B_{x2} =: B_x \quad , \tag{10.1.6}$$

and from Faraday's law (using $\nabla \times \mathbf{E} = 0$) combined with $\mathbf{E} = -(\mathbf{V} \times \mathbf{B})/c$ one derives

$$V_1 B_{z1} = V_{x2} B_{z2} - V_{z2} B_x \quad . \tag{10.1.7}$$

Equations (10.1.2) – (10.1.7) relate the downstream parameters to the upstream state. They are generally called *Rankine-Hugoniot relations*. There is a general solution for each downstream MHD parameter in terms of upstream values. Since it is rather intricate and not instructive, we shall consider here three special cases.

If the shock speed is *extremely high*, the upstream energy density is dominated by the kinetic energy. Neglecting upstream magnetic and thermal energy density in Equation (10.1.3), the Rankine-Hugoniot relations reduce to

$$\frac{\mathbf{V}_{x2}}{\mathbf{V}_1} = \frac{\rho_1}{\rho_2} \approx \frac{1}{4} \tag{10.1.8}$$

and

$$p_2 \approx 3\rho_2 V_{x2}^2 \tag{10.1.9}$$

(Exercise 10.2). Surprisingly, Equation (10.1.8) states a maximum compression ratio for strong shocks. Equation (10.1.9) expresses that the fraction of initial energy converted into thermal energy is 9/16 and is independent of the shock velocity. This heating is independent of the processes taking place in the shock layer as long as they yield Maxwellians and $T_e = T_i$. Heating is one of the prime reasons for interest in shock waves, both in the laboratory and in the universe, as it provides a controlled means of producing high-temperature plasmas.

For *perpendicular* fast-mode shocks ($B_{x1} = B_{x2} = 0$), the Rankine-Hugoniot relations simplify to

$$\frac{\mathbf{V}_{x2}}{\mathbf{V}_1} = \frac{n_1}{n_2} = \frac{B_{z1}}{B_{z2}} = \frac{1}{8}\left[1 + \frac{5p_1}{\rho_1 V_1^2} + \frac{5B_{z1}^2}{8\pi\rho_1 V_1^2}\right] + \frac{1}{8}\left[\left(1 + \frac{5p_1}{\rho_1 V_1^2} + \frac{5B_{z1}^2}{8\pi\rho_1 V_1^2}\right)^2 + \frac{2B_{z1}^2}{\pi\rho_1 V_1^2}\right]^{1/2} . \quad (10.1.10)$$

Equation (10.1.4) requires that $V_{z2} = 0$. A compression shock only exists if $V_{x2}/V_1 < 1$, thus from Equation (10.1.10)

$$M := \frac{V_1}{\sqrt{c_{s1}^2 + c_{A1}^2}} > 1 \quad . \quad (10.1.11)$$

M is called the *magnetoacoustic Mach number*. The sound velocity, c_s, and the Alfvén velocity, c_A, have been defined previously (Eqs. 3.2.12 and 3.2.13). Their combination in the denominator is the magnetoacoustic speed (Eq. 3.2.24). Equation (10.1.11) expresses that the shock must move faster than the magnetoacoustic speed in the upstream plasma. Condition (10.1.11) can be inverted to give $V_{x2} < \sqrt{(c_{s2}^2 + c_{A2}^2)}$. The physics of these conditions is that fast-mode disturbances in the downstream region reach the front and pile up; the front, however, cannot discharge into the region ahead. The strong shock jump relations (10.1.8 and 10.1.9) are valid for $M \gg 1$.

The ratio, β, of thermal to magnetic pressure (Eq. 3.1.51) controls the influence of the terms in Equation (10.1.10). For $\beta \ll 1$ – as in the solar corona – the magnetic pressure dominates, and the relevant Mach number is approximately the Alfvénic Mach number, $M \approx M_A := V_1/c_{A1}$.

Shocks propagating *parallel* to the magnetic field are particularly simple as the magnetic field terms drop out of the Rankine-Hugoniot relation (10.1.10). The jump conditions are purely hydrodynamic and identical to compression shocks of the sound mode. The shock condition (10.1.11) becomes $M = V_1/c_{s1} > 1$.

10.2. Collisionless Shocks in the Solar System

Cosmic shock theories were historically derived for hydrodynamic shocks in neutral gases. When the first evidence of shocks in the solar corona was identified in coronal radio type II bursts by R. Payne-Scott and coworkers in 1947, it was not clear that collisionless plasma waves were the source of the radio emission, nor was it known that they constitute an important part of the shock phenomenon. Soon after, T. Gold postulated shock waves propagating into interplanetary space and causing aurorae at Earth. Ten years later the IMP-1 satellite detected the bow shock of the Earth in the solar wind and opened a nearby testing ground for theories of collisionless shocks. Since then the fields of solar, interplanetary and planetary shocks have interacted fruitfully and are now growing together. Here we focus on fast-mode shocks, associated turbulent waves, and particle acceleration.

For all known shocks in the interplanetary medium, the population of accelerated particles carries relatively little mass, momentum, and energy, and therefore has a negligible impact on the large-scale dynamics. The macrostructure can be described approximately by single-fluid MHD equations (Section 10.1.2). Nevertheless, the single particle physics and collisionless processes – although of second order in the solar system – are important in astrophysics (in particular for cosmic ray acceleration). The kinetic microphysics has received great observational and theoretical attention for this reason. This overview of observations of shocks aims at a basic understanding of the most important kinetic processes.

10.2.1. Planetary and Cometary Bow Shocks

A collisionless shock forms as the super-Alfvénic solar wind hits a planet's magnetosphere. It is called *bow shock* in analogy to the non-linear waves forming at the bow of a boat moving faster than the surface gravity waves of water. The analogy should not be taken too literally since the wind flowing from the Sun carries the interplanetary magnetic field. This super-Alfvénic flow hitting an obstacle yields an illustrative example of a magnetic piston-driven shock. The kinetic shock phenomena are controlled by the magnetic field and strongly depend on the direction of the upstream field.

Figure 10.3 outlines the geometry of a bow shock for a planet with its own magnetic field. As qualitatively expected from MHD considerations (Section 10.1.1), the shock is much thinner where it is quasi-perpendicular (near A) than where it is quasi-parallel. Intense magnetic field fluctuations (indicated by wavy lines) are ubiquitous in the region ahead of the quasi-parallel shock. The various regions in the Earth's bow shock have been investigated extensively by orbiting satellites. Other spacecraft have flown through the bow shocks of nearly all other planets with similar results. Marked differences have only been found in bodies without magnetospheres (such as Venus, the Moon, and comets).

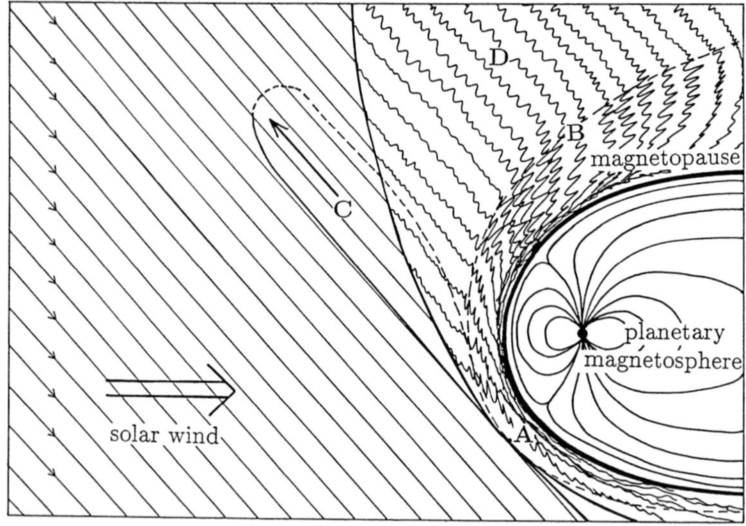

Fig. 10.3. A collisionless shock (dashed curve) forms as the solar wind impinges on the magnetosphere of a planet such as the Earth. The two plasmas and magnetic fields are separated by the magnetopause (thick curve). The points A and B mark the locations where the magnetic field relative to the shock normal is perpendicular and parallel, respectively. The magnetic field is compressed and variable in the shock. Energetic electrons and electron driven waves are observed in the electron foreshock C. Accelerated ions are unstable to magnetoacoustic waves forming the foot of the shock (or 'ion foreshock') D.

A. *Non-Thermal Particles*

Typically, 1% of the solar wind energy impinging on the Earth's magnetosphere is transferred to super-thermal particles in the upstream region. Most of the energy is taken up by ions. There are several particle populations distinct by energy and angular distributions.

- The population of reflected ions is most important: they form a steady field-aligned beam in the sunward direction with a velocity of about $-V_1$ in the frame of the shock (or a few keV). The ions are found on interplanetary field lines that are nearly tangent to the shock surface and connected to the bow shock at point A in Figure 10.3. (In other words, they are a result of the quasi-perpendicular shock.)

- An ion population having a broad energy distribution (with energies beyond 100 keV) is also observed. It is nearly isotropic and highly variable in density. These *diffuse ions* may (at least partially) originate from the above ion beam by quasi-linear diffusion (Section 7.3.3). Their energy density is comparable to that of the reflected ions. As an alternative, the theory of acceleration by quasi-parallel shock processes has also received wide support. The diffuse ions are observed downstream (anti-Sunward) of the reflected ions (in the upper right corner of Fig. 10.3), almost filling the entire ion foreshock region. In terms of total energy, they are the dominant non-thermal particle population.

- *Electron beams* of 1 – 2 keV per particle are observed mainly on interplanetary magnetic field lines that are newly connected to the bow shock (electron foreshock, region C in Fig. 10.3). These field lines are nearly tangential to the shock front (point A in Fig. 10.3). Presumably, the electrons are accelerated where the field lines meet the shock in the quasi-perpendicular region. Because the particle velocities for these energies are much higher than the solar wind speed, energetic electrons can be found far upstream.

- The background electron distribution has a *super-thermal tail* in the sunward direction between the shock and the tangential field line.

- Electrons and protons with energies > 100 keV have been reported. They may also be a part of the electron beam and the diffusive ion populations, respectively. Compared to the other non-thermal particles, they are energetically unimportant.

B. *Upstream Waves*

The various non-thermal particle populations in the upstream region represent sources of free energy. They can drive many types of waves. The relation between observed waves and driving particles is not always clear.

Figure 10.4 displays the waves measured by a spacecraft traversing the Earth's bow shock from the solar wind (upstream) into the magnetosheath (region between shock front and magnetopause, cf. Fig. 10.3). The ordinate axis is observing time. The satellite moves at 83 km per minute. The amplitude of the oscillating electric and magnetic fields are shown in many channels at a large range of frequencies (top and middle). The magnetic field (given in Fig. 10.4, bottom) jumps by a factor of 3 from upstream to downstream. In the shock region, the overshoot even reaches 4.8 times the upstream value.

Most energetic in the absolute scale are large amplitude, low-frequency waves of the *fast magnetoacoustic* type (i.e. combining magnetic and density compressions like the MHD mode derived in Section 3.2.3). They are most prominent in the shock and downstream regions of Figure 10.4. Nevertheless, they also dominate energetically in the upstream plasma. As the waves are strongly correlated in time and space with the diffusive ions, the ions seem to be the source of the waves. An attractive possibility is the electromagnetic instability of ion beams (Section 7.3.1). As the solar wind moves faster than these waves, the waves are convected into the shock front and influence its nature. Also prominent (see Figure 10.4) but much weaker, are *whistler* waves, which may have several sources, including streaming ions and transverse electrons (Section 8.2.1).

Electron plasma waves near the plasma frequency are detected throughout the foreshock region. Their source is at the boundary of the electron foreshock. The waves correlate with the electron beams and have phase velocities of $1 - 2 \cdot 10^9$

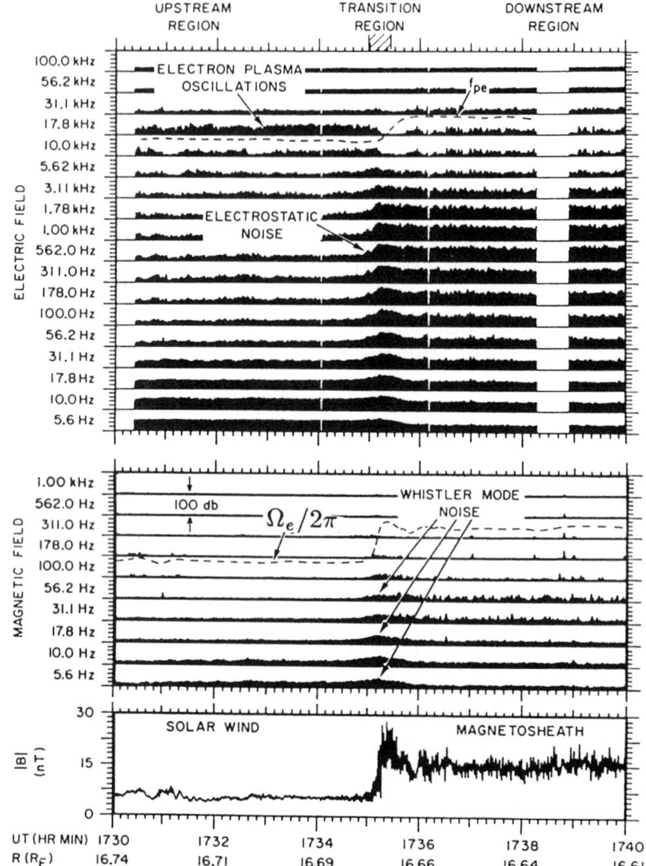

Fig. 10.4. The Earth's bow shock excites many wave modes. The wave electric (*top*) and magnetic (*middle*) fields are shown versus time as the ISEE–1 satellite traverses from the upstream into the downstream regions. The distance R from Earth center is also given in units of Earth radii. The spacing between frequencies is logarithmic. The plasma frequency and gyrofrequency are indicated by dashed curves. The amplitude scale is also logarithmic in each channel, ranging from the background to a factor of 10^{10} (100 dB). The background (DC) magnetic field strength is presented in the bottom panel in units of nanotesla ($= 10^{-5}$G). (After D.A. Gurnett, published in Tsurutani and Stone (eds.), 1985.)

cm s^{-1} corresponding to the observed electron energies of a few keV. A likely mechanism is the bump-on-tail instability (Section 5.2.4). In addition, *ion acoustic* waves, caused apparently by the diffusive ions, are frequently observed in the range 0.1–10 kHz (Fig. 10.4, labelled 'electrostatic noise'), upstream of the quasi-parallel shock region. At even lower frequencies ($\lesssim$ 50 Hz), Figure 10.4 shows unidentified broadband and impulsive electrostatic emissions. In observations near the quasi-perpendicular shock region, *lower hybrid* waves have been identified. They may be driven by reflected ions.

Distant spacecraft in the upstream solar wind have observed propagating *electromagnetic radiation* originating from the bow shock. Both the fundamental (at about the plasma frequency) and the harmonic emission have been detected. The bandwidth of the harmonic is very narrow, sometimes less than 3% of the center frequency. Two emission mechanisms have been proposed: (*i*) Langmuir waves may be driven unstable by the electron beams accelerated by the quasi-perpendicular shock. They could generate radio emission much like a solar type III burst (Section 6.3). (*ii*) Alternatively, a likely electron acceleration mechanism (to be discussed in Section 10.3.1) produces a loss-cone distribution. As in the solar wind $\omega_p/\Omega_e \approx 40$, the theory presented in Section 8.2 predicts instability for z-mode and upper hybrid waves (Eq. 8.2.23). The coupling of such waves, or an upper hybrid wave with a low-frequency wave then could produce radio emissions as in solar type IV bursts (Section 8.4.2), but from a small, narrowband source.

Finally, *Alfvén waves* have been discovered far upstream of bow shocks. They play a particularly prominent role in the shocks forming ahead of comets. As cometary nuclei are small and have a negligible magnetic field, comets generate shocks in a different way than planets. Sunlight evaporates atoms and molecules from the comet's surface, and solar UV radiation ionizes them. A plasma forms (called ionosphere), that cannot be penetrated by the solar wind (property of frozen in flux, Section 3.1.3). A few neutral atoms escape beyond the shock front between the cometary ionosphere and the solar wind. Ultimately ionized, they form a counter-streaming ion beam in the solar wind. Such a beam is prone to the classic electromagnetic ion beam instability of Alfvén waves (Section 7.3).

10.2.2. Interplanetary shocks

Interplanetary space is furrowed by shocks, in particular at distances beyond 1 AU from the Sun. Here we concentrate on the most abundant shocks, the fast-mode shocks produced by coronal mass ejections (CME). CMEs are observed by white-light coronagraphs and will be discussed in the next section.

The phenomena found in interplanetary travelling shocks are similar to the bow shocks ahead of planets. Not every interplanetary shock has all the bow shock features, however. The bow shocks comprise quasi-perpendicular as well as quasi-parallel sections; interplanetary shocks may have smaller curvature with less variation in shock angle. Generally, the density of non-thermal particles accelerated by interplanetary shocks is smaller, but the particles reach higher energy. Interplanetary shocks generally have a lower Mach number and a much greater spatial extent.

Interplanetary shocks are usually supercritical. They reflect and accelerate ions. The number of accelerated ions increases with Mach number; for quasi-parallel shocks, it was found to increase also with the upstream wave level. These relations are suggestive of first-order Fermi acceleration to be discussed in Section 10.3.2 (scattering between converging waves or shocks). Quasi-parallel shocks cause non-thermal ions and ion acoustic turbulence millions of kilometers upstream.

Radio emissions are of particular interest as they yield information by remote sensing. Interplanetary shocks with detectable radio emission belong to the most energetic shocks produced by the Sun. They are associated with powerful flares and with the most massive and energetic CMEs. The radio emission decreases with distance from the Sun and usually disappears before about 0.7 AU in the background noise. Some shock-associated radio emission is produced by electron beams escaping far into the upstream region. Other emissions seem to originate near the shock. All radio sources, however, are located ahead of the interplanetary shock.

10.2.3. Coronal shocks

In the corona we can study much more powerful shocks, in which certain features are more pronounced. The radio emission alone (energetically a very minor phenomenon) emitted by a strong coronal shock sometimes exceeds the energy flux of all non-thermal particles upstream of the Earth's bow shock. Coronal shocks are also optically observable in Thomson scattered white light and by their footprints in the chromosphere (Moreton wave). Optical images give a global view of the phenomenon (Fig. 10.5), complementary to spacecraft observations probing the details along their trajectory.

A. *Coronal Mass Ejections*

Thomson scattering of electromagnetic radiation on free electrons is proportional to electron density. Density enhancements in the corona can be observed by coronagraphs (optical telescopes with an occulting device blocking the direct sunlight from the photosphere). Figure 10.5 shows material ejected by a coronal process. It produces a density enhancement that is observable on its way through the upper corona into interplanetary space. Such an event is called *coronal mass ejection* (CME).

In periods of high solar activity, more than one CME per day on average is observable with current techniques. They are related to eruptive prominences (also visible in Fig. 10.5) or flares. The speed of the leading edge ranges from less than 100 km s^{-1} to well beyond 1000 km s^{-1} at an estimated Alfvén velocity of 400 km s^{-1}. Apparently, not all CMEs are fast enough to be shocks. The flare-associated CMEs have generally a higher speed, and the fastest ones often propagate as far as 1 AU to be registered by spacecraft.

Curiously, detailed investigations have revealed that the relation between flare associated CMEs and flares are not as straightforward as one may think. CMEs can start up to 30 minutes before the flare. Thus the flare cannot be directly responsible for driving the CME. The CME seems to have its own driver that, under certain conditions, triggers a flare. CMEs are far more numerous than

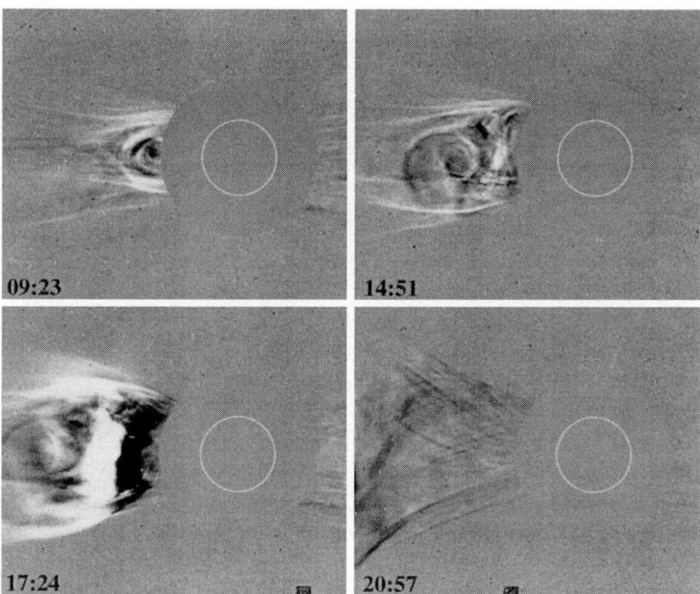

Fig. 10.5. The white light image of a coronal mass ejection (observed by the LASCO C2 coronagraph on the SoHO satellite) shows a bright arc moving out. The images are running differences to the preceding image. The white circle is the size of the photospheric disk (from Dere et al, 1999).

interplanetary shocks. For further details and references, the reader is referred to the reviews by Kahler (1987) and Hundhausen (1988).

B. *Type II Radio Bursts*

Extremely intense, narrow bands of radio emission caught the attention of the earliest observers. The bands are at harmonic frequencies having a ratio of slightly less than 1 : 2 and usually drift to lower frequency. The drift rate is typically two orders of magnitude smaller than for type III bursts produced by electron beams. The phenomenon has been named type II burst.

As an example, Figure 10.6 shows the start of a type II burst. After an interruption at 14:24:40, the event continues for more than 15 minutes. The observational characteristics are summarized in the following paragraphs. The references are listed in the reviews given at the end of the chapter.

The emission in two harmonic frequency bands is similar to the Earth's bow shock and to interplanetary shocks. It is suggestive of plasma emission near the plasma frequency and twice its value. The solar rate of frequency drift, combined with a density model, suggests radial velocities between 200 and 2000 km s^{-1}, about 2 orders of magnitude lower than type III bursts. Such outward source motions have been confirmed by multi-frequency radioheliographs. The enlargements

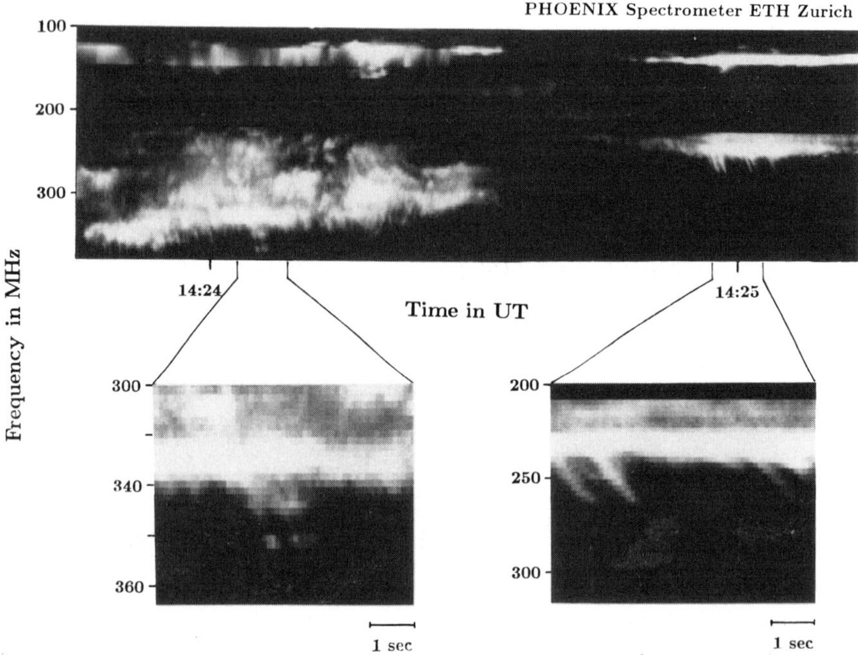

Fig. 10.6. Spectrogram of a section of a solar type II radio burst. The bright features indicate enhanced radio flux in logarithmic scale. *Bottom*: Two enlargements of the above overview showing horizontal 'backbone' and skew 'herringbone' emissions (from Benz and Thejappa, 1988).

in Figure 10.6 demonstrate that the radio emission has two components. A variable, narrow band drifts slowly to lower frequency at the same rate as the whole phenomenon. It is metaphorically termed the *backbone.* In addition, rapidly drifting, broadband structures shoot out of the backbone to higher and (not visible in Fig. 10.6) lower frequency. They are called *herringbones.* The two components are obviously different in their intrinsic time scale and bandwidth. Note that the backbone has a bandwidth of only $2 - 3\%$ at some time. If interpreted in terms of plasma emission ($\omega \approx \omega_p$), the bandwidth requires a source density homogeneity better than 5%, suggesting a small, stationary source moving with the shock. The herringbones resemble type III bursts and are generally interpreted as signatures of a beam of energetic electrons accelerated near the source of the backbone and escaping upstream from the shock. It is very likely that the herringbone emission is due to electron plasma waves excited by the bump-on-tail instability (Section 5.2.4) and scattered into radio waves by means similar to type III bursts.

Herringbone emission, and thus appreciable electron acceleration, occurs only in about 20% of all type II bursts. The stronger the backbone emission, the more likely are herringbones. The backbone flux density correlates with shock velocity (e.g. Cane and White, 1989).

Only 65% of the shocks observed as a fast ($v > 500$ km s^{-1}) coronal mass ejection radiate type II emission. On the other hand, CMEs slower than 200 km s^{-1} are occasionally accompanied by type II bursts (e.g. Kahler *et al.*, 1985). Furthermore, type II bursts often end around 20 MHz (about 2 $R_\odot$ from the center of the Sun) while the shocks apparently continue. There seem to be additional conditions on electron acceleration and radio emission.

Type II associated shocks are generally productive in particle acceleration, manifest in significant associations with interplanetary proton streams and coronal type IV radio emission. The causal relation to flares is clear for type II bursts. They start about one minute after the peak of the flare hard X-rays and are more delayed the lower the starting frequency.

A tentative and popular scenario proposes CME to be flare independent shocks, piston-driven by rising magnetic loops. If a large flare is associated, the flare blast wave – starting later and catching up from behind – causes a type II-emitting, high Mach number shock. A blast wave interpretation is attractive due to the property of all waves (including shocks) to refract toward regions of low phase velocity (see for example Fig. 6.4). As the energy of a shock is approximately conserved, the magnetoacoustic Mach number increases in such a region. For the same reason, the shock would develop into a more quasi-perpendicular type, prone to electron acceleration (Section 10.3.1). This scenario is awaiting observational confirmation.

10.3. Particle Acceleration and Heating by Shocks

Particle acceleration is a characteristic property of collisionless shocks. There are two types of acceleration processes. (*i*) Most important are mechanisms reflecting upstream particles by the moving front. There are several derivatives of this idea, two will be presented in Sections 10.3.1 (for electrons) and 10.3.2 (for ions). (*ii*) In addition, the upstream wave turbulence can heat or – more generally – energize background particles (primarily electrons). It will be discussed in Section 10.3.3. These types of acceleration should not be considered a complete list of all possible acceleration processes at shocks.

10.3.1. ELECTRON ACCELERATION AT QUASI-PERPENDICULAR SHOCKS

In 1949 E. Fermi proposed that 'magnetic clouds' moving in the Galaxy could accelerate charged particles to cosmic ray energies. He did not use the word 'shock', but considered magnetic mirrors (Section 2.1.2) embedded in the clouds that elastically reflect particles. In the observer's frame of reference, the particle gains energy in a head-on collision (opposite initial velocities of particle and mirror), and it loses energy in a catch-up collision, when the particle approaches the mirror from behind.

Fermi later developed the idea into two types of acceleration mechanisms. In the *first-order* Fermi process, two mirrors approach each other and the particles in between collide many times, gaining energy at each reflection. The *second* type assumes clouds moving in random directions. Since head-on collisions are statistically more likely than catch-up ones, the colliding particles gain energy on the average. Both historic concepts are still in use, although modified in details. The first-order Fermi process, being simpler to realize and more efficient, will be studied first.

In this section we concentrate on a single reflection with high shock velocity. The process is referred to as *shock drift acceleration* or fast-Fermi acceleration. Assume that the gyroradius of a charged particle is much smaller than the shock thickness. Then a shock can act like a magnetic mirror, as there is an increase in magnetic field strength across the front for fast-mode shocks (Eqs. 10.1.2 – 10.1.7). The magnetic moment of the particle is conserved in the shock-particle interaction. According to Equation (2.2.3) and if there were no electric field, a particle with velocity v is reflected under the condition

$$B_m > B_1 \left(\frac{v}{v_\perp^1} \right)^2 \quad , \tag{10.3.1}$$

where the superscript 1 refers to the values before the encounter, and B_m is the mirroring magnetic field in the shock front. However, Equation (10.3.1) is not generally valid, since $\mathbf{E}$ is usually not zero – neither in the observer's frame nor the normal incidence frame defined in Section 10.1 – as $\mathbf{E} = -(\mathbf{V} \times \mathbf{B})/c$. For this reason, we have to study a special frame of reference in the following subsection, where Equation (10.3.1) can be used.

A. *De Hoffmann-Teller Frame*

Equation (10.3.1) only holds in the particular reference frame where the electric field due to the upstream plasma motion vanishes. This is known as the *de Hoffmann-Teller (HT) frame*, in reference to early workers in collisionless shock theory. In this frame the shock front is at rest and the upstream flow is along the magnetic field, thus $\mathbf{V}_1^{\rm HT} \times \mathbf{B}_1 = 0$. The HT frame moves along the shock front relative to the normal incidence frame with a velocity $V_{\rm HT}$ so that in the HT frame one would see the upstream plasma flowing in along the magnetic field lines. This is the geometry of somebody watching rain from a moving car. The speed, $V_{\rm HT}$, is chosen to see the flow coming at the given angle θ_1 from zenith, where θ_1 is the magnetic field angle to the shock normal, $\mathbf{n}$ (Fig. 10.7), namely

$$\mathbf{V}_{\rm HT} = \frac{\mathbf{n} \times (\mathbf{V}_1 \times \mathbf{B}_1)}{\mathbf{B}_1 \cdot \mathbf{n}} = V_1 \tan \theta_1 \quad . \tag{10.3.2}$$

As observed in the normal incidence frame, $\mathbf{V}_{\rm HT}$ is simply the velocity at which the intersection point of a given field line and the shock surface moves along the shock front. Faraday's law requires the flow behind the shock to be also aligned with the magnetic field in the HT frame.

The upstream inflow velocity in the HT frame, $V_1^{\rm HT}$ (not to be confused with $V_{\rm HT}$, see Fig. 10.7), is along the magnetic field by definition. It amounts to

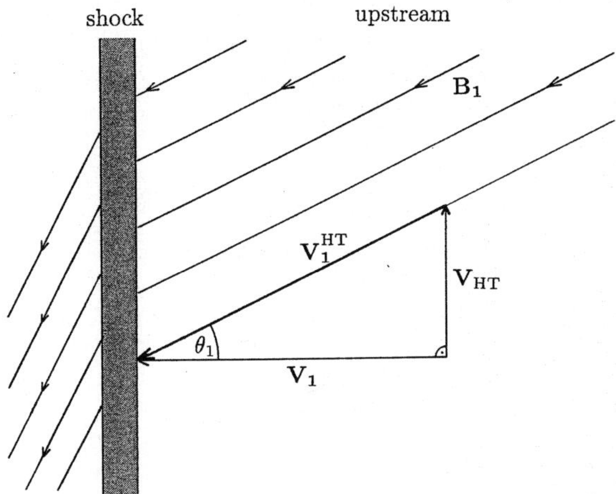

Fig. 10.7. Upstream flow velocities in the normal incidence frame of reference, $\mathbf{V}_1$, and in the de Hoffmann-Teller frame, $\mathbf{V}_1^{\mathrm{HT}}$. The transformation velocity, $\mathbf{V}_{\mathrm{HT}}$, is parallel to the shock front.

$$V_1^{\mathrm{HT}} = \frac{V_1}{\cos\theta_1} \tag{10.3.3}$$

from simple geometry. In the HT frame, the upstream flow of a quasi-perpendicular shock hits the front at a high velocity as $\cos\theta_1 \ll 1$. In the observer's frame, where the intersection with the shock moves along a given field line at a high velocity, one anticipates that the reflected particles gain considerable energy. We shall investigate this in the following section.

B. *Electron Acceleration*

Since there is no electric field, the magnetic moment of a charged particle is conserved in the HT frame of reference. According to Equation (10.3.1), particles with a pitch angle $\alpha > \alpha_c := \arcsin(B_1/B_2)^{1/2}$ – i.e. those with high transverse velocities – are reflected. Figure 10.8, drawn in the HT frame, displays the velocity distribution. Incoming particles are found at an average parallel velocity of $-V_1^{\mathrm{HT}}$, and reflected particles are shifted to $+V_1^{\mathrm{HT}}$. Reflection is limited to particles with large pitch angle in the HT frame. As the condition for reflection (Eq. 10.3.1) is independent of charge and mass, and as electrons have a much faster transverse motion, there are more electrons reflected than ions. Shocks with high inflow speed accelerate electrons to the highest energies and will be of primary interest. If V_1^{HT} is large – that is, in nearly perpendicular shocks – reflection is limited to the electrons with the largest pitch angle (i.e. to a fraction of the halo population in Fig. 10.8).

We have neglected any electric fields in the shock layer building up by preferential electron acceleration or by local field fluctuations. They can be included easily into the orbit calculation through a potential. Its effect is to further reduce the population of reflected electrons particularly at low parallel velocity.

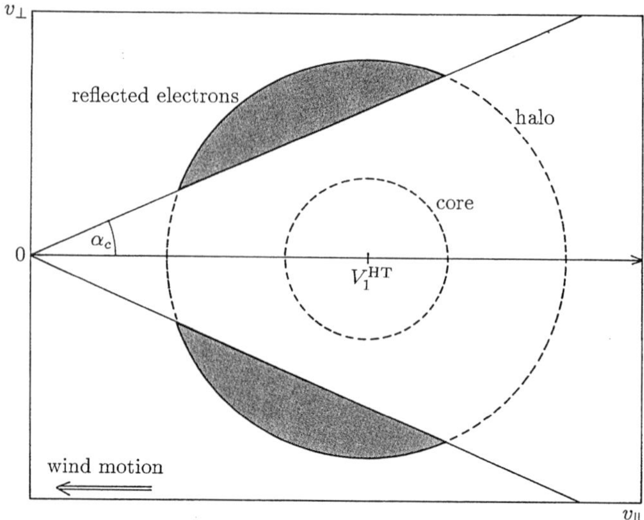

Fig. 10.8. Schematic velocity distribution of upstream electrons in the de Hoffmann-Teller frame of reference. The reflected electrons are shaded. The coordinates are chosen parallel (sunward is positive) and perpendicular to the magnetic field. The initial distribution is assumed to have two Maxwellian populations forming a core and a halo. The halo corresponds to a tail in the upstream velocity distribution providing the seed particles for acceleration.

In the HT frame, the particles are reflected at the same energy, but negative velocity. The Lorentz transformation by $\mathbf{V}_1^{\mathrm{HT}}$ into the normal incidence frame, reveals that the reflected particles gain energy. A particle with a velocity $v_\parallel^{\mathrm{HT}}$ parallel to the magnetic field in the HT frame has a larger velocity in the normal incidence frame (ni) by the transformation velocity. The y-component (Fig. 10.2) is of primary interest and is given by

$$v_y^{ni} = \frac{v_\parallel^{\mathrm{HT}} \sin\theta_1 + V_{\mathrm{HT}}}{1 + v_\parallel^{\mathrm{HT}} \sin\theta_1 V_{\mathrm{HT}}/c^2} \quad . \tag{10.3.4}$$

The denominator in Equation (10.3.4) is required by the theorem of velocity addition in special relativity. For nearly perpendicular shocks ($V_{\mathrm{HT}} \gg v_\parallel^{\mathrm{HT}}$), the particles are reflected with about V_{HT}, which was given by Equation (10.3.2) and is limited by the speed of light.

The distribution of reflected particles (Fig. 10.8) may be called a displaced loss-cone. It resembles the distributions of trapped particles in its inverted distribution of transverse energy. Additionally, it has the properties of a stream. The two positive gradients in velocity space – in perpendicular as well as parallel directions – drive instabilities similar to trapped particles and beams. In an initial phase, still close to the shock, the distribution will relax to a plateau-like distribution, releasing its transverse free energy and a part of its parallel free energy. This process may be the source of the radiation seen in the backbone of solar type II radio bursts (Section 10.2.3). The plateau-distribution propagates subsequently away from the shock, producing electrostatic plasma waves at its front. It is very likely the source of the herringbones in type II bursts.

Drift (or fast-Fermi) acceleration can explain the observed energetic electrons at the Earth's bow shock. It is likely to take place in the solar corona and possibly in supernova remnants. Furthermore, the process is also proposed to accelerate intermittent ion beams observed from the quasi-perpendicular region of the Earth's bow shock.

10.3.2. Ion Acceleration at Quasi-Parallel Shocks

Shock drift acceleration discussed in the previous section is primarily effective for a halo electron population. Now we study an acceleration process that is well suited for ions. If some ions are reflected on a shock front, they carry much free energy into the upstream region and can drive low-frequency, electromagnetic beam instabilities according to Section 7.3. Strong magnetoacoustic and Alfvén waves have been observed and interpreted this way. As the ions excite waves, they diffuse in velocity space into a more isotropic distribution (Section 6.2.1). The average ion velocity is drastically reduced. Since the shock speed exceeds the magnetoacoustic velocity, the inflow drives eventually the ions – trapped in their self-excited waves – back to the shock front. The particles experience repeated reflections between the waves and the front. The process is known as *diffusive shock acceleration*. It is important to note that the particles are not accelerated in a single step, but by multiple scatterings off a single shock. The upstream waves are overtaken by the shock, both together provide for a set of converging mirrors, characteristic for first-order Fermi acceleration. If electrons and ions get accelerated to similar velocities, the ions gain more energy due to their larger mass.

Diffusive acceleration can be modelled in a simple way, following the approach of Bell (1978). Let χ be the ratio of particle energy increase per reflection in the observer's frame. In a one-dimensional geometry (head-on collisions only), a particle with an initial velocity v_i reflected by a shock having a velocity in the observer's frame of $V_1^{\rm obs} \ll v_i$ gains kinetic energy by a factor of

$$\chi := \frac{\varepsilon + \Delta\varepsilon}{\varepsilon} \approx 1 + 4\frac{V_1^{\rm obs}}{v_i} \quad . \tag{10.3.5}$$

Equation (10.3.5) is non-relativistic. The relativistic energy gain is calculated from energy and momentum conservation in the HT frame. Without loss of generality, we assume one spatial dimension and Lorentz transform to the observer's frame. One finds that the particle energy changes in the observer's frame by

$$\Delta\varepsilon = 2\varepsilon_i\gamma_1^2 \frac{V_1^{\rm obs}(v_i + V_1^{\rm obs})}{c^2} \quad , \tag{10.3.6}$$

where ε_i is the total initial energy, $m\gamma_i c^2$, and γ_1 is the Lorentz factor of the shock velocity, both in the observer's frame (Exercise 10.5). The energy gain is reduced by a factor of 2 when averaged over angles in three dimensions. For $v_i \approx c$, and $V_1^{\rm obs} \ll c$ (thus $\gamma_1 \approx 1$),

$$\chi \approx 1 + \frac{V_1^{\rm obs}}{c} \quad . \tag{10.3.7}$$

Furthermore, let P be the probability that the particle remains in the upstream region and will have another reflection ($P \leq 1$). Let us start with a density n_0 of particles with an initial energy ε_0. After j reflections there is a density of $n = n_0 P^j$ particles with energies $\varepsilon = \varepsilon_0 \chi^j$. After taking the logarithm and eliminating j, the two relations combine to

$$n(\geq \varepsilon) = n_0 \left(\frac{\varepsilon}{\varepsilon_0} \right)^{\ln P / \ln \chi} \quad , \tag{10.3.8}$$

a power-law distribution. As n refers to the particles that have reached the energy ε and are further accelerated, it is related to the spectral energy density by $n(\geq \varepsilon) = \int_\varepsilon^\infty f(\varepsilon) d\varepsilon$. The energy distribution, $f(\varepsilon)$, is the derivative of Equation (10.3.8). It is also a power law of the form $f(\varepsilon) \propto \varepsilon^{-\delta}$, where δ is the power-law index of the energy distribution, amounting to

$$\delta = 1 - \frac{\ln P}{\ln \chi} \quad . \tag{10.3.9}$$

There are several approaches to estimate P, the fraction of particles of an isotropic distribution that is reflected. In the simplest version, often used for cosmic rays, it is assumed that the shock front is completely permeable to energetic ions from both sides. The ions ($v \approx c$ and trapped in magnetic field fluctuations) have the same density n on both sides and move through the front at a rate of $nc/4$. Let us assume that particles are only lost by convecting away in the downstream region. The loss rate then is nV_2 or about $nV_1/4$ in the strong shock limit (Eq. 10.1.8). The probability of escape is V_1/c, thus $P \approx 1 - V_1/c$ and $\delta \approx 2$. This value becomes larger if the average compression factor (Eq. 10.1.8) is below 4, as often observed at the bow shock. The predicted power-law exponent thus is close to the one observed in the cosmic ray spectrum (Section 7.1.2).

10.3.3. Resonant Acceleration and Heating

The entropy jumps in the plasma flowing through the shock front. Thus the temperature increases non-reversibly. In collisionless shocks, microscopic electric and magnetic fields are the source of the entropy increase. The interaction of particles with these field fluctuations drains energy from the shock and heats the plasma.

An example is heating of electrons by magnetic compression. It is an intrinsically adiabatic process, but can be made irreversible by wave turbulence. If the shock thickness is larger than the gyroradius, particles penetrating the shock conserve their magnetic moment. Equation (2.1.7) in relativistic form reads as

$$\frac{(p_\perp^1)^2}{B_1} = \frac{(p_\perp^2)^2}{B_2} \quad , \tag{10.3.10}$$

where $p_\perp = \gamma m v_\perp$ is the perpendicular particle momentum, and the superscripts 1 and 2 refer to the upstream and downstream plasma, respectively. The process is also known as *betatron acceleration* in analogy to the laboratory application. The maximum increase of the magnetic field predicted by MHD theory (Eq. 10.1.10) for strong perpendicular shocks is $B_2/B_1 = 4$. The magnetic field in the Earth's bow shock is often observed to overshoot four times the upstream value (Fig. 10.4). Nevertheless, very high energies cannot be achieved through this simple version of the betatron process. Even worse, in a subsequent expansion of the magnetic field the particle loses all the energy gain since the process is reversible.

A betatron-accelerated population is anisotropic in velocity space because the particles gain only transverse energy. If pitch angles are isotropized in a random way before expansion, some heating is irreversible. The cycle may repeat many times. In each cycle, the particles experience a net energy gain. This process has been called *magnetic pumping*.

Magnetic pumping could be particularly effective in the strong low-frequency turbulence ahead of quasi-parallel shocks (ion foreshock) where the waves take the role of magnetic pumps. It has been proposed that electrons could be accelerated this way and be subsequently scattered in pitch angle by whistler waves.

Another form of heating is *resonance acceleration*. An example has already been given in Section 9.4.3 on particle acceleration by currents. The essence of the mechanism is wave turbulence created by one particle population that stochastically accelerates another population. The process is a diffusion in velocity space. As pointed out in Section 9.4.3, lower hybrid waves are an attractive medium to transfer energy from ions to electrons. Perpendicular currents in the form of moving ions exist in oblique shock fronts. Perpendicular ion flows are unstable toward growing lower hybrid waves. The scenario is supported by observations at the Earth's bow shock, where lower hybrid waves have been discovered that may occasionally accelerate electrons in parallel velocity.

10.4. Stochastic Particle Acceleration

Concerning Fermi acceleration on magnetic mirros we have so far considered only first-order models. These processes, however, are not consistent with observations of the prompt acceleration of electrons and ions in flares, where acceleration of large numbers of particles is required preceding the observed shocks in the high corona and interplanetary space. Second-order Fermi acceleration is most attractive where shocks or large amplitude waves propagate in a limited volume such as the reconnection outflows and energize some or most of its particles. Charged particles move along magnetic field lines and thus get accelerated in the direction parallel to the magnetic field in head-on collisions. The particles lose energy in the slightly less frequent catch-up encounters. Thus their energy varies stochastically in both directions, but has a trend to increase.

If waves act as the mirrors, only a small fraction of the particles is affected. The interaction requires a resonance condition of the type

$$\omega - k_z v_z = l\Omega_\alpha \quad . \tag{10.4.1}$$

Contrary to (non-resonant) Fermi acceleration, a spectrum of waves is now necessary for substantial acceleration. A particle having gained energy then can interact with another wave meeting its new resonance condition for further energy gain. Attractive for particle acceleration are waves that resonate with particles in the bulk part of the thermal velocity distribution. Such waves can act as primary accelerators without needing already fast seed particles. In addition, waves are required that are naturally formed in a situation of magnetic reconnection.

Of particular interest for flare electron acceleration are oblique fast magnetoacoustic waves (also called compressional Alfvén waves, see Section 3.2.4). These waves have a dispersion relation $\omega \approx kc_A$ (assuming $c_A \gg c_s$) from the MHD range up to about the proton gyrofrequency. Above about $10\Omega_p$, the branch enters the whistler regime having enhanced phase velocity. As $c_A < v_{te}$, the waves are in $l = 0$ (Čerenkov) resonance with the bulk part of the thermal electrons, but generally not with the ions.

Fast magnetoacoustic waves have an oscillating wave magnetic field component parallel to the undisturbed field. It produces a series of compressive and rarefactive perturbations moving in parallel direction with a velocity of about $v_A/\cos\theta$, where θ is the angle between $\mathbf{k}$ and B_0. If the distribution of waves is isotropic, the range of phase velocity projections in z-direction is continuous from v_A to infinity.

This characteristic property makes it easy to envisage the interaction with electrons. The oscillations act as magnetic mirrors for charged particles like converging fields (Section 2.2). If the gyroradius is much smaller than the scale of contraction, the particles experience a mirror force $-\mu\partial B_z/\partial z$ (Eq. 2.1.12), and those with pitch angles above a critical value are reflected. The critical value depends on the wave amplitude. In small amplitude waves, only particles moving nearly in phase with the wave are reflected. Then the resonance condition (10.4.1) must be closely satisfied. In the frame moving with the wave, reflected particles have large pitch angles as the loss-cone is nearly 90°. For larger amplitudes, the velocity range of interaction becomes larger and more particles interact. In reality, the waves should not be envisaged as infinite sinusoidals, but distributions in frequency, angle, and amplitude of finite wave packets. Then it becomes easy to conceptualize a transition to (non-resonant) shocks.

If more interacting particles are slightly slower than the parallel wave phase velocity, the particles gain on the average and the wave is damped by the interaction. It is thus the magnetic analog to Landau damping of a wave with parallel electric field (Section 5.2.3). The resonance condition (Eq. 10.4.1, with $l = 0$) can be rewritten as $\lambda_z/v_z \approx \tau$, where τ is the wave period and $\lambda_z = 2\pi/k\cos\theta$ is the parallel wavelength. A small amplitude wave and a particle interact if the particle transit time across the wavelength is approximately equal to the period.

This property is the origin of the name *transit-time damping* for this acceleration process. It is basically resonant second-order Fermi acceleration.

Electron acceleration by transit-time damping has a major drawback. The acceleration is in parallel direction only. As v_z increases, the pitch angle $\alpha = \arctan(v_\perp/v_z)$ therefore decreases. The requirement on the wave amplitude for reflection increases until it cannot be satisfied anymore and acceleration stops. Several processes have been proposed that may scatter the accelerated electrons into more transverse orbits, such as collisions or a pre-existing population of whistler waves. Most likely, however, is that the velocity distribution stretches out into parallel direction until it becomes unstable to L-mode waves by anomalous Doppler resonance. This is named the *electron firehose instability*.

Fast magnetoacoustic waves may originate in the rearrangement of the large-scale magnetic field by reconnection jets or their shear flow instabilities. Long wavelength modes cascade to smaller wavelengths to reach eventually waves with a sufficiently high k_z to interact with thermal electrons, when they are transit-time damped. The electron acceleration process can then be described by quasi-linear diffusion (Section 6.2.1). If the pitch angle distribution remains more or less isotropic, electron acceleration by transit-time damping can proceed to ultrarelativistic velocities within a fraction of a second.

Large amplitude, fast magnetoacoustic waves preferentially accelerate electrons by the $l = 0$ resonance. The $l \neq 0$ resonances in Eq. (10.4.1)are negligible as long as the conditions for transit-time acceleration are satisfied. A similar scenario has also been proposed for flare proton acceleration, assuming Alfvén waves and $l \neq 0$ resonances. Stochastic acceleration is a simple and most promising process for flares if the required input wave turbulence is given.

Exercises

10.1: The thickness Δx of a shock front is given by the scale over which the inflowing energy can be dissipated. Do a thought experiment to study whether this may occur by Ohmic dissipation. An MHD shock is assumed to move in x-direction. B_{1y} and B_{2y} are the transverse field components ahead and behind the shock in the rest frame of the shock, respectively. For a strong shock, the magnetic field compression is $B_{2y} \approx 4B_{1y}$. The kinetic energy density deposited by the inflowing plasma in the shock layer is $\Delta\mathcal{E} \approx \frac{1}{2}\rho_1 V_1^2$. This must occur within a time $\Delta t \approx \Delta x/V_1$. Use Coulomb conductivity and, as an alternative, the minimum anomalous conductivity (Eq. 9.4.3); let $T = 10^6$ K, $c_A = 10^8$ cm s^{-1}, $M_A := V_1/c_A = 3$, and $n_e = 10^9$ cm^{-3}. Apply Equation (10.1.1) and

$$J \approx \frac{c}{4\pi} \frac{B_{1y} - B_{2y}}{\Delta x} \tag{10.4.2}$$

to calculate the shock thickness Δx.

10.2: For high Mach number shocks, the kinetic energy inflow largely exceeds the thermal energy and magnetic energy. The inflowing energy then is partitioned between kinetic, thermal, and magnetic energy density in the downstream medium. The division is independent of Mach number as stated by Equations (10.1.8) and (10.1.9). Prove these equations!

10.3: A gradient in magnetic field strength perpendicular to the field causes a gradient current. Calculate the scale length of the field in perpendicular direction required to cause a current drift velocity, V_d, equal to the ion sound velocity, c_{is}. Take for example $T_l = T_i = 10^7$ K, $n_e = 10^{10}$ cm^{-3}, $B = 100$ G.

10.4: A bow shock forms where the solar wind hits the magnetosphere of the Earth. Calculate the distance r of the plasma pause (the surface separating the two plasmas) from the center of the Earth in the direction toward the Sun. Use momentum conservation at the shock front and assume constant pressure between front and pause. Let the solar wind have a proton density of 5 cm^{-3}, magnetic field of $2 \cdot 10^{-4}$ G, temperature 10^5 K, and a velocity of 400 km s^{-1}, and assume a terrestrial magnetic field in the equatorial plane of $B(r) = 0.3(r/R_\oplus)^3$ G. Why does MHD give the correct answer?

10.5: A particle with initial velocity v_i and initial energy ε_i gains energy when reflected by a shock moving at $V_1^{\rm obs}$ in the observers frame of reference. Assume a head-on collision and a one-dimensional geometry. Energy ε and three-momentum $\mathbf{p}$ are Lorentz transformed into the frame moving with the shock, where

$$\varepsilon' = \gamma_1(\varepsilon + pV_1^{\rm obs}) \quad , \tag{10.4.3}$$

$$\mathbf{p}' = \gamma_1(\mathbf{p} + \frac{V_1^{\rm obs}\varepsilon}{c^2}) \quad , \tag{10.4.4}$$

where $\gamma_1 = (1 - (V_1^{\rm obs}/c)^2)^{-1/2}$ is the Lorentz factor of the shock. What is the energy gain per reflection? Prove Equations (10.3.5) and (10.3.6).

Further Reading and References

General reviews

Tidman, D.A. and Krall, N.A.: 1971, *Shock Waves in Collisionless Plasmas*, Wiley-Interscience, New York.

Tsurutani, B.T. and Stone, R.G. (eds.): 1985, 'Collisionless Shocks in the Heliosphere', *Geophysical Monograph* **35**, American Geophysical Union, Washington.

Priest, E.R.: 1982, *Solar Magnetohydrodynamics*, D. Reidel Publishing Comp., Dordrecht, Holland, Chapter 5 on MHD shocks.

Observations of shocks

Aurass, H.: 1992, 'Radio Observations of Coronal and Interplanetary Type II Bursts', *Am. Geophys.* **10**, 359.

Bavassano-Cattaneo, M.B., Tsurutani, B.T., and Smith, E.J.: 1986, 'Subcritical and Supercritical Interplanetary Shocks: Magnetic Field and Energetic Particle Observations', *J. Geophys. Res.* **91**, 11929.

Hundhausen, A.J.: 1988, 'The Origin and Propagation of Coronal Mass Ejections', *Proc. Sixth Int. Solar Wind Conf.* (V.J. Pizzo, T.E. Holzer, and D.G. Sime, eds.), NCAR Tech. Note, TN-**306**, Vol. 1, p. 181.

Kahler, S.W.: 1987, 'Coronal Mass Ejections', *Rev. Geophys.* **25**, 663.

Mann, G.: 1995, 'Theory and Observations of Coronal Shock Waves', in *Coronal Magnetic Energy Releases,* (A.O. Benz and A. Krüger, eds.), Lecture Notes in Physics **444**, 183.

Nelson, G.J. and Melrose, D.B.: 1985, in *Solar Radiophysics* (D.J. McLean and N.R. Labrum, eds.), Cambridge University Press, Chapter 13 on type II radio bursts.

Russel, C.T. and Hoppe, M.M.: 1983, 'Upstream Waves and Particles', *Space Science Rev.* **34**, 155.

Particle acceleration

Decker, R.B.: 1988, 'Computer Modelling of Test Particle Acceleration at Oblique Shocks', *Space Sci. Rev.* **48**, 195.

Jones, F.C. and Ellison, D.C.: 1991, 'The Plasma Physics of Shock Acceleration', *Space Sci. Rev.* **58**, 259.

Holman, G.D. and Pesses, M. E.: 1983, 'Solar Type II Radio Emission and the Shock Drift Acceleration of Electrons', *Astrophys. J.* **267**, 837.

Lee, M.A.: 1983, 'Coupled Hydromagnetic Wave Excitation and Ion Acceleration at Interplanetary Traveling Shocks', *J. Geophys. Res.* **88**, 6109.

Miller, J.A.: 1997, 'Electron Acceleration in Solar Flares by Fast Mode Waves: Quasi-linear Theory and Pitch-angle Scattering', *Astrophys. J.* **491**, 939.

Schlickeiser, R. and Steinacker, J.: 1989, 'Particle Acceleration in Impulsive Solar Flares II. Nonrelativistic Protons and Ions', *Solar Phys.* **122**, 29.

References

Bell, A.R.: 1978, 'The Acceleration of Cosmic Rays in Shock Fronts', MNRAS **182**, 147.

Benz, A.O. and Thejappa, G.: 1988, 'Radio Emission of Coronal Shock Waves', *Astron. Astrophys.* **202**, 267.

Cane, H.V. and White, S.M.: 1989, 'On the Source Conditions for Herringbone Structure in Type II Solar Radio Bursts', *Solar Phys.* **120**, 137.

Dere, K. P., Brueckner, G. E., Howard, R. A., Michels, D. J. and Delaboudiniere, J.: 1991, 'LASCO and EIT Observations of Helical Structure in Coronal Mass Ejections', *Astrophys. J.* **516**, 465.

Kahler, S.W., Reames, D.V., Sheeley, N.R.Jr., Howard, R.A., Koomen, M.J., and Michels, D.J.: 1985, 'A Comparison of Solar ^{3}Helium-Rich Events with Type II Bursts and Coronal Mass Ejections', *Astrophys. J.* **290**, 742.

CHAPTER 11

PROPAGATION OF RADIATION

In the preceding chapters we have studied a number of ways high-frequency, transverse electromagnetic waves – generally called *radiation* for short – are produced in coronal plasmas. How the wave energy builds up in the source and propagates to the observer also requires scrutiny. Most of the necessary elements have already been developed in previous chapters. In particular, the dispersion relation of electromagnetic waves has been discussed in Chapter 4. The concepts of wave intensity and photon temperature have been introduced in Sections 5.1.1 and 6.3.2, respectively.

Since each emission process operates in reverse as an absorption, a good emitter is also a good absorber. The book-keeping within the source is accomplished by the transfer equation (Section 11.1).

Radiation is modified as it propagates from the source to Earth in a number of ways. We limit ourselves to the effects of the plasma component of the propagation medium. Its free electric charges react to the electric and magnetic wave fields, and partake in the oscillation. The wave energy is partitioned between oscillating fields and particles. Moreover, the wave energy in the particles can be randomized by particle collisions, causing wave absorption (Section 11.2). The oscillating free electrons also alter the dispersion – the function $\omega(\mathbf{k})$ – and modify the wave's phase and group velocities. The basic effects of dispersion are discussed in Section 11.3. They include refraction caused by spatial variations of ω_p or Ω_e. If these variations are of small scale and can be analysed statistically, the effect is termed *scattering* (Section 11.4). Inhomogeneities in coronae are usually not random, but structures aligned along the magnetic field, requiring specialized concepts to be discussed in Section 11.5.

We have seen in Section 4.3 that the more the wave frequency exceeds the plasma frequency and the gyrofrequencies, the smaller the influence of the plasma on the wave and the more the propagation medium resembles a vacuum (e.g. Eq. 4.3.16). Therefore, plasma propagation effects in coronae are most severe for radiation at radio frequencies. Although they are often a nuisance for the investigation of a source, propagation effects yield also rewarding information on the intervening medium.

11.1. Transfer Equation

The observable parameter of main interest is the intensity, I, in units of energy of electromagnetic radiation per collector area, time, frequency, and solid angle [erg s^{-1} Hz^{-1} cm^{-2} $sterad^{-1}$]. This is related to the brightness temperature T_b (Eq. 5.1.1), and its measurement requires spatial resolution of the source. Note that the intensity, and thus the brightness temperature, do not change in a vacuum as the radiation travels. Hence they are independent of the distance between the source and the observer. This is simply a consequence of energy conservation. In a semi-classical way, it can also be viewed as a result of photon conservation. In contrast, the flux density F in [erg s^{-1} Hz^{-1} cm^{-2}] is proportional to the angular source size and, in vacuum, falls off inversely as the square of the distance.

In a plasma without radiation sources and sinks, the photon number is conserved. However, the photon propagation paths may not be straight lines, and the photon *density* can be modified. In the wave approach generally followed in this book we describe the flow of photons by a wave intensity. Its complex behavior in an anisotropic medium whose properties are variable in space forces us to make approximations. In this section we use *geometric optics* originally developed for the practical needs in constructing optical instruments. Its main assumption is that the medium varies slowly with position or, more precisely, that the scale length of the variation is much longer than the wavelength in the medium. In particular, we exclude diffraction effects at structures or scale lengths of the size of a wavelength or smaller.

If geometric optic applies, radiation propagates according to the extremal principle of Fermat, and the radiation energy can be considered being transported along curves or *rays*. Different rays do not interact with one another. Propagation along rays is reversible. In an isotropic medium – neglecting the magnetic field – the propagation path of the radiation in geometric optics is simply determined by Snell's law,

$$\mathcal{N} \sin \alpha = \text{constant} \quad , \tag{11.1.1}$$

where α is the angle of propagation relative to the gradient in the refraction index $\mathcal{N}$. Equation (11.1.1) defines the rays that represent the propagation paths of waves in a medium.

If $\mathcal{N}$ increases, Equation (11.1.1) requires a decrease in α. In a non-magnetized plasma, $\mathcal{N} = (1 - \omega_p^2/\omega^2)^{1/2}$. The radiation is refracted toward the direction of the gradient. The net effect – sketched in Figure 11.1 – collimates radiation into regions of low phase velocity. In a horizontally stratified corona, it is refracted toward the radially outward direction. Observations from two spacecraft viewing a radio source in the solar corona from different angles have confirmed the beaming of the radiation in vertical.

Since intensity is per solid angle, energy conservation and Snell's law of refraction require $I \sin^2 \alpha \propto I/\mathcal{N}^2 = \text{constant}$ along the line of sight. The intensity can therefore change, even in the absence of absorption and emission.

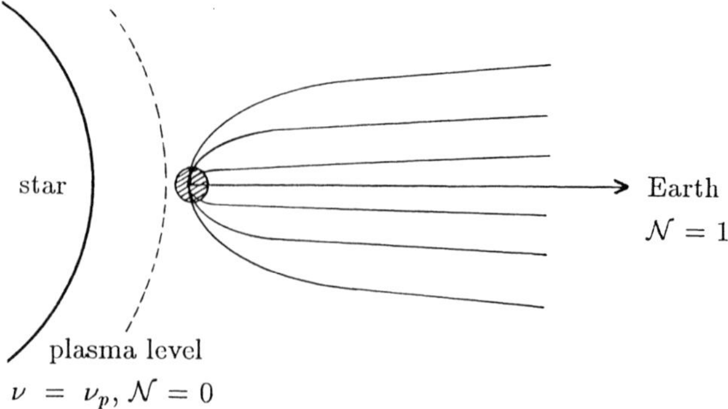

Fig. 11.1. Radiation escaping from near the plasma level is beamed toward regions of larger refractive index, $\mathcal{N}$, thus lower phase velocity.

If the radiation is emitted in a plasma, the refractive index $\mathcal{N}$ at the source is smaller than unity (Eq. 4.3.21, resp. 4.5.1), and increases on the way to Earth (where $\mathcal{N} = 1$). Therefore, the intensity is higher at Earth than in the source by the square of the refractive index in the source. This is a result of the reduced range in the angle of propagation (and presumes that the source is visible from Earth). The effect is particularly strong for emission near the cutoff frequency ($\omega \approx \omega_p$ for o-mode, $\omega \approx \omega_x$ for x-mode, given by Eq. 4.4.10).

A general expression for $\mathcal{N}$ in a magnetized plasma has been given in Equation (4.5.1). It can be approximated for frequencies well above the cutoff frequency (for o-mode $\omega^2 \gg \omega_p^2$; for x-mode $\omega^2 \gg \omega_x^2$ and propagation away from perpendicular ($|\cos\theta|/\sin^2\theta \gg \Omega_e/2\omega$) by

$$\left(\mathcal{N}_{\substack{o\\x}}\right)^2 \approx 1 - \left(\frac{\omega_p}{\omega}\right)^2 \left(1 \pm \frac{\Omega_e}{\omega}|\cos\theta|\right)^{-1} \quad , \tag{11.1.2}$$

where θ is the angle between wave propagation and magnetic field (Exercise 11.1). This approximation is termed *quasi-longitudinal.* The word 'longitudinal', meaning here propagation parallel to the magnetic field, is a legacy of early magnetoionic terminology. The quasi-longitudinal approximation is valid in a large part of the solar corona, where generally $\omega_p \gg \Omega_e$ (Exercise 11.1).

The angle θ makes the refraction index anisotropic. Particularly at frequencies close to the plasma frequency, the refractive index depends strongly on propagation angle to the magnetic field (Eq. 4.5.1). As a consequence, the group velocity of the radiation is not parallel to the phase velocity, and Snell's law is not valid or must be generalized. We shall not consider this case, and the interested reader is adverted to the references at the end of this chapter. Nevertheless, Figure 11.1 is still a good representation of propagation for weakly magnetized plasmas.

We have introduced the emissivity, $\eta(\omega)$, in Section 6.3.3 and defined by the energy radiated per unit of time, volume, frequency, and solid angle. Furthermore, absorption attenuates I proportionally to path length. We denote it by the absorption coefficient, κ, referred to unit length. The *transfer equation* describes the evolution of I as the difference between a source term (emission) and a loss term (absorption),

$$\frac{d}{ds}\left(\frac{I}{\mathcal{N}^2}\right) = \frac{\eta}{\mathcal{N}^2} - \kappa\left(\frac{I}{\mathcal{N}^2}\right) \quad , \tag{11.1.3}$$

where s is distance along the ray path from the star to the observer. With the following definitions for the source function $\mathcal{J}$ and optical depth τ:

$$\mathcal{J} := \frac{\eta}{\kappa \mathcal{N}^2} \quad [\text{erg s}^{-1}\text{ cm}^{-2}\text{ sterad}^{-1}] \ , \tag{11.1.4}$$

$$\tau(s) := \int_s^\infty \kappa\, ds \quad , \tag{11.1.5}$$

the transfer equation becomes

$$\frac{d}{d\tau}\left(\frac{I}{\mathcal{N}^2}\right) = \left(\frac{I}{\mathcal{N}^2}\right) - \mathcal{J} \quad , \tag{11.1.6}$$

a first-order differential equation. When the plasma in the source is in thermodynamic equilibrium, the source function (11.1.4) equals the Planck function (5.1.1) evaluated at the source temperature T_e (Kirchhoff's law). Optical depth is a measure of the transparency of a medium. Note that it is defined as zero at infinity (that is at the observer) and that it increases into the source (see Fig. 11.2). It is an obvious misnomer since it can refer to any kind of radiation, not just optical.

Let the boundary values for the integration of Equation (11.1.6) be τ_1 (far end of the source), $\tau_2 = 0$ (observer), and let $\mathcal{N}(\tau_2) = 1$. The general solution of Equation (11.1.6) is the observed intensity,

$$I(\tau = 0) = \frac{I(\tau_1)}{\mathcal{N}^2(\tau_1)} e^{-\tau_1} + \int_0^{\tau_1} \mathcal{J}(\tau') e^{-\tau'} d\tau' \quad . \tag{11.1.7}$$

The first term on the right (mathematically speaking the general solution of the homogeneous part of Eq. 11.1.6) is the radiation entering the source from behind, and being attenuated by the source and the propagation thereafter (Fig. 11.2). The second term is the actual source term (a particular solution of the inhomogeneous equation) and is here of main interest. $\mathcal{J}$ and τ depend strongly on ω. Assuming no radiation from beyond the source (i.e. $I(\tau_1) = 0$) and $\mathcal{J}(\tau') =$ constant in space within the source,

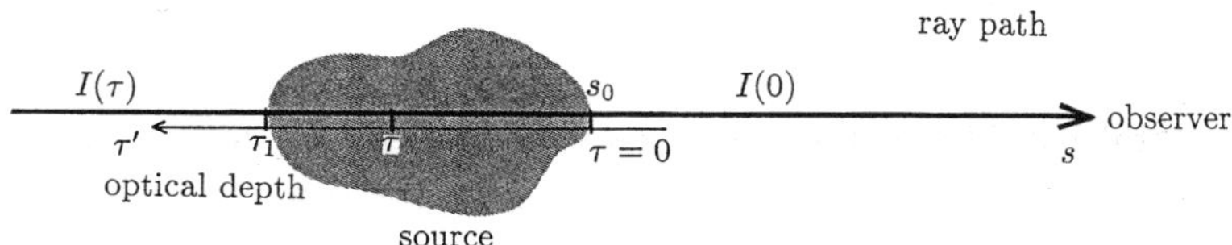

Fig. 11.2. The intensity of the radiation observed at Earth is propagated along the path s through a source region.

$$I(\omega) = \mathcal{J}(\omega)(1 - e^{-\tau_1(\omega)}) \quad , \tag{11.1.8}$$

where $\tau_1 \approx \kappa L$ and L is the source depth. For an *optically thick* source, by definition $\tau_1(\omega) \gg 1$, and

$$I(\omega) \approx \mathcal{J}(\omega) \quad . \tag{11.1.9}$$

In the *optically thin* case ($\tau_1(\omega) \ll 1$),

$$I(\omega) \approx \mathcal{J}(\omega)\tau_1(\omega) \approx \frac{\eta(\tau_1)L}{\mathcal{N}^2} \quad . \tag{11.1.10}$$

Imagine the source surrounded by a cavity and in thermodynamic equilibrium with it. A source that only emitted would cool at the expense of the cavity. The second law of thermodynamics requires that the source also absorbs radiation. The inverse of an emission process is called *self-absorption*. Moreover, as the Maxwell-Vlasov equations are time reversible, the process absorbing energy from a wave field must also be able to emit in the presence of a wave field. This is called *stimulated emission*. In thermal equilibrium, it limits the intensity to that of the Planck function.

During propagation from the source to Earth, the radiation is far from thermodynamic equilibrium with the background plasma. At this stage, absorption is usually a thermal process, in which background particles extract energy from the wave and lose it either by collisional interactions (Section 11.2) or gyromagnetic resonance.

11.2. Collisional Absorption

Electromagnetic waves are absorbed in a plasma, since electrons oscillating in the wave fields have collisions and lose a part of the wave's energy. The absorption is sometimes also called *free-free absorption* following the notation of atomic physics. It is the inverse process of bremsstrahlung (or free-free) emission. Radiation near the plasma frequency is strongly affected by absorption for two reasons. (*i*) The fraction of wave energy in oscillatory electron motion is $(\omega_p/\omega)^2$ (Exercise 5.3). This is the fraction of the wave energy that can be damped by collisions. Once it starts to be damped, the wave converts electromagnetic to kinetic energy, so that the energy ratio is conserved. Therefore, particle collisions can effectively thermalize the entire wave energy. (*ii*) The group velocity – about $\mathcal{N}c$ in the non-magnetic approximation – is small, and the waves stay in the region for a long time.

The absorption coefficient of transverse waves is given by

$$\kappa = \frac{1}{t_{e,i}\mathcal{N}c}\left(\frac{\omega_p}{\omega}\right)^2 \quad , \tag{11.2.1}$$

where $t_{e,i}$ is the thermal electron-ion collision time (Eq. 2.6.32). The product $t_{e,i}\mathcal{N}c$ is the propagation length after which collisions randomize the momentum of particle oscillations. Having travelled this distance, the wave energy residing in particle energy is lost and must be replenished. The slower the wave, the shorter this absorption length. Putting in the numbers, we get

$$\kappa \approx \frac{9.88 \cdot 10^{-3} n_e \sum_i Z_i^2 n_i \ln\Lambda_t}{\nu^2 \mathcal{N} T^{3/2}} \quad [\mathrm{cm}^{-1}] \ , \tag{11.2.2}$$

where the observing frequency ν ($\gtrsim \nu_p$) is in Hz, T in K, n_e in cm^{-3}, and $\mathcal{N}$ is given by Equation (11.1.2). For a fully ionized plasma with solar abundances $\sum_i Z_i^2 n_i \approx 1.16 n_e$. The factor $\ln\Lambda_t$ – often referred to as the Gaunt factor – is slightly different from the case of a test particle studied in Section 2.6. For waves with $\nu > \nu_p$, the time to build up Debye shielding of a charge is limited to half a wave period. Therefore, the maximum distance for binary interactions is $v_{te}/\nu < \lambda_D$. For $\nu \gtrsim \nu_p$, Bekefi (1966), in an extensive and more accurate approach, has found

$$\ln\Lambda_t = \begin{cases} 17.6 + \frac{3}{2}\ln T - \ln\nu & \text{for } T \lesssim 8.2 \cdot 10^5 \langle Z_i^2 \rangle \ [\mathrm{K}] \\ 24.6 + \ln T - \ln\nu & \text{for } T \gtrsim 8.2 \cdot 10^5 \langle Z_i^2 \rangle \ [\mathrm{K}] \end{cases} \quad . \tag{11.2.3}$$

The frequency ν in Equation (11.2.3) is in Hz, and $\langle Z_i^2 \rangle$ is the abundance weighted average ion charge. (For solar abundances and complete ionization $\langle Z_i^2 \rangle = 1.16$.)

Collisional damping reduces the intensity of transverse waves particularly if radiated near the plasma frequency. Let I_0 be the intensity just when the radiation leaves the source (at s_0 in Fig. 11.2). The intensity observed at $\mathcal{N} = 1$ is then

$$I = \frac{I_0}{(\mathcal{N}_0)^2} e^{-\tau} \quad . \tag{11.2.4}$$

The optical depth τ here refers to collisional damping during propagation to Earth. It has been defined in general form by Equation (11.1.5) and can be evaluated easily for any given coronal model. For example, assume an exponentially decreasing density with scale height H_n (implying constant temperature and gravity), $T > 10^6$ K, solar abundances, and neglect the magnetic field. The optical depth of collisional absorption then amounts to

$$\tau = \begin{cases} 464 \ \nu_9^2 \ T_6^{-3/2} (H_{10}/\cos\phi)(\ln\Lambda_t/20) & \text{for } \nu = \nu_p^0 \\ 12.0 \ \nu_9^2 \ T_6^{-3/2} (H_{10}/\cos\phi)(\ln\Lambda_t/20) & \text{for } \nu = 2\nu_p^0 \\ 141 \ n_{10} \ (\nu_p^0/\nu)^2 T_6^{-3/2} (H_{10}/\cos\phi) \ln\Lambda_t/20 & \text{for } \nu \gg \nu_p^0 \end{cases} \tag{11.2.5}$$

for radiation emitted at the fundamental, at the harmonic, or far above the plasma frequency of the source, respectively. The plasma frequency in the source is ν_p^0, and ν_9 is the emitted frequency in units of 10^9 Hz (= 1 GHz). H_{10} is the density scale height in units of 10^{10} cm, T_6 the temperature in 10^6 K, n_{10} the electron density in 10^{10} cm^{-3}, and ϕ (assumed constant and evaluated for $\mathcal{N} = 1$) is the angle between propagation and the density gradient. The third line shows the original dependence on n and ν in Equation (11.2.2). The specified relation between ν and ν_p^0 reduces it to the forms of the first and second line.

Equation (11.2.5) shows that collisional absorption depends strongly on emission frequency relative to ν_p^0. Little radiation emerges from $\tau \gtrsim 3$. Harmonic emission is much less affected than fundamental radiation. If the density scale height is linearly proportional to temperature – as would be the case in an unmagnetized, isothermal atmosphere (Eq. 3.1.54) – absorption depends only on the inverse square-root of the temperature.

Plasma emission at $\nu \approx \nu_p^0$ is more attenuated at high frequency than at low frequency (because of the higher density at the source). For interplanetary type III bursts (ν_p a few tens of kHz), attenuation even at the fundamental can be neglected. In fundamental metric bursts (ν_p a few tens of MHz), the effects of collisional attenuation and of refractive focussing are about equal, but of opposite sign. They are usually neglected. For $\nu \gtrsim$ 0.5 GHz and solar coronal conditions ($H_n \approx 10^{10}$ cm, $T \approx 10^6$ K) no fundamental emission was expected. Observations such as Figure 6.5 indicating fundamental radiation, however, suggest that the solar corona is transparent even for microwave emission propagating near the plasma frequency. The density scale height in Equation (11.2.5) seems to be reduced to an effective value about a factor 20 smaller than estimated from an isothermal, horizontally stratified atmosphere. It will be interpreted in Section 11.5 by inhomogeneity due to density filamentation along the magnetic field, allowing harmonic plasma emission to escape up to about 7 GHz.

11.3. Dispersion Effects

11.3.1. Geometric Optics

Geometric optics assumes a medium of propagation in which the waves propagate independently of each other, following the laws of refraction and reflection. The waves can be represented by rays. The conditions for geometric optics have been discussed in Section 11.1. If these assumptions apply, propagation in a plasma is fully determined by the refractive indices of the two transverse modes.

Equation (11.1.1) can be integrated numerically for a given coronal model. Figure 11.3 illustrates the result in the form of rays having their end points at Earth. We note in particular:

- Refraction always shifts the apparent source of the rays *toward* the star. If a source is visible for an observer, refraction makes it appear at lower altitude than in reality.

- The minimum height of all rays through Earth forms an envelope below which a source is invisible from Earth at the given frequency. This surface may be called the *horizon of refraction*. It is always at or above the plasma layer.

- From each source above the horizon, there are two rays to Earth – a *direct* ray and a *refracted* ray – and therefore two images. In practice, the refracted

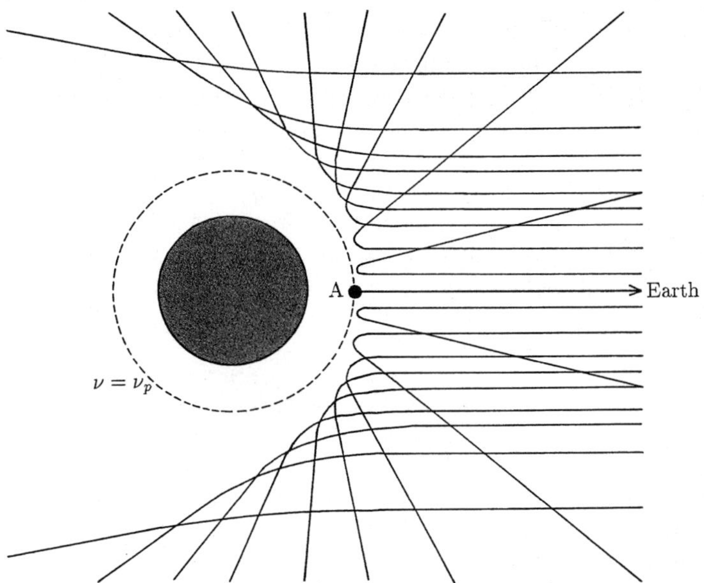

Fig. 11.3. Ray paths in a spherically symmetric corona for radio waves at a given frequency ω. Magnetic fields have been neglected. Only rays propagating into the direction of Earth are plotted. The full circle indicates the photosphere; the dashed circle is the plasma level, where $\omega_p = \omega$. Note that this figure is highly idealized as in reality coronae are extremely structured (cf. Fig. 1.1).

image is strongly absorbed, as it propagates close to the plasma level, and has not yet been observed.

- At the subearth point (A) a ray penetrates to the cutoff level, where $\mathcal{N} = 0$. Detailed investigation shows that waves approaching the cutoff level couple to waves propagating away so that the radiation is reflected. Figure 11.3 suggests that this case is of minor importance, as most rays are prevented by refraction from reaching the cutoff level.

There are important exceptions to geometric optics in coronae, in particular if the wave frequency is close to the plasma frequency (see above) or if it is nearly perpendicular to the background magnetic field (Section 11.3.4).

We note finally that the presence of coronal structure on all scales greatly complicates the actual picture. In reality, the apparent position of solar radio sources is *higher* than their actual position, and sources well below the theoretical horizon of refraction – even behind the Sun – are sometimes visible. These discrepant findings will be discussed and interpreted in Sections 11.4 and 11.5.

11.3.2. PLASMA DISPERSION

The group velocity of radiation depends on the electron density and on the magnetic field in the medium of propagation. A radiation pulse that has travelled through a plasma arrives later than a pulse that has passed the same distance in vacuum. The delay, τ, is called *group delay* and amounts to

$$\tau = \int \left(\frac{1}{v_{gr}} - \frac{1}{c} \right) ds \quad , \tag{11.3.1}$$

integrated from the source to the observer. The group velocity, v_{gr}, is calculated from the dispersion relation (Eq. 11.1.2). We simplify for $\omega \gg \Omega_e$ to obtain the non-magnetic case,

$$v_{gr} = \frac{\partial \omega}{\partial k} \approx c(1 - \omega_p^2/\omega^2)^{1/2} < c \quad . \tag{11.3.2}$$

Inserting this into Equation (11.3.1), one gets, in first order of $(\omega_p/\omega)^2$:

$$\tau \approx \frac{1}{2c\omega^2} \int \omega_p^2 \, ds = \frac{2\pi e^2}{c \, m_e} \frac{1}{\omega^2} \int n_e \, ds \quad . \tag{11.3.3}$$

The integral over electron density along the ray path is termed the *dispersion measure* or electron column density. The delay is an observable quantity when a broadband pulse, emitted over a wide frequency range at the same time, is observed at two frequencies separated by $\Delta\omega$. The dispersion measure can be evaluated from the observed difference in arrival time, $\Delta\tau$,

$$\int n_e \, ds \approx - \frac{c m_e \omega^3}{4\pi e^2} \frac{\Delta\tau}{\Delta\omega} \quad . \tag{11.3.4}$$

The negative sign indicates that a pulse arrives first at the higher frequency. The effect is observable if the pulse is short and emitted simultaneously over a wide bandwidth. The pulse is then observed to drift from high to low frequency and can be used to measure the electron column depth. The method is practicable for pulsars, where it is called *interstellar dispersion* and has been applied successfully to measure the average interstellar electron density or, in some cases, the distance of the pulsar.

Dispersion also causes a polarization effect. Neglecting terms of order ω/Ω_e in Equation (11.3.2) has eliminated the effects of the magnetic field in the plasma of propagation and of the gyromotion of the oscillating electrons. Under the influence of the background magnetic field, the electrons are constrained to move in spiral orbits. Therefore, they interact differently with left and right-hand polarized waves (o and x mode, respectively).

The unpolarized or linearly polarized radiation contains waves of both modes propagating independently. The two modes are circularly polarized in opposite senses if they propagate parallel to the magnetic field (Section 4.3). They are elliptically polarized in the general case. If the magnetic field is included in $\mathcal{N}$, the two modes propagate at different speeds. Two fully polarized pulses of opposite polarity emitted simultaneously arrive at different times (Exercise 11.2), the o-mode travelling faster at a given frequency. The effect is observable in solar narrowband spikes (Section 8.1.3).

11.3.3. Faraday rotation

Let us consider the radiation of a linearly polarized, stationary source. Emission processes producing linear polarization include, for example, the synchrotron mechanism and the coupling of perpendicular electrostatic waves. We shall show in this section that dispersion rotates the position angle of the electric field of the wave.

The oscillatory motion of a wave is generally described by a phase

$$\psi = ks - \omega t + \psi_0 \quad . \tag{11.3.5}$$

Linearly polarized waves propagating through a magnetized plasma must be viewed as a combination of equal components of oppositely handed elliptically polarized waves. Their electric field vectors add to the electric field of the wave. In the quasi-longitudinal limit (Eq. 11.1.2) and for identical frequencies, the two modes differ in refractive index by

$$\Delta\mathcal{N} := \mathcal{N}_o - \mathcal{N}_x \approx \frac{\omega_p^2 \Omega_e}{\omega^3} |\cos\theta| \quad . \tag{11.3.6}$$

Since $v_{ph} = c/\mathcal{N}$, the phase velocities of the two modes are different. The phase velocity of the o-mode is slower than the x-mode. On propagating a distance ds the phases of the two modes are shifted in relation to each other by

$$\begin{aligned} d\psi = (k_o - k_x)ds &= \left(\frac{1}{v_{ph}^0} - \frac{1}{v_{ph}^x} \right) \omega \ ds \\ &= \frac{\Delta\mathcal{N} \ \omega \ ds}{c} \quad . \end{aligned} \tag{11.3.7}$$

In the frame moving with the o-mode, the phase of the x-mode moves ahead. The direction of the electric field vector of the combined waves is the bisector of the angle between the electric field vectors of the two elliptically polarized waves. Therefore, it rotates by an angle $d\phi = d\psi/2$, or

$$d\phi = \frac{\omega_p^2 \Omega_e \cos\theta}{2\omega^2 c} dx \quad . \tag{11.3.8}$$

The disappearance of the absolute marks for $\cos\theta$ subtly takes care of all angles between the direction of propagation and the magnetic field. As a wave propagates, the component of the magnetic field parallel to the direction of propagation may reverse sign. Remember that the sense of rotation of the x-mode is the same as the sense of the electron gyration in the magnetic field; it is right-hand (relative to $\mathbf{k}$) for $\theta < \pi/2$ and left-hand for $\theta > \pi/2$. A normal wave mode with sufficiently high frequency (defined in Eq. 11.3.16) keeps its sense of rotation when propagating from a region with $\theta < \pi/2$ to a region with $\theta > \pi/2$. The wave changes its mode, and what was a faster x-mode before, becomes a slower o-mode. The electric field vector of a linear wave therefore changes its sense of rotation. If the magnetic field is directed toward the observer, $d\phi$ is positive; if the field is directed away from the observer, $d\phi < 0$.

The total *Faraday rotation* is the integral along the ray from the source to the observer,

$$\phi = \frac{2\pi e^3}{m_e^2 c^2 \omega^2} \int n_e B \cos\theta \; ds \quad . \tag{11.3.9}$$

The integral in Equation (11.3.9) is called the *rotation measure.* It can readily be determined by two measurements of the direction of linear polarization at adjacent frequencies (Exercise 11.3, see also Fig. 11.4), similar to the dispersion measure (Eq. 11.3.4). Thus, $\Delta\phi/\Delta\omega$ gives the integral of $n_e B \cos\theta$ along the ray path. It yields an average magnetic field – more precisely, its longitudinal component to the line of sight – weighted by the electron density. The sign of ϕ indicates whether the magnetic field is directed predominantly toward the observer (if $\phi > 0$) or away from it.

The combination of rotation measure and dispersion measure is a successfully applied method for estimating the magnetic field strength of an intervening plasma. If the rotation measure is dominated by a roughly uniform region, the *weighted* (!) magnetic field strength along the ray path is

$$\langle B \cos\theta \rangle = \frac{\int n_e B \cos\theta \; ds}{\int n_e \; ds} \quad . \tag{11.3.10}$$

The Faraday rotation of pulsar radio emission has been used to measure the magnetic field of the interstellar medium. The method has also been applied successfully to the solar wind using extragalactic background sources. Figure 11.4 shows the effect of Faraday rotation in the interplanetary medium. It rotates the polarization angle of radiation proportional to the square of the wavelength as predicted by Equation (11.3.9). Two linearly polarized quasars are observed at different angles from the Sun. The quasar 3C353 has a larger rotation measure since its radiation passes through interplanetary plasma of higher density and magnetic field.

Faraday rotation is rapid for radiation frequencies within, say, an order of magnitude of the plasma frequency, and a meaningful rotation measure would have to be determined at extremely close frequencies (Exercise 11.3). This is the case of radio waves emitted in coronae. The electric vector rotates many times within a source, destroying any initial linear polarization. No linear radio emission has ever been observed conclusively from the Sun and the stars, because any initial linear polarization is washed out by different rotation angles from different parts of a source.

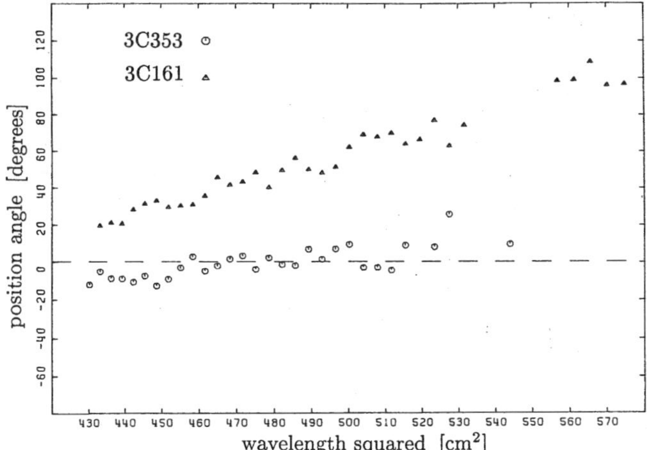

Fig. 11.4. The position angle of the electric field of the linearly polarized radiation is plotted against λ^2. The estimated rms error for each point is $\pm 6°$. The observations of two quasars, 3C353 and 3C161, are shown, located at 32.6 and 83.1 degrees from the Sun, respectively (from Gauss and Goldstein, 1973).

11.3.4. Quasi-Transverse Regions

Electromagnetic waves in a homogeneous medium can be considered as eigenmodes. What happens if they travel along a ray where the magnetic field changes direction, as in Figure 11.5 (top)? In an inhomogeneous medium there are no eigenmodes. In first approximation, a real wave can be described as being composed of the two eigenmodes of the homogeneous case. However, this is a deliberate choice, and the ratio between the amplitudes varies along the path. The two modes are no longer independent; in other words, the amplitude of one cannot be chosen without specifying the other. This is called *mode coupling*.

The freely moving charged particles in a plasma – in particular the electrons – oscillate simultaneously with both modes of radiation. The oscillating particles couple the two modes to each other. In contrast to wave scattering (Section 6.3.2) mode coupling is a linear process. The amplitude of one mode is proportional to the amplitude of the other, and is not multiplied by the amplitude of density fluctuations or of a third wave. Remember that the two coupled modes are in reality only the components of one wave.

The coupling is particularly notable

(1) near the plasma frequency ($\omega \approx \omega_p$), where a large fraction of the wave energy resides in particle motion,

(2) in quasi-transverse regions, where propagation is nearly perpendicular to the magnetic field and the wave modes are substantially elliptically or linearly polarized (Section 4.4).

There is no coupling in vacuum and in a homogeneous medium. Furthermore, there is no mode coupling for waves travelling parallel ($\theta = 0$) to the magnetic field and as long as the quasi-longitudinal approximation (11.1.2) holds.

A. *Mode Coupling in Quasi-Transverse Regions*

The quasi-longitudinal approximation breaks down in the range of quasi-transverse behavior. This range is limited by $\pi/2 - \Delta\theta \lesssim \theta \lesssim \pi/2 + \Delta\theta$, with

$$\Delta\theta \approx \frac{\Omega_e}{2\omega(1-\omega_p^2/\omega^2)} \quad , \tag{11.3.11}$$

as can be derived from the dispersion relation (4.5.1) assuming $(\Omega_e/\omega)^2 \ll (1 - \omega_p^2/\omega^2)^2$ (Exercise 11.1). In Figure 11.5 a ray propagates along a path s. The path length within the quasi-transverse region becomes

$$\Delta s \approx \Delta\theta \, \frac{ds}{d\theta} = \frac{\Omega_e \, H_B}{2\omega(1-\omega_p^2/\omega^2)^{3/2}} \quad . \tag{11.3.12}$$

We have defined $H_B := dx/d\theta$, the scale length in transverse direction over which the angle of **B** changes. The x-axis is assumed parallel to the ray at the point of transverse propagation (Fig. 11.5, top). The ray path is affected by refraction, introducing a further factor $(1-\omega_p^2/\omega^2)^{-1/2}$ by Snell's law: $ds/dx \approx \sin\theta \approx 1/\mathcal{N}$.

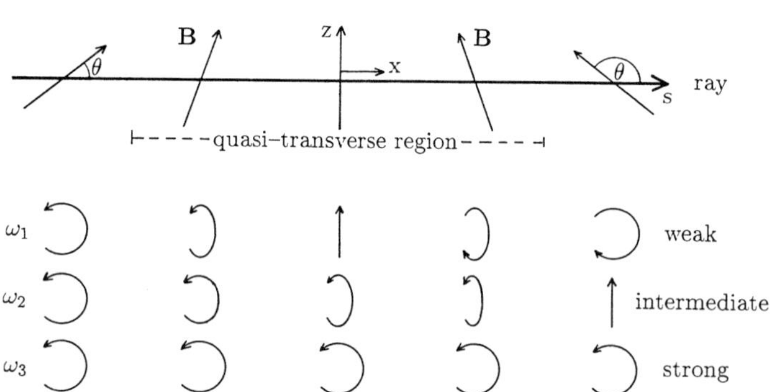

Fig. 11.5. *Top*: A ray traverses a region of quasi-transverse magnetic field. *Bottom*: Each horizontal row of five polarization ellipses corresponds to a different physical situation. The first row corresponds to the behavior of a low-frequency wave, $\omega_1^4 \ll \omega_t^4$, the second row to the intermediate case, $\omega_2 \approx \omega_t$, and the bottom row to a high-frequency wave, $\omega_3^4 \gg \omega_t^4$. The coupling of these waves is weak, intermediate, and strong, respectively.

Two oppositely polarized modes interact efficiently as long as their phases remain approximately equal. Therefore, the condition for severe coupling is that the change in phase difference, $\Delta\psi$, is small during the interaction over a length Δs;

$$\Delta\psi = (k_o - k_x)\Delta s \ll 1 \quad . \tag{11.3.13}$$

We use the dispersion relation (Eq. 4.5.1) in the quasi-transverse limit to derive

$$k_o - k_x \approx \frac{\Omega_e^2 \omega_p^2}{2\omega^3 c(1 - \omega_p^2/\omega^2)} \quad . \tag{11.3.14}$$

The smaller the phase difference, $\Delta\psi$, through the quasi-transverse region, the better the coupling. The degree of coupling is described by the *coupling constant*,

$$\begin{aligned} Q := \frac{1}{\Delta\psi} &= \frac{4\, c\, \omega^4}{\Omega_e^3 \omega_p^2 H_B} \left(1 - \omega_p^2/\omega^2\right)^{5/2} \\ &=: \left(\frac{\omega}{\omega_t}\right)^4 \left(1 - \omega_p^2/\omega^2\right)^{5/2} \quad . \end{aligned} \tag{11.3.15}$$

The transition frequency, ω_t, has been defined with

$$\omega_t := \left(\frac{\Omega_e^3 \omega_p^2 H_B}{4c}\right)^{1/4} \quad . \tag{11.3.16}$$

Q vanishes in the absence of a gradient ($H_B = \infty$). Small Q means slow change of mode ratio. Except for $\omega \approx \omega_p$, the ratio ω/ω_t determines the strength of the coupling. We may estimate H_B in a corona to be of the order of 10^9 cm and $\Omega_e/\omega_p \approx 0.1$. The transition frequency is then typically $\omega_t \approx 10\ \omega_p$. We shall compare these expectations with observation in the following subsection.

The effect of a quasi-transverse region depends on the region in which it is traversed: this is sketched in Figure 11.5 (bottom). The polarization ellipse indicates the sense and orientation of elliptical polarization. For $\omega^4 \ll \omega_t^4$ and thus $Q \ll 1$, the phase difference changes too much to transfer energy efficiently; mode coupling is said to be *weak*. If a wave is in o-mode before a quasi-transverse region with weak coupling, it remains in o-mode throughout it. This means that, if initially $\theta < \pi/2$ and the E-vector rotates counterclockwise, the wave becomes elliptically polarized in the quasi-transverse region and linearly polarized at $\theta = \pi/2$. It leaves the quasi-transverse region in o-mode, but the E-vector rotates clockwise because now $\theta > \pi/2$ (Figure 11.5, ω_1). The emerging mode is the same as the emitted mode, but the sense of the observed circular polarization is opposite to that in the source. Weak coupling changes the sense of circular polarization.

Intermediate coupling requires $Q \approx 1$. About half of the energy is transferred to the other mode, so that the final radiation is elliptically or – in the ideal case – linearly polarized. Faraday rotation (Section 11.2.4) quickly reduces the polarization to the circular part, diminishing the degree of polarization.

Mode coupling is *strong* if $\omega^4 \gg \omega_t^4$ and $Q \gg 1$. Energy is exchanged effectively between modes so that a wave arriving in pure o-mode is decomposed into o-mode and x-mode. The circular polarization remains as if the magnetic field did not change. The rotation is counterclockwise before the quasi-transverse region (assuming o-mode and $\theta < \pi/2$), in the region and thereafter. The handedness (screw sense) of the rotation does not change when viewed in forward direction of propagation, but the name of the mode changes. Since $\theta > \pi/2$ behind the quasi-transverse region, the o-mode wave turns into x-mode. The observed sense of circular polarization then is identical to the sense of polarization in the source. Strong coupling preserves the sense of polarization.

The deduction of the original mode and of the field direction in the source requires information on the mode coupling. Then, if we know the mode of emission, we can deduce the orientation of the magnetic field in the source, and *vice versa*.

B. *Confrontation with Observations*

If a coronal source emits radiation at ω between one and two times ω_p, the radiation first passes through a region with $\omega < \omega_t$ for seemingly plausible values of the relevant parameters. Thus mode coupling in a quasi-transverse region near the source is expected to be weak for plasma emission.

The observational evidence is controversial. The strength of mode coupling can be tested by two sources located in the legs of a loop and emitting the same mode. Taking o-mode, for example, the right ray in Figure 11.6 (left) is emitted with counterclockwise polarization and the left ray clockwise. The left ray passes through a quasi-transverse region. If mode coupling is weak, the sense of rotation flips to clockwise, becoming identical to the right ray. Solar type III radio bursts travelling in loops (called U-bursts, Fig. 5.8) have been studied for this effect. At frequencies above 200 MHz the sources in the rising and descending phase have usually the same polarization confirming the expectation. However, at lower frequencies the two sources show different polarization indicating strong coupling throughout the high corona (beyond about 10^5km).

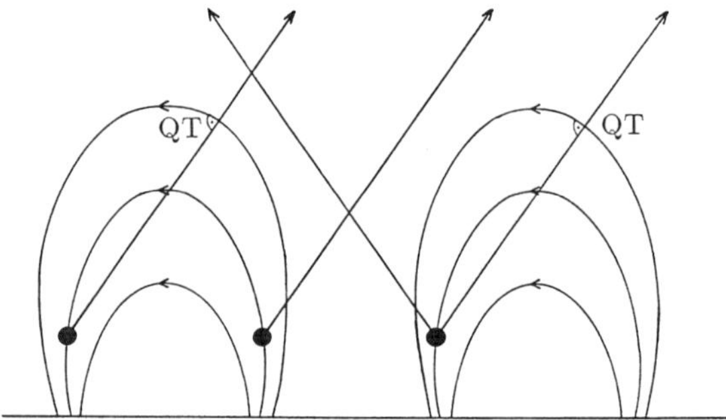

Fig. 11.6. Illustrations of two tests of the strength of mode coupling in quasi-transverse regions (QT). *Left*: Two sources with opposite longitudinal magnetic field relative to the observer. *Right*: One source seen from two directions.

In another test, a radio source is observed from two directions (Fig. 11.6, right). In practice, one uses stationary sources and observes in two phases of rotation. For example, noise storms (Section 9.5.2) are highly polarized, stationary sources in the high corona at $\omega \approx \omega_p$. When the solar rotation moves them to different aspect angles, their sense of polarization is not observed to change, suggesting strong coupling.

Strong coupling in high coronal plasma emission is generally interpreted by a small transition frequency, reduced by small-scale fluctuations in the magnetic field in the quasi-transverse region. They may be caused, for example, by the continuous passage of Alfvén waves through the corona. The ripples effectively reduce H_B in Equation (11.3.16). The angle between the ray and the magnetic field may change significantly on scales as small as 10^5 cm, making $\omega_t \approx \omega_p$. All emissions then are strongly coupled in any quasi-transverse region they encounter in propagating, and the observed sense of circular polarization is the same as the sense emitted.

C. *Depolarization*

Several emission mechanisms proposed for solar radio bursts produce predominantly one mode of radiation. Yet the observed bursts are only weakly polarized. This is the case for type II and type III bursts in general, and noise storms, narrowband spikes and decimetric pulsations near the limb. On the other hand, coherent radio bursts of dMe stars are highly circularly polarized (Section 8.4.4).

The observational discrepancies may be interpreted by depolarization in quasi-transverse regions. In a quasi-transverse region where $\omega \approx \omega_t$ within about a factor of two, highly elliptical or linear polarization results from incident circularly polarized radiation (Fig. 11.5, ω_2). Faraday rotation eliminates this linear polarization and the emerging radiation is partially circularly polarized. If the magnetic field is rippled in the quasi-transverse region, the ray passes effectively through many quasi-transverse regions and loses most of its initial polarization.

The coupling constant increases rapidly for frequencies between ω_p and $2\omega_p$ (Eq. 11.3.15). Therefore, plasma emission at the fundamental has a high probability to encounter a quasi-transverse region with $Q \approx 1$. It is most likely to be affected by depolarization.

In the coronae of dwarf M-stars the ratio Ω_e/ω_p is probably close to unity, thus ω_t is larger and coupling near the source is weak throughout the corona. The radiations from sources with different signs of the longitudinal magnetic field combine to radiation with the same sense of rotation after passing a quasi-transverse region near the star ($\omega \ll \omega_t$). The domain where $Q \approx 1$ is far from the star, and the probability to cross a quasi-transverse region there is small. This may be the reason why the observed radiation is not depolarized.

11.4. Scattering at Plasma Inhomogeneities

In an irregular propagation medium, a wave cannot be represented by a single ray. Small fluctuations in density or magnetic field distort an incident plane wave as the wave phase propagates at different speeds. The wave breaks up into parts that are deflected, focussed, and defocussed in a random way. A wave then resembles a bundle of interfering rays traversing slightly different paths.

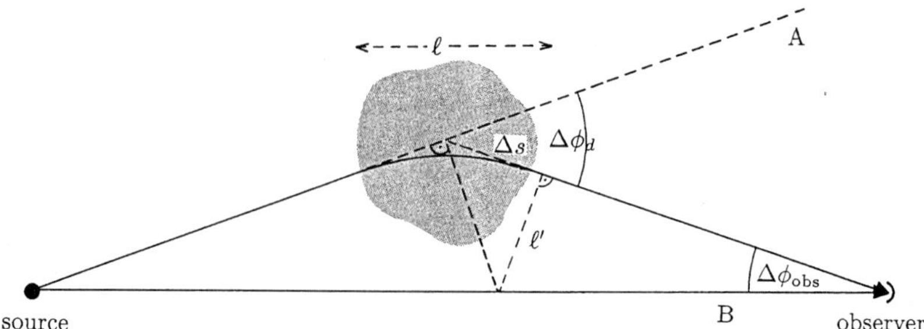

Fig. 11.7. A wave front (ray A) is distorted by an inhomogeneity (say a density fluctuation) in the medium of propagation and is deflected by an angle $\Delta\phi_d$. Ray B propagates to the observer in a straight line.

Consider a single scattering process as illustrated in Figure 11.7. Ray A passes through an inhomogeneity of size ℓ assumed to be surrounded by a homogeneous background. In the limit of very small deflection angle $\Delta\phi_d$, we can neglect the difference in the path length of rays A and B for the following discussion. The wave phase (Eq. 11.3.5) is shifted relative to ray B in the undisturbed medium by

$$\Delta\psi = \Delta k\ \ell = \left(\frac{1}{v_{ph}^1} - \frac{1}{v_{ph}^0}\right)\omega\ell \quad . \tag{11.4.1}$$

The superscripts 0 and 1 refer to the ambient medium and the irregularity, respectively. Scattering is called *strong*, if $|\Delta\psi| > \pi$. Let us neglect magnetic effects and assume that the irregularity is a density fluctuation of Δn. The phase velocity, derived from Equation (11.1.2) for $\omega \gg \Omega_e$, is

$$v_{ph} = \frac{\omega}{k} \approx \frac{c}{\sqrt{1 - \omega_p^2/\omega^2}} \quad . \tag{11.4.2}$$

For $\omega^2 \gg \omega_p^2$, one derives

$$\Delta\psi \approx -\frac{\Delta n}{n}\left(\frac{\omega_p}{\omega}\right)^2 \frac{\ell\omega}{2c} \quad . \tag{11.4.3}$$

The negative sign in Equation (11.4.3) indicates that the phase moves faster for $\Delta n > 0$. When the ray leaves the density fluctuation, its phase is ahead by Δs compared to the ray in the homogeneous medium. It continues to propagate as a plane wave, but arrives at the observer from a different angle. The deflection angle, $\Delta\phi_d$ (see Fig. 11.7), is

$$\Delta\phi_d \approx \frac{\Delta s}{\ell'} = \frac{\Delta\psi}{k_1\ell'} = -\frac{1}{2}\,\frac{\Delta n}{n}\left(\frac{\omega_p}{\omega}\right)^2 \tag{11.4.4}$$

We have used $v_{ph}^1 \approx c$, $\ell' \approx \ell$, and Equation (11.4.3). Let us now study the effect of many scattering processes. The total deflection angle of multiple scatterings is the sum of a random walk in two dimensions, on average

$$\langle \phi_d \rangle \;\approx\; \sqrt{N}\,\langle|\,\Delta\phi_d\,|\rangle \approx \frac{1}{2}\sqrt{\frac{L}{\ell}}\,\frac{\langle \Delta n^2 \rangle^{1/2}}{n}\,\left(\frac{\omega_p}{\omega}\right)^2 \quad , \tag{11.4.5}$$

where N is the number of scatterings approximated by L/ℓ, and L is the size of the scattering medium. The observer sees a multitude of rays from the scattering region. If the scattering occurs in a thin region midway between the source and the observer, a point source appears smeared out by half of the rms deflection angle, thus $\langle \phi_{\rm obs} \rangle \approx \langle \phi_d \rangle/2$. In practice, the diameters ℓ and ℓ' can be defined by the autocorrelation lengths of the density distribution, and $\langle \Delta n^2 \rangle^{1/2}$ by its standard deviation.

Scattering broadens the image of a source. Assuming Gaussian statistics with $\langle \phi_{\rm obs} \rangle$ being the angular radius at half power of the scattered (apparent) image of a point source, a flux density F is smeared out to an extended image with an intensity distribution

$$I(\phi) = \frac{F}{2\pi \langle \phi_{\rm obs} \rangle^2} \exp\left[-\frac{\phi^2}{2\langle \phi_{\rm obs} \rangle^2}\right] \quad . \tag{11.4.6}$$

In addition to angular broadening, the scattering process washes out the time profile of an impulsive signal. A scattered ray is delayed relative to the direct ray for the simple reason of its longer path. For a Gaussian distribution of irregularities and hence a Gaussian distribution of scattering angles, a pulse is convolved with a truncated exponential,

$$g(t) = \begin{cases} \exp(-2t/\tau_s) & t \geq 0 \\ 0 & t < 0 \end{cases} \quad . \tag{11.4.7}$$

The time constant,

$$\tau_s \approx \frac{D_2}{D_1}\,\frac{D\,\langle \phi_{\rm obs} \rangle^2}{2c} \quad , \tag{11.4.8}$$

can be derived as an exercise (11.4) from a geometrical argument. D is the distance to the source, D_1 denotes the distance from the source to the scattering region assumed to have a thickness much smaller than D (thin screen model), and $D_2 = D - D_1$. In the derivation of Equation (11.4.8), the approximations $(D_2)^2 \langle \phi_{\rm obs} \rangle^2 \ll (D_1)^2, (D_2)^2$ have been used. A group velocity of c have been assumed, and ϕ is in units of radians.

Strong and multiple scattering are the two conditions for a phenomenon called *scintillation*. It is caused by interference between the direct ray and scattered rays. It can be constructive – then the signal is enhanced – or destructive, and the signal weakens. If the source or the scattering medium move in relation to the observer, the observed flux varies, since the phase relations of the two rays – and thus their interference – change rapidly. Another characteristic of scintillations is that interference remains constructive only over a limited bandwidth. The variations of scintillation therefore are not correlated if observed at two frequencies separated more than $\Delta\nu$, usually a very small value (e.g. about $10^{-33} \cdot \nu^4$ in the interstellar medium, $\Delta\nu$ and ν in units of Hz).

In conclusion, we summarize the key effects of scattering by random fluctuations on radiation from sources in coronae:

(1) The apparent size of a point source is bigger the closer the frequency to the plasma frequency and the thicker the scattering region (Eq. 11.4.5).

(2) The apparent source height near the limb increases with scattering, since rays passing through the corona at low altitude are strongly absorbed.

(3) Large-scale refraction makes radiation more directive (Section 11.2.2), scattering makes it less directive.

(4) Angular broadening by scattering implies temporal smoothing (Eq. 11.4.8). For short pulses both effects of scattering are observable and related.

The scattering hypothesis has been successful for pulsars and irregularities of the interstellar medium. The evidence of scattering of solar radio bursts is ambiguous. In agreement with the scattering hypotheses, the diameter of the observed images increases from the center of the disk to the limb, compatible with the longer propagation path in the corona. Solar radio emissions from near the plasma level have apparent diameters of several arcminutes, apparently broadened by scattering. However, the observed shift of the apparent sources near the limb to higher altitude is too large to be interpreted by scattering on random fluctuations, as they would introduce a range of time delays in excess of the observed decay time of the short bursts (e.g. solar type I bursts). Furthermore, the observed source size of fundamental emission would be larger than for harmonic emission at a given frequency, contrary to observations. And finally, the observed directivity of some radiation (particularly type I bursts) excludes the scattering predicted from the apparent source size. It is the focus of the next section to interpret the propagation effects of solar radio bursts by irregularities that are not isotropic and have not a random shape.

11.5. Propagation in a Fibrous Medium

In the previous section we have found discrepancies between observed solar radio emission and the prediction of both refraction models in a spherically symmetric atmosphere and models involving scattering on random irregularities. The discrepancies disappear if the corona is modelled as a fibrous medium. The ratio β between thermal plasma pressure and magnetic field pressure is generally much smaller than unity in the corona of the Sun and other late-type stars. The magnetic field dominates (Section 3.1.3) and arranges itself into a smooth configuration. The density can then vary perpendicular to the magnetic field with little effect on the pressure equilibrium (Eq. 3.1.50). On the other hand, the density fluctuations are smoothed in the third dimension along the field lines. The coronal density may thus be highly structured in flux tubes of high and low density.

This is confirmed at the arcsecond scale by solar UV and soft X-ray images and in interplanetary space by *in situ* observations. Figure 11.8 shows that surfaces of constant density in interplanetary space are highly corrugated. The curves were derived from solar wind density measurement near Earth, and using the assumptions $n_e \propto r^{-2}$ and that the solar wind motion with $V_{sw} = 400$ km s^{-1} forms an Archimedean spiral magnetic field. The figure illustrates the concept of a plasma structured as *fibers* of high and low density.

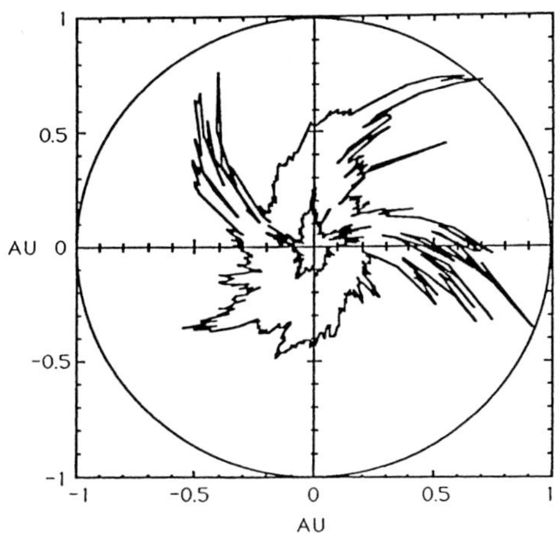

Fig. 11.8. Curves of constant density in the ecliptic plane, the inner one corresponding to $n_e = 7.1 \cdot 10^2$ cm^{-3} and the outer one to $n_e = 44$ cm^{-3}. The curves were constructed from 27 days of ion density measurements made once per hour near Earth (from Lecacheux *et al.*, 1989).

White-light and soft X-ray observations suggest coronal density variations across the magnetic field of an order of magnitude at a given height over distances of 10^4km or less (Fig. 1.2). The electron density in the interplanetary medium is known to vary over small scales transverse to the magnetic field with a root-mean-square factor of about 2.

There are two models of the effect of a fibrous medium on propagation. (*i*) *Ducting* by low density, elongated structures collimates and guides the radiation. (*ii*) Deflection by high density fibers causes *anisotropic scattering*. The two concepts are limiting cases. In reality, they may occur simultaneously.

11.5.1. Ducting

Figure 11.9 depicts radiation originating near the plasma level ($\omega \approx \omega_p$) in a fibrous source at left. Radiation propagating along the magnetic field lines (i.e. perpendicular to the density gradient and horizontal in the figure) is absorbed by electron collisions, regardless of whether it has originated in the high density region or in the low density region (Section 11.2.1, Eq. 11.2.5a).

Since refraction bends radiation toward regions with lower phase velocity (higher $\mathcal{N}$), implying lower density, radiation from the high density part may be refracted to the low density regions, where its frequency is now well above the

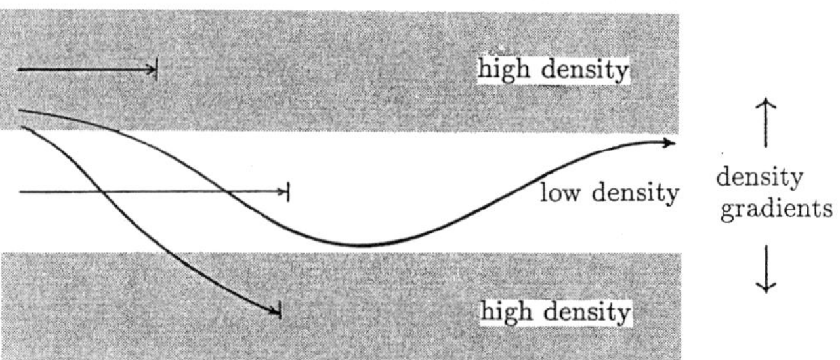

Fig. 11.9. Propagation of radiation emitted near the local plasma frequency in a dense, fibrous corona. Rays perpendicular to the density gradient are absorbed. Only radiation refracted and scattered from a high density source into a low density trough can escape and is ducted.

local plasma frequency. The collisional absorption is then reduced according to Equation (11.2.5.C). A part of the radiation is scattered into a direction nearly parallel to the density structures. Refraction will then guide the radiation along the low density troughs (called *ducts*) as light waves are in an optical fiber. When the radiation reaches the higher corona, the plasma frequency of even the high-density regions falls below the frequency of the radiation and the radiation escapes freely.

In a conical duct expanding in outward direction, guiding becomes even more efficient. Each refraction at the wall decreases the angle between the ray and the axis of the duct by 2ϕ, where ϕ is the half-angle of the cone. The ducted rays are collimated into the direction of the cone axis.

Ducting has been proposed (by Duncan, 1977) to account for the apparent positions of the fundamental ($\omega \approx \omega_p$) and harmonic ($\omega \approx 2\omega_p$) components of radiation caused by beams and shock waves (solar type III and type II radio bursts, respectively). Fundamental and harmonic radiation can be identified from spectral observations. The positions of fundamental-harmonic pairs – observed at the same frequency – nearly coincide, although the source of the harmonic emission should be at lower density by a factor of four. Correspondingly, the height of a fundamental source observed at frequency ν is always greater than the height of its harmonic emission observed at 2ν, despite the fact that the source is the same.

A likely situation is sketched in Figure 11.10. The fundamental emission – and probably the harmonic emission as well – first propagates in field aligned ducts at a low density, until the high density filaments become transparent for the observing frequency. Thus, fundamental and harmonic emission may appear at the same location for a given observing frequency. The condition for this to occur is simply that a duct is surrounded by plasma having at least twice the plasma frequency (i.e. the surrounding density exceeds four times the value in the duct). The harmonic emission is ducted along with the fundamental of the same radiation frequency originating farther down. The two emissions escape from the duct at the same position.

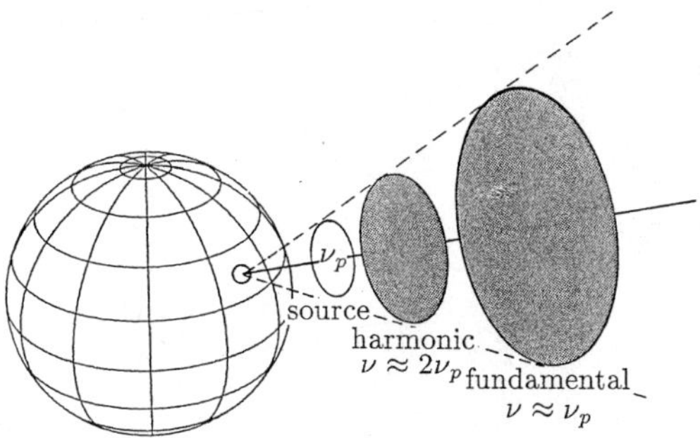

Fig. 11.10. Schematic drawing of plasma radiation originating at a source with a plasma frequency ν_p. The radiation is observed at the apparent sources (shaded) in fundamental ($\nu \approx \nu_p$) and in harmonic ($\nu \approx 2\nu_p$) emissions.

The ducting model helps to reconcile the reported discrepancy between coronal densities derived from the observed radio frequency (Section 5.3) and the white-light intensity. The radio measurements systematically show densities that are higher by a factor of 10 to 20. Ducted radio emission refers to the peak density at the apparent source position. On the other hand, white-light measurements yield average values along the line of sight. This can explain at least a factor of 4. The preferential escape of emission from high density structures (Fig. 11.9) may contribute a similar factor.

Ducting explains how a radio source appears at greater heights without excessively increasing its apparent size. It gives a natural interpretation of the observed similarity of source sizes of fundamental and harmonic emissions.

In conclusion, ducting may interpret many propagation phenomena of waves having frequencies near the local plasma frequency. However, it highly collimates the escaping radiation in the radially outward direction, contradicting the observation of limb events. Obviously, at higher altitude – where $\omega \gg \omega_p$ – some form of scattering will take over. This is the topic of the next section.

11.5.2. Anisotropic Scattering

Ducting is important as long as the radiation frequency is close to the average plasma frequency in the plasma of propagation. When the radiation frequency exceeds the plasma frequency of most flux tubes, the radiation is refracted away from the direction of the duct. Nevertheless, the radiation may still be reflected by the most dense fibers (Fig. 11.11). A refracted ray conserves the angle to the fibers at each reflection. The alignment of the flux tubes imposes an anisotropy into the scattering process.

In the two-dimensional geometry of Figure 11.11, the radiation of a highly directive source is reflected into the two directions with angle θ to $\mathbf{B}$. The apparent

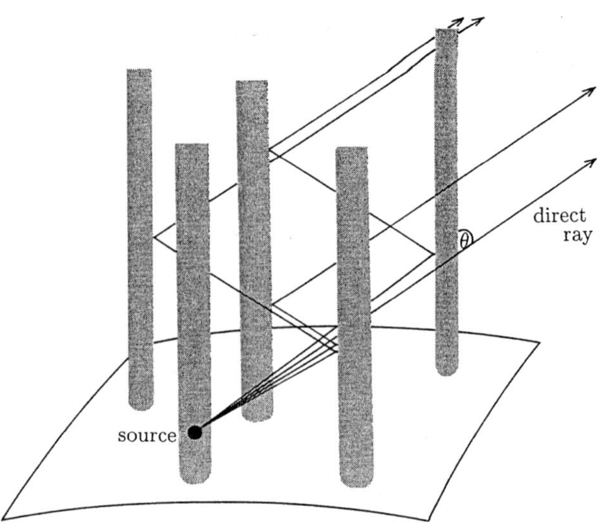

Fig. 11.11. Overdense flux tubes (shaded) scatter the radiation to a larger apparent source size without reducing its directivity. The rays cannot penetrate the overdense flux tubes. They are either reflected, pass in front or behind the tubes.

size of the source increases, and its average position is apparently at higher altitude. The directivity of the source remains high.

Anisotropic scattering has been proposed to explain the observed directivity in solar type I radio bursts associated with noise storms. They appear to have a larger source size L than their short rise time Δt allows ($L > \Delta t\ v_{gr}$, where the group velocity for fundamental emission ($\omega \approx \omega_p$) is about $0.2c$). This seems to be the result of scattering. Anisotropic scattering increases the source size with a smaller loss of directivity than isotropic scattering. Since anisotropic scattering occurs at propagation speeds of nearly c, the extra path for reflections does not efficiently delay the radiation.

All propagation effects intensify the closer the wave frequency is to the plasma frequency of the medium traversed. The densest region in the propagation path is therefore most influential. For emissions from a corona it is usually the coronal plasma overlaying the source that modifies the radiation.

The fibrous structure of coronae becomes the dominant effect at $\omega \approx \omega_p$. It not only distorts the emission, but allows radiation to escape even from the densest part of a corona where a horizontally stratified atmosphere would be opaque. Propagation effects thus contain interesting information about density inhomogeneity in magnetically structured coronae.

Exercises

11.1: The quasi-longitudinal approximation (11.1.2) is often used for estimating propagation effects. What is the limiting angle θ where the longitudinal and transverse terms of Equation (4.5.1) are equal? Evaluate the angle for $\Omega_e/\omega = 0.1$ in the two cases $\omega_p^2/\omega^2 = 0.01$ and 0.8.

11.2: The two modes of radiation usually propagate independently. Consider a radiation pulse that is initially linearly polarized and that propagates quasi-longitudinally through a cloud of plasma having $\omega_p/\omega = 0.5$, $\Omega_e/\omega = 0.1$, and a length $L = 3 \cdot 10^{10}$ cm. What is the difference in arrival time of the o-mode and x-mode parts of the radiation?

11.3: The direction of the electric field vector of a linearly polarized wave rotates around **k** as the wave propagates through a magnetized plasma. This Faraday rotation angle depends on frequency. Calculate the frequency difference at which the rotation differs by 2π. How much is it for linear polarization emitted in a corona? Use $\nu = 10^{10}$ Hz, $n_e = 10^9$ cm^{-3}, $B\cos\theta = 100$ G, and $L = 10^{10}$ cm (about one scale height in the solar corona). How many times does the electric vector rotate in this distance?

11.4: Scattering smears out the time profile of an impulsive signal. Calculate the delay (Eq. 11.4.8) of a ray scattered in a thin screen between the source and the observer. Assume that the group velocity outside the screen is c and that the observer sees the ray arriving at an angle $\langle\phi_{\rm obs}\rangle$ from the direct ray. What is the time constant due to scattering of solar type I bursts having an apparent diameter of $2'$ and $D_1 \approx 10^{10}$ cm?

Further Reading and References

General treatises including anisotropic refraction

Bekefi, G.: 1966, *Radiation Processes in Plasmas*, J. Wiley and Sons, Inc., New York.

Melrose, D.B.: 1980, *Plasma Astrophysics – Non-thermal Processes in Diffuse Magnetized Plasmas*, Vol.1, Gordon and Breach, New York.

Chandrasekhar, S.: 1955, *Radiative Transfer*, Oxford University Press.

Dispersion and scattering

Gauss, F.S., and Goldstein, S.J.Jr.: 1973, 'Multifrequency Polarization Observations of Eight Extragalactic Sources', *Astrophys. J.* **179**, 439.

Manchester, R.W. and Taylor, J.H.: 1977, *Pulsars*, W.H. Freeman, San Francisco.

Melrose, D.B.: 1988, 'Depolarization of Solar Bursts due to Scattering by Low-Frequency Waves', *Solar Phys.* **119**, 143.

Wentzel, D.G.: 1984, 'Polarization of Fundamental Type III Radio Bursts', *Solar Phys.* **90**, 139.

White, S.M., Thejappa, G., and Kundu, M.R.: 1992, 'Observations of Mode Coupling in the Solar Corona and Bipolar Noise Storms', *Solar Phys.* **138**, 163.

Propagation in a fibrous medium

Bougeret, J.L. and Steinberg, J.L.: 1977, 'A New Scattering Process above Solar Active Regions: Propagation in a Fibrous Medium' *Astron. Astrophys.* **61**, 777.

Duncan, R.A.: 1977, 'Wave Ducting of Solar Meter-Wave Radio Emission as an Explanation of Fundamental/Harmonic Sources Coincidence and other Anomalies', *Solar Phys.* **63**, 389.

Lecacheux, A., Steinberg, J.-L., Hoang, S., and Dulk, G.A.: 1989, 'Characteristics of Type III Bursts in the Solar Wind from Simultaneous Observations on Board ISEE-3 and Voyager', *Astron. Astrophys.* **217**, 237.

Roelof, E.C. and Pick, M.: 1989, 'Type III Radio Bursts in a Fibrous Corona', *Astron. Astrophys.* **210**, 417.

MATHEMATICAL EXPRESSIONS

Definitions of Vector and Tensor Operations

Scalar product:

$$\mathbf{A} \cdot \mathbf{B} := A_x B_x + A_y B_y + A_z B_z \tag{A.1}$$

Vector product:

$$\mathbf{A} \times \mathbf{B} := (A_y B_z - A_z B_y,\ A_z B_x - A_x B_z,\ A_x B_y - A_y B_x) \tag{A.2}$$

Tensor product:

$$\mathbf{A} \circ \mathbf{B} := \begin{pmatrix} A_x B_x & A_x B_y & A_x B_z \\ A_y B_x & A_y B_y & A_y B_z \\ A_z B_x & A_z B_y & A_z B_z \end{pmatrix} \tag{A.3}$$

Divergence:

$$\nabla \cdot \mathbf{A} := \frac{\partial A_x}{\partial x} + \frac{\partial A_y}{\partial y} + \frac{\partial A_z}{\partial z} \tag{A.4}$$

Rotation (curl):

$$\nabla \times \mathbf{A} := \left(\frac{\partial A_z}{\partial y} - \frac{\partial A_y}{\partial z}, \frac{\partial A_x}{\partial z} - \frac{\partial A_z}{\partial x}, \frac{\partial A_y}{\partial x} - \frac{\partial A_x}{\partial y} \right) \tag{A.5}$$

Laplace operator:

$$\nabla^2 |\mathbf{A}| = \frac{\partial^2}{\partial x^2} |\mathbf{A}| + \frac{\partial^2}{\partial y^2} |\mathbf{A}| + \frac{\partial^2}{\partial z^2} |\mathbf{A}| \tag{A.6}$$

Vector-tensor products:

$$\mathbf{A} * \hat{\mathcal{B}} := (\sum_i A_i B_{ix} \ , \ \sum_i A_i B_{iy} \ , \ \sum_i A_i B_{iz}) \tag{A.7}$$

$$\hat{\mathcal{B}} * \mathbf{A} := (\sum_i B_{xi} A_i \ , \ \sum_i B_{yi} A_i \ , \ \sum_i B_{zi} A_i) \tag{A.8}$$

Vector Relations

$$(\mathbf{A} \times \mathbf{B}) \times \mathbf{C} \equiv -\mathbf{C} \times (\mathbf{A} \times \mathbf{B}) \equiv (\mathbf{A} \cdot \mathbf{C})\mathbf{B} - (\mathbf{B} \cdot \mathbf{C})\mathbf{A} \tag{A. 9}$$

$$\nabla \times (\nabla \times \mathbf{A}) \equiv \nabla(\nabla \cdot \mathbf{A}) - \nabla^2 \mathbf{A} \tag{A.10}$$

$$\mathbf{A} \times (\nabla \times \mathbf{A}) \equiv \nabla(\frac{|\mathbf{A}|^2}{2}) - (\mathbf{A}\nabla)\mathbf{A} \tag{A.11}$$

$$\nabla \cdot (\nabla \times \mathbf{A}) \equiv 0 \tag{A.12}$$

Frequently Used Integrals

For $a > 0$ and n a natural number (n = 1,2,3,...),

$$\int_0^\infty x^n e^{-\frac{x}{a}} dx = a^{n+1} n! \quad . \tag{A.13}$$

For $n = 0, 1, 2, \ldots$

$$\int_0^\infty x^{2n+1} e^{-\frac{x^2}{a^2}} dx = \frac{n!}{2} a^{2n+2} \quad , \tag{A.14}$$

$$\int_0^\infty x^{2n} e^{-\frac{x^2}{a^2}} dx = \frac{1 \cdot 3 \ldots (2n-1)\sqrt{\pi}}{2^{n+1}} a^{2n+1} \quad . \tag{A.15}$$

In particular,

$$\int_0^\infty e^{-\frac{x^2}{a^2}} dx = \frac{\sqrt{\pi}}{2} a \quad . \tag{A.16}$$

We write the *Fourier transformation* in space and the *Laplace transformation* in time as

$$\mathbf{A}(\mathbf{k}, \omega) := \int_0^\infty dt \int_{-\infty}^\infty d^3x \mathbf{A}(\mathbf{x}, t) \; e^{-i[\mathbf{kx} - \omega t]} \quad , \tag{A.17}$$

yielding a back transformation

$$\mathbf{A}(\mathbf{x}, t) = \int_{\mathcal{L}} \frac{d\omega}{2\pi} \int_{-\infty}^\infty \frac{d^3k}{(2\pi)^3} \mathbf{A}(\mathbf{k}, \omega) \; e^{i[\mathbf{kx} - \omega t]} \quad . \tag{A.18}$$

$\mathcal{L}$ indicates the Landau prescription on the contour integral in the complex ω-plane (see Fig. 5.5).

UNITS

We are using *Gaussian* (cgs) units throughout this book, since it is the most widely spread system in the field. The equations in the Gaussian system can easily be transformed into the *mks* system by replacing each symbol in the equation with the mks symbol given in Table B.1. Symbols for length and time, and their derivatives remain unchanged. When an equation is transformed to mks symbols, mks units (cf. Table B.2) have to be used.

Table B.1. Transformation of symbols in equations from Gaussian (cgs) units to mks units.

Quantity	Gaussian	Rationalised mks
speed of light	c	$(\varepsilon_0\mu_0)^{-1/2}$
charge	e	$e(4\pi\varepsilon_0)^{-1/2}$
current	j	$j(4\pi\varepsilon_0)^{-1/2}$
electric field	E	$E(4\pi\varepsilon_0)^{1/2}$
magnetic field	B	$B(4\pi/\mu_0)^{1/2}$
dielectric constant	ε	$\varepsilon/\varepsilon_0$
magnetic permeability	μ	μ/μ_0
electrical conductivity	σ	$\sigma(4\pi\varepsilon_0)^{-1}$

Table B.2. Units of symbols in Gaussian (cgs) and mks systems.

Quantity	Gaussian	Rationalized mks
length	1 cm =	10^{-2} m
mass	1 g =	10^{-3} kg
time	1 s =	1 s
force	1 dyne =	10^{-5} N
energy	1 erg =	10^{-7} joule (J)
power	1 erg s^{-1} =	10^{-7} watt (W)
charge	1 statcoul =	$\frac{1}{3} \cdot 10^{-9}$ coulomb (C)
electric field	1 statvolt cm^{-1} =	$3 \cdot 10^{4}$ V m^{-1}
current	1 statamp =	$\frac{1}{3} \cdot 10^{-9}$ ampère (A)
current density	1 statamp cm^{-2} =	$\frac{1}{3} \cdot 10^{-5}$ A m^{-2}
electrical conductivity	1 s^{-1} =	$\frac{1}{9} \cdot 10^{-9}$ siemens m^{-1}
magnetic field	1 gauss (G) =	10^{-4} tesla (T)

FREQUENTLY USED EXPRESSIONS

We assume a fully ionized plasma with solar abundances (27% helium by mass) and use cgs units (B in gauss, T in kelvin, $\ln\Lambda = 20$). The subscript $\odot$ refers to parameters in units of the solar values, and the subscript α denotes a general particle species.

Physical Constants

Speed of light	$c = 2.99792458 \cdot 10^{10}$	cm s^{-1}
Electron charge	$e = 4.803206 \cdot 10^{-10}$	esu
Electron mass	$m_e = 9.1093898 \cdot 10^{-28}$	g
Proton mass	$m_p = 1.6726231 \cdot 10^{-24}$	g
Mass ratio	$m_p/m_e = 1836.152735$	
Electron volt	$1 \text{ eV} = 1.602177 \cdot 10^{-12}$	erg
Boltzmann constant	$k_B = 1.38065 \cdot 10^{-16}$	erg deg^{-1}
Gravitational constant	$G = 6.6726 \cdot 10^{-8}$	cm^3 $g^{-1}s^{-2}$
Planck constant	$h = 6.62607 \cdot 10^{-27}$	erg s

Plasma Properties

Gyroradius	$R_\alpha = 5.69 \cdot 10^{-8} v_\perp B^{-1}(m_\alpha/m_e)$	cm
Debye length	$\lambda_D = 6.65\, T^{1/2} n_e^{-1/2}$	cm
Gyrofrequency	$\Omega_\alpha/2\pi = 2.80 \cdot 10^6 B\gamma^{-1}(m_e/m_\alpha)$	Hz
Plasma frequency	$\nu_p = 8.98 \cdot 10^3 \sqrt{n_e}$	Hz
Self-collision time	$t_{t\alpha} = 1.34 \cdot 10^{-2}\, T_\alpha^{3/2} Z_\alpha^{-4} n_\alpha^{-1} (m_\alpha/m_e)^{1/2}$	
Thermal electron-ion		s
collision time	$t_{e,i} = 1.65 \cdot 10^{-2}\, T^{3/2} n^{-1}$	s
Coulomb mean free path	$\ell_{c\alpha} = v_{t_\alpha} t_{t_\alpha} = 5.21 \cdot 10^3\, T_\alpha^2 n_\alpha^{-1} Z_\alpha^{-4}$	cm
Coulomb conductivity	$\sigma = 3.22 \cdot 10^6\, T_e^{3/2}$	s^{-1}
Dreicer field	$E_D = 2.02 \cdot 10^{-10}\, Z_i n_e/T$	statvolts cm^{-1}
Deflection time of non-thermal electron	$t_d^{tot} = 1.44 \cdot 10^{-20} v^3/n_e$	s
Mean thermal velocity	$v_{t\alpha} = 3.89 \cdot 10^5 T^{1/2}(m_e/m_\alpha)^{1/2}$	cm s^{-1}
Sound velocity	$c_s = 1.51 \cdot 10^4 \sqrt{T}$	cm s^{-1}
Alfvén velocity	$c_A = 2.03 \cdot 10^{11} B/\sqrt{n_e}$	cm s^{-1}
Pressure ratio	$\beta = 3.47 \cdot 10^{-15}\, n_e(T_e + T_p)/B^2$	

Solar Properties

Age	$4.65 \cdot 10^9$	y
Mass	$1.99 \cdot 10^{33}$	g
Radius	$6.96 \cdot 10^{10}$	cm
Luminosity	$3.83 \cdot 10^{33}$	erg s^{-1}
Surface gravity	$2.74 \cdot 10^4$	cm s^{-2}
Escape velocity	$6.18 \cdot 10^7$	cm s^{-1}
Density scale height	$5.00 \cdot 10^3\ T$	cm

Stellar Characteristics

Brightness temperature	$T_b = 1.77 \cdot 10^{27}\ \nu^{-2} F_{mJy} \ell_{mas}^{-2}$	K
(unpolarized)	$= 2.04 \cdot 10^{25} \nu^{-2} F_{mJy} D_{pc}^2 \ell_{\odot}^{-2}$	K
Surface gravity	$g = 2.74 \cdot 10^4\ M_{\odot} R_{\odot}^{-2}$	cm s^{-2}
Escape velocity	$v_{\rm esc} = 6.18 \cdot 10^7\ M_{\odot}^{1/2} R_{\odot}^{-1/2}$	cm s^{-1}
Density scale height	$H_n = 5.00 \cdot 10^3\ T\ g_{\odot}^{-1}$	cm

Propagation

Free-free absorption	$\kappa \approx 0.2\ n_e^2\ T_e^{-3/2} \nu^{-2}$	cm^{-1}
Group velocity delay	$\tau = 1.345 \cdot 10^{-3}\ \nu^{-2} \int n_e ds$	s
Faraday rotation	$\phi = 2.365 \cdot 10^4\ \nu^{-2} \int n_e B \cos\theta ds$	rad
Transition frequency	$\nu_t = 1.745 \cdot 10^4\ B^{3/4} n_e^{1/4} H_B^{1/4}$	Hz

Astronomical Constants and Units

Year	$y = 3.156 \cdot 10^7$	s
Parsec	$pc = 3.086 \cdot 10^{18}$	cm
Astronomical unit	$AU = 1.496 \cdot 10^{13}$	cm
Angstrom	$\text{Å} = 10^{-8}$	cm
Jansky	$Jy = 10^{-23}$	erg s^{-1} cm^{-2} Hz^{-1}
Solar flux unit	$sfu = 10^{-19}$	erg s^{-1} cm^{-2} Hz^{-1}

NOTATION

LATIN ALPHABET

A	general parameter or variable
$\bar{A}$	amplitude of a general oscillating parameter A
a	constant
$\mathbf{B}$	magnetic field
c	speed of light
c_A	Alfvén velocity (Eq. 3.2.13)
c_s	sound velocity (Eq. 3.2.12)
D	source distance
$\hat{\mathcal{D}}$	diffusion tensor
d	thickness of current sheet
$\mathbf{E}$	electric field
$\mathbf{E}_D$	Dreicer electric field (Eq. 9.2.7)
e	charge of an electron (modulus), or subscript for electron
$\mathbf{e_z}$	unity vector in z-direction
$\mathbf{F}_L$	Lorentz force
$\mathbf{F}$	general force
F	flux density of radiation, or subscript for field particle
f	distribution function of particles (Eq. 1.4.8)
G	various functions
g	gravitational acceleration
H	scale height or scale length
$H(\omega, \omega/k)$	part of dielectric function
h	height, or Planck's constant
I	intensity of radiation (Eq. 5.1.1), or total current
i	subscript for ion, or summation index
$\mathcal{J}$	source function of radiation (Eq. 11.1.4), or action integral
$\mathbf{J}$	current density
$\mathbf{k}$	wave vector ($\|k\| = 2\pi/\lambda$)
k_B	Boltzmann's constant
k_D	inverse Debye length ($= 1/\lambda_D$)
L	source size or source depth, or subscript for Langmuir wave
$L(\mathbf{k})$	interaction length
$\mathcal{L}$	longitudinal invariant
ℓ	length or diameter
l	harmonic number

M	mirror ratio, or magnetoacoustic Mach number (Eq. 10.1.11)
m	particle mass
N	total particle number, or quantum density (Eq. 6.3.3)
$\mathcal{N}$	refractive index
$\mathbf{n}$	unit vector normal to surface
n_α	particle density of species α
n_b	beam density
P	polarization, or probability
$\hat{\mathbf{P}}$	stress tensor (Eq. 3.1.10)
$\hat{\mathbf{p}}$	pressure tensor (Eq. 3.1.11)
p	isotropic pressure, or subscript for proton, or $i\omega$
$\mathbf{p}$	momentum
q	particle charge ($:= Ze$)
R	gyroradius (Eq. 2.1.2)
R_m	magnetic Reynolds number (Eq. 3.1.49)
r	radial coordinate, or impact parameter
r_c	impact parameter for 90° deflection (Eq. 2.6.1)
s	path length, or subscript for ion acoustic wave
T	temperature (Eq. 1.4.10), or subscript for test particle
T_b	brightness temperature (Eq. 5.1.1)
t	time, or subscript for transverse wave, or for thermal population
t_d	collisional deflection time (Eq. 2.6.19)
t_s	collisional slowing-down time (Eq. 2.6.25)
t_t	thermal collision time (Eq. 2.6.31)
t_ε	collisional energy-loss time (Eq. 2.6.29)
u	relative velocity
$\mathbf{V}$	mean velocity (Eq. 1.4.9)
$\mathcal{V}$	volume
V_d	drift velocity of bulk electrons vs. ions in a current
$\mathbf{v}$	particle velocity
v_r	runaway velocity (Eq. 9.2.10)
v_{gr}	group velocity of wave
v_{par}	velocity component parallel to initial velocity
v_{perp}	velocity component perpendicular to initial velocity
v_{ph}	phase velocity of wave
$v_{t\alpha}$	mean thermal velocity of species α
v_T	velocity of test particle
v_z	velocity component parallel to $\mathbf{B}$
$v_\perp$	velocity component perpendicular to $\mathbf{B}$
$W_{\rm tot}$	total wave energy density
$W(\mathbf{k})$	spectral wave energy density
w	total scattering rate
w_T	Thomson scattering rate
$w(\mathbf{v}, \mathbf{k_L}, \mathbf{k_t})$	differential scattering rate
$\mathbf{w^d}$	particle drift velocity

x	coordinate
y	coordinate
Z	electric charge in units of e
z	coordinate in the direction of $\mathbf{B}$

GREEK ALPHABET

α	particle species, or pitch angle
α_c	loss-cone half-angle
β	v/c, or plasma beta (Eq. 3.1.51)
γ	Lorentz factor, $(1-(v/c)^2)^{-1/2}$
γ_k	imaginary part of frequency (growth rate)
δ	δ-function, or small number, or exponent of energy distribution
$\mathcal{E}$	mean thermal energy density
ε	particle energy
$\hat{\epsilon}$	dielectric tensor
η	resistivity $(=1/\sigma)$, or emissivity
θ	angle between $\mathbf{B}$ and $\mathbf{k}$
κ	absorption coefficient
Λ	λ_D/r_c and approximate number of particles in a Debye sphere
λ	wavelength
λ_D	Debye length (Eq. 2.4.10)
μ	magnetic moment (Eq. 2.1.6)
ν	frequency
ξ	exponent of particle-flux distribution in energy
ρ	mass density
ρ^*	charge density
σ	conductivity
τ	period, or optical depth (Eq. 11.1.5)
τ_e	escape time from trap (Eq. 8.3.1)
Φ	electric potential
ϕ	angle
Ψ	error function, scalar magnetic potential
ψ	phase angle
$\mathit{\Omega}$	solid angle
Ω_α	gyrofrequency of species α (Eq. 2.1.4)
Ω_z^α	scalar gyrofrequency with sign of charge
ω	angular frequency $(=2\pi\nu)$
ω_p	plasma frequency $(=2\pi\nu_p)$ (Eq. 4.2.24)
ω_r	real part of frequency
ω_t	transition frequency for mode coupling (Eq. 11.3.16)

AUTHOR INDEX

SUBJECT INDEX

Previously published in Astrophysics and Space Science Library book series:

- **Volume 273**: **Lunar Gravimetry**
  Author: Rune Floberghagen
  Hardbound, ISBN 1-4020-0544-X, April 2002
- **Volume 271**: **Astronomy-inspired Atomic and Molecular Physics**
  Author: A.R.P. Rau
  Hardbound, ISBN 1-4020-0467-2, March 2002
- **Volume 269**:**Mechanics of Turbulence of Multicomponent Gases**
  Authors: Mikhail Ya. Marov, Aleksander V. Kolesnichenko
  Hardbound, ISBN 1-4020-0103-7, December 2001
- **Volume 268**:**Multielement System Design in Astronomy and Radio Science**
  Authors: Lazarus E. Kopilovich, Leonid G. Sodin
  Hardbound, ISBN 1-4020-0069-3, November 2001
- **Volume 267: The Nature of Unidentified Galactic High-Energy Gamma-Ray Sources**
  Editors: Alberto Carramiñana, Olaf Reimer, David J. Thompson
  Hardbound, ISBN 1-4020-0010-3, October 2001
- **Volume 266: Organizations and Strategies in Astronomy II**
  Editor: André Heck
  Hardbound, ISBN 0-7923-7172-0, October 2001
- **Volume 265: Post-AGB Objects as a Phase of Stellar Evolution**
  Editors: R. Szczerba, S.K. Górny
  Hardbound, ISBN 0-7923-7145-3, July 2001
- **Volume 264: The Influence of Binaries on Stellar Population Studies**
  Editor: Dany Vanbeveren
  Hardbound, ISBN 0-7923-7104-6, July 2001
- **Volume 262: Whistler Phenomena**
  Short Impulse Propagation
  Authors: Csaba Ferencz, Orsolya E. Ferencz, Dániel Hamar, János Lichtenberger
  Hardbound, ISBN 0-7923-6995-5, June 2001
- **Volume 261: Collisional Processes in the Solar System**
  Editors: Mikhail Ya. Marov, Hans Rickman
  Hardbound, ISBN 0-7923-6946-7, May 2001
- **Volume 260: Solar Cosmic Rays**
  Author: Leonty I. Miroshnichenko
  Hardbound, ISBN 0-7923-6928-9, May 2001
- **Volume 259: The Dynamic Sun**
  Editors: Arnold Hanslmeier, Mauro Messerotti, Astrid Veronig
  Hardbound, ISBN 0-7923-6915-7, May 2001
- **Volume 258: Electrohydrodynamics in Dusty and Dirty Plasmas**
  Gravito-Electrodynamics and EHD
  Author: Hiroshi Kikuchi
  Hardbound, ISBN 0-7923-6822-3, June 2001
- **Volume 257: Stellar Pulsation - Nonlinear Studies**
  Editors: Mine Takeuti, Dimitar D. Sasselov
  Hardbound, ISBN 0-7923-6818-5, March 2001

- **Volume 256: Organizations and Strategies in Astronomy**
  Editor: André Heck
  Hardbound, ISBN 0-7923-6671-9, November 2000
- **Volume 255: The Evolution of the Milky Way**
  Stars versus Clusters
  Editors: Francesca Matteucci, Franco Giovannelli
  Hardbound, ISBN 0-7923-6679-4, January 2001
- **Volume 254: Stellar Astrophysics**
  Editors: K.S. Cheng, Hoi Fung Chau, Kwing Lam Chan, Kam Ching Leung
  Hardbound, ISBN 0-7923-6659-X, November 2000
- **Volume 253: The Chemical Evolution of the Galaxy**
  Author: Francesca Matteucci
  Hardbound, ISBN 0-7923-6552-6, May 2001
- **Volume 252: Optical Detectors for Astronomy II**
  State-of-the-art at the Turn of the Millennium
  Editors: Paola Amico, James W. Beletic
  Hardbound, ISBN 0-7923-6536-4, December 2000
- **Volume 251: Cosmic Plasma Physics**
  Author: Boris V. Somov
  Hardbound, ISBN 0-7923-6512-7, September 2000
- **Volume 250: Information Handling in Astronomy**
  Editor: André Heck
  Hardbound, ISBN 0-7923-6494-5, October 2000
- **Volume 249: The Neutral Upper Atmosphere**
  Author: S.N. Ghosh
  Hardbound, ISBN 0-7923-6434-1, (in production)
- **Volume 247: Large Scale Structure Formation**
  Editors: Reza Mansouri, Robert Brandenberger
  Hardbound, ISBN 0-7923-6411-2, August 2000
- **Volume 246: The Legacy of J.C. Kapteyn**
  Studies on Kapteyn and the Development of Modern Astronomy
  Editors: Piet C. van der Kruit, Klaas van Berkel
  Hardbound, ISBN 0-7923-6393-0, August 2000
- **Volume 245: Waves in Dusty Space Plasmas**
  Author: Frank Verheest
  Hardbound, ISBN 0-7923-6232-2, April 2000
- **Volume 244: The Universe**
  Visions and Perspectives
  Editors: Naresh Dadhich, Ajit Kembhavi
  Hardbound, ISBN 0-7923-6210-1, August 2000
- **Volume 243: Solar Polarization**
  Editors: K.N. Nagendra, Jan Olof Stenflo
  Hardbound, ISBN 0-7923-5814-7, July 1999
- **Volume 242: Cosmic Perspectives in Space Physics**
  Author: Sukumar Biswas
  Hardbound, ISBN 0-7923-5813-9, June 2000
- **Volume 241: Millimeter-Wave Astronomy: Molecular Chemistry & Physics in Space**

Editors: W.F. Wall, Alberto Carramiñana, Luis Carrasco, P.F. Goldsmith
Hardbound, ISBN 0-7923-5581-4, May 1999

- **Volume 240: Numerical Astrophysics**
Editors: Shoken M. Miyama, Kohji Tomisaka, Tomoyuki Hanawa
Hardbound, ISBN 0-7923-5566-0, March 1999
- **Volume 239: Motions in the Solar Atmosphere**
Editors: Arnold Hanslmeier, Mauro Messerotti
Hardbound, ISBN 0-7923-5507-5, February 1999
- **Volume 238: Substorms-4**
Editors: S. Kokubun, Y. Kamide
Hardbound, ISBN 0-7923-5465-6, March 1999
- **Volume 237: Post-Hipparcos Cosmic Candles**
Editors: André Heck, Filippina Caputo
Hardbound, ISBN 0-7923-5348-X, December 1998
- **Volume 236: Laboratory Astrophysics and Space Research**
Editors: P. Ehrenfreund, C. Krafft, H. Kochan, V. Pirronello
Hardbound, ISBN 0-7923-5338-2, December 1998
- **Volume 235: Astrophysical Plasmas and Fluids**
Author: Vinod Krishan
Hardbound, ISBN 0-7923-5312-9, January 1999
Paperback, ISBN 0-7923-5490-7, January 1999
- **Volume 234: Observational Evidence for Black Holes in the Universe**
Editor: Sandip K. Chakrabarti
Hardbound, ISBN 0-7923-5298-X, November 1998
- **Volume 233: B[e] Stars**
Editors: Anne Marie Hubert, Carlos Jaschek
Hardbound, ISBN 0-7923-5208-4, September 1998
- **Volume 232: The Brightest Binaries**
Authors: Dany Vanbeveren, W. van Rensbergen, C.W.H. de Loore
Hardbound, ISBN 0-7923-5155-X, July 1998
- **Volume 231: The Evolving Universe**
Selected Topics on Large-Scale Structure and on the Properties of Galaxies
Editor: Donald Hamilton
Hardbound, ISBN 0-7923-5074-X, July 1998
- **Volume 230: The Impact of Near-Infrared Sky Surveys on Galactic and Extragalactic Astronomy**
Editor: N. Epchtein
Hardbound, ISBN 0-7923-5025-1, June 1998
- **Volume 229: Observational Plasma Astrophysics: Five Years of Yohkoh and Beyond**
Editors: Tetsuya Watanabe, Takeo Kosugi, Alphonse C. Sterling
Hardbound, ISBN 0-7923-4985-7, March 1998
- **Volume 228: Optical Detectors for Astronomy**
Editors: James W. Beletic, Paola Amico
Hardbound, ISBN 0-7923-4925-3, April 1998
- **Volume 227: Solar System Ices**
Editors: B. Schmitt, C. de Bergh, M. Festou
Hardbound, ISBN 0-7923-4902-4, January 1998

- **Volume 226: Observational Cosmology with the New Radio Surveys**
  Editors: M.N. Bremer, N. Jackson, I. Pérez-Fournon
  Hardbound, ISBN 0-7923-4885-0, February 1998
- **Volume 225: SCORe'96: Solar Convection and Oscillations and their Relationship**
  Editors: F.P. Pijpers, Jørgen Christensen-Dalsgaard, C.S. Rosenthal
  Hardbound, ISBN 0-7923-4852-4, January 1998
- **Volume 224: Electronic Publishing for Physics and Astronomy**
  Editor: André Heck
  Hardbound, ISBN 0-7923-4820-6, September 1997
- **Volume 223: Visual Double Stars: Formation, Dynamics and Evolutionary Tracks**
  Editors: J.A. Docobo, A. Elipe, H. McAlister
  Hardbound, ISBN 0-7923-4793-5, November 1997
- **Volume 222: Remembering Edith Alice Müller**
  Editors: Immo Appenzeller, Yves Chmielewski, Jean-Claude Pecker, Ramiro de la Reza, Gustav Tammann, Patrick A. Wayman
  Hardbound, ISBN 0-7923-4789-7, February 1998
- **Volume 220: The Three Galileos: The Man, The Spacecraft, The Telescope**
  Editors: Cesare Barbieri, Jürgen H. Rahe†, Torrence V. Johnson, Anita M. Sohus
  Hardbound, ISBN 0-7923-4861-3, December 1997
- **Volume 219: The Interstellar Medium in Galaxies**
  Editor: J.M. van der Hulst
  Hardbound, ISBN 0-7923-4676-9, October 1997
- **Volume 218: Astronomical Time Series**
  Editors: Dan Maoz, Amiel Sternberg, Elia M. Leibowitz
  Hardbound, ISBN 0-7923-4706-4, August 1997
- **Volume 217: Nonequilibrium Processes in the Planetary and Cometary Atmospheres: Theory and Applications**
  Authors: Mikhail Ya. Marov, Valery I. Shematovich, Dmitry V. Bisikalo, Jean-Claude Gérard
  Hardbound, ISBN 0-7923-4686-6, September 1997
- **Volume 216: Magnetohydrodynamics in Binary Stars**
  Author: C.G. Campbell
  Hardbound, ISBN 0-7923-4606-8, August 1997
- **Volume 215: Infrared Space Interferometry: Astrophysics & the Study of Earth-like Planets**
  Editors: C. Eiroa, A. Alberdi, Harley A. Thronson Jr., T. de Graauw, C.J. Schalinski
  Hardbound, ISBN 0-7923-4598-3, July 1997
- **Volume 214: White Dwarfs**
  Editors: J. Isern, M. Hernanz, E. García-Berro
  Hardbound, ISBN 0-7923-4585-1, May 1997
- **Volume 213: The Letters and Papers of Jan Hendrik Oort as archived in the University Library, Leiden**
  Author: J.K. Katgert-Merkelijn
  Hardbound, ISBN 0-7923-4542-8, May 1997

- **Volume 212: Wide-Field Spectroscopy**
  Editors: E. Kontizas, M. Kontizas, D.H. Morgan, G.P. Vettolani
  Hardbound, ISBN 0-7923-4518-5, April 1997
- **Volume 211: Gravitation and Cosmology**
  Editors: Sanjeev Dhurandhar, Thanu Padmanabhan
  Hardbound, ISBN 0-7923-4478-2, April 1997
- **Volume 210: The Impact of Large Scale Near-IR Sky Surveys**
  Editors: F. Garzón, N. Epchtein, A. Omont, B. Burton, P. Persi
  Hardbound, ISBN 0-7923-4434-0, February 1997
- **Volume 209: New Extragalactic Perspectives in the New South Africa**
  Editors: David L. Block, J. Mayo Greenberg
  Hardbound, ISBN 0-7923-4223-2, October 1996
- **Volume 208: Cataclysmic Variables and Related Objects**
  Editors: A. Evans, Janet H. Wood
  Hardbound, ISBN 0-7923-4195-3, September 1996
- **Volume 207: The Westerbork Observatory, Continuing Adventure in Radio Astronomy**
  Editors: Ernst Raimond, René Genee
  Hardbound, ISBN 0-7923-4150-3, September 1996
- **Volume 206: Cold Gas at High Redshift**
  Editors: M.N. Bremer, P.P. van der Werf, H.J.A. Röttgering, C.L. Carilli
  Hardbound, ISBN 0-7923-4135-X, August 1996
- **Volume 205: Cataclysmic Variables**
  Editors: A. Bianchini, M. Della Valle, M. Orio
  Hardbound, ISBN 0-7923-3676-3, November 1995
- **Volume 204: Radiation in Astrophysical Plasmas**
  Author: V.V. Zheleznyakov
  Hardbound, ISBN 0-7923-3907-X, February 1996
- **Volume 203: Information & On-Line Data in Astronomy**
  Editors: Daniel Egret, Miguel A. Albrecht
  Hardbound, ISBN 0-7923-3659-3, September 1995
- **Volume 202: The Diffuse Interstellar Bands**
  Editors: A.G.G.M. Tielens, T.P. Snow
  Hardbound, ISBN 0-7923-3629-1, October 1995
- **Volume 201: Modulational Interactions in Plasmas**
  Authors: Sergey V. Vladimirov, Vadim N. Tsytovich, Sergey I. Popel, Fotekh Kh. Khakimov
  Hardbound, ISBN 0-7923-3487-6, June 1995
- **Volume 200: Polarization Spectroscopy of Ionized Gases**
  Authors: Sergei A. Kazantsev, Jean-Claude Henoux
  Hardbound, ISBN 0-7923-3474-4, June 1995
- **Volume 199: The Nature of Solar Prominences**
  Author: Einar Tandberg-Hanssen
  Hardbound, ISBN 0-7923-3374-8, February 1995
- **Volume 198: Magnetic Fields of Celestial Bodies**
  Author: Ye Shi-hui
  Hardbound, ISBN 0-7923-3028-5, July 1994

- **Volume 193: Dusty and Self-Gravitational Plasmas in Space**
  Authors: Pavel Bliokh, Victor Sinitsin, Victoria Yaroshenko
  Hardbound, ISBN 0-7923-3022-6, September 1995
- **Volume 191: Fundamentals of Cosmic Electrodynamics**
  Author: Boris V. Somov
  Hardbound, ISBN 0-7923-2919-8, July 1994
- **Volume 190: Infrared Astronomy with Arrays**
  The Next Generation
  Editor: Ian S. McLean
  Hardbound, ISBN 0-7923-2778-0, April 1994
- **Volume 189: Solar Magnetic Fields**
  Polarized Radiation Diagnostics
  Author: Jan Olof Stenflo
  Hardbound, ISBN 0-7923-2793-4, March 1994
- **Volume 188: The Environment and Evolution of Galaxies**
  Authors: J. Michael Shull, Harley A. Thronson Jr.
  Hardbound, ISBN 0-7923-2541-9, October 1993
  Paperback, ISBN 0-7923-2542-7, October 1993
- **Volume 187: Frontiers of Space and Ground–Based Astronomy**
  The Astrophysics of the 21st Century
  Editors: Willem Wamsteker, Malcolm S. Longair, Yoji Kondo
  Hardbound, ISBN 0-7923-2527-3, August 1994
- **Volume 186: Stellar Jets and Bipolar Outflows**
  Editors: L. Errico, Alberto A. Vittone
  Hardbound, ISBN 0-7923-2521-4, October 1993
- **Volume 185: Stability of Collisionless Stellar Systems**
  Mechanisms for the Dynamical Structure of Galaxies
  Author: P.L. Palmer
  Hardbound, ISBN 0-7923-2455-2, October 1994
- **Volume 184: Plasma Astrophysics**
  Kinetic Processes in Solar and Stellar Coronae
  Author: Arnold O. Benz
  Hardbound, ISBN 0-7923-2429-3, September 1993
- **Volume 183: Physics of Solar and Stellar Coronae:**
  G.S. Vaiana Memorial Symposium
  Editors: Jeffrey L. Linsky, Salvatore Serio
  Hardbound, ISBN 0-7923-2346-7, August 1993
- **Volume 182: Intelligent Information Retrieval: The Case of Astronomy and Related Space Science**
  Editors: André Heck, Fionn Murtagh
  Hardbound, ISBN 0-7923-2295-9, June 1993
- **Volume 181: Extraterrestrial Dust**
  Laboratory Studies of Interplanetary Dust
  Author: Kazuo Yamakoshi
  Hardbound, ISBN 0-7923-2294-0, February 1995
- **Volume 180: The Center, Bulge, and Disk of the Milky Way**
  Editor: Leo Blitz
  Hardbound, ISBN 0-7923-1913-3, August 1992

- **Volume 179: Structure and Evolution of Single and Binary Stars**
  Authors: C.W.H. de Loore, C. Doom
  Hardbound, ISBN 0-7923-1768-8, May 1992
  Paperback, ISBN 0-7923-1844-7, May 1992
- **Volume 178: Morphological and Physical Classification of Galaxies**
  Editors: G. Longo, M. Capaccioli, G. Busarello
  Hardbound, ISBN 0-7923-1712-2, May 1992
- **Volume 177: The Realm of Interacting Binary Stars**
  Editors: J. Sahade, G.E. McCluskey, Yoji Kondo
  Hardbound, ISBN 0-7923-1675-4, December 1992
- **Volume 176: The Andromeda Galaxy**
  Author: Paul Hodge
  Hardbound, ISBN 0-7923-1654-1, June 1992
- **Volume 175: Astronomical Photometry, A Guide**
  Authors: Christiaan Sterken, J. Manfroid
  Hardbound, ISBN 0-7923-1653-3, April 1992
  Paperback, ISBN 0-7923-1776-9, April 1992
- **Volume 174: Digitised Optical Sky Surveys**
  Editors: Harvey T. MacGillivray, Eve B. Thomson
  Hardbound, ISBN 0-7923-1642-8, March 1992
- **Volume 173: Origin and Evolution of Interplanetary Dust**
  Editors: A.C. Levasseur-Regourd, H. Hasegawa
  Hardbound, ISBN 0-7923-1365-8, March 1992
- **Volume 172: Physical Processes in Solar Flares**
  Author: Boris V. Somov
  Hardbound, ISBN 0-7923-1261-9, December 1991
- **Volume 171: Databases and On-line Data in Astronomy**
  Editors: Miguel A. Albrecht, Daniel Egret
  Hardbound, ISBN 0-7923-1247-3, May 1991
- **Volume 170: Astronomical Masers**
  Author: Moshe Elitzur
  Hardbound, ISBN 0-7923-1216-3, February 1992
  Paperback, ISBN 0-7923-1217-1, February 1992
- **Volume 169: Primordial Nucleosynthesis and Evolution of the Early Universe**
  Editors: Katsuhiko Sato, J. Audouze
  Hardbound, ISBN 0-7923-1193-0, August 1991

Missing volume numbers have not yet been published.
For further information about this book series we refer you to the following web site: http://www.wkap.nl/prod/s/ASSL

To contact the Publishing Editor for new book proposals:
Dr. Harry (J.J.) Blom: harry.blom@wkap.nl